T0279352

AEROSPACE PROPULSION

Aerospace Series List

Aerospace Propulsion	Lee	October 2013
Aircraft Flight Dynamics and Control	Durham	August 2013
Civil Avionics Systems, Second Edition	Moir, Seabridge and Jukes	August 2013
Modelling and Managing Airport Performance	Zografos	July 2013
Advanced Aircraft Design: Conceptual Design, Analysis and Optimization of Subsonic Civil Airplanes	Torenbeek	June 2013
Design and Analysis of Composite Structures: With applications to aerospace Structures, Second Edition	Kassapoglou	April 2013
Aircraft Systems Integration of Air-Launched Weapons	Rigby	April 2013
Design and Development of Aircraft Systems, Second Edition	Moir and Seabridge	November 2012
Understanding Aerodynamics: Arguing from the Real Physics	McLean	November 2012
Aircraft Design: A Systems Engineering Approach	Sadraey	October 2012
Introduction to UAV Systems, Fourth Edition	Fahlstrom and Gleason	August 2012
Theory of Lift: Introductory Computational Aerodynamics with MATLAB and Octave	McBain	August 2012
Sense and Avoid in UAS: Research and Applications	Angelov	April 2012
Morphing Aerospace Vehicles and Structures	Valasek	April 2012
Gas Turbine Propulsion Systems	MacIsaac and Langton	July 2011
Basic Helicopter Aerodynamics, Third Edition	Seddon and Newman	July 2011
Advanced Control of Aircraft, Spacecraft and Rockets	Tewari	July 2011
Cooperative Path Planning of Unmanned Aerial Vehicles	Tsourdos et al	November 2010
Principles of Flight for Pilots	Swatton	October 2010
Air Travel and Health: A Systems Perspective	Seabridge et al	September 2010
Unmanned Aircraft Systems: UAVS Design, Development and Deployment	Austin	April 2010
Introduction to Antenna Placement & Installations	Macnamara	April 2010
Principles of Flight Simulation	Allerton	October 2009
Aircraft Fuel Systems	Langton et al	May 2009
The Global Airline Industry	Belobaba	April 2009
Computational Modelling and Simulation of Aircraft and the Environment: Volume 1 – Platform Kinematics and Synthetic Environment	Diston	April 2009
Handbook of Space Technology	Ley, Wittmann Hallmann	April 2009
Aircraft Performance Theory and Practice for Pilots	Swatton	August 2008
Aircraft Systems, Third Edition	Moir & Seabridge	March 2008
Introduction to Aircraft Aeroelasticity And Loads	Wright & Cooper	December 2007
Stability and Control of Aircraft Systems	Langton	September 2006
Military Avionics Systems	Moir & Seabridge	February 2006
Design and Development of Aircraft Systems	Moir & Seabridge	June 2004
Aircraft Loading and Structural Layout	Howe	May 2004
Aircraft Display Systems	Jukes	December 2003
Civil Avionics Systems	Moir & Seabridge	December 2002

AEROSPACE PROPULSION

T.-W. Lee
Arizona State University, USA

This edition first published 2014
© 2014, John Wiley & Sons, Ltd

Registered office

John Wiley & Sons Ltd, The Atrium, Southern Gate, Chichester, West Sussex, PO19 8SQ, United Kingdom

For details of our global editorial offices, for customer services and for information about how to apply for permission to reuse the copyright material in this book please see our website at www.wiley.com.

The right of the author to be identified as the author of this work has been asserted in accordance with the Copyright, Designs and Patents Act 1988.

All rights reserved. No part of this publication may be reproduced, stored in a retrieval system, or transmitted, in any form or by any means, electronic, mechanical, photocopying, recording or otherwise, except as permitted by the UK Copyright, Designs and Patents Act 1988, without the prior permission of the publisher.

Wiley also publishes its books in a variety of electronic formats. Some content that appears in print may not be available in electronic books.

Designations used by companies to distinguish their products are often claimed as trademarks. All brand names and product names used in this book are trade names, service marks, trademarks or registered trademarks of their respective owners. The publisher is not associated with any product or vendor mentioned in this book.

Limit of Liability/Disclaimer of Warranty: While the publisher and author have used their best efforts in preparing this book, they make no representations or warranties with respect to the accuracy or completeness of the contents of this book and specifically disclaim any implied warranties of merchantability or fitness for a particular purpose. It is sold on the understanding that the publisher is not engaged in rendering professional services and neither the publisher nor the author shall be liable for damages arising herefrom. If professional advice or other expert assistance is required, the services of a competent professional should be sought.

MATLAB® is a trademark of The MathWorks, Inc. and is used with permission. The MathWorks does not warrant the accuracy of the text or exercises in this book. This book's use or discussion of MATLAB® software or related products does not constitute endorsement or sponsorship by The MathWorks of a particular pedagogical approach or particular use of the MATLAB® software.

Library of Congress Cataloging-in-Publication Data

Lee, T.-W. (Tae-Woo)
Aerospace propulsion / TW Lee.
1 online resource.
Includes bibliographical references and index.
Description based on print version record and CIP data provided by publisher; resource not viewed.
ISBN 978-1-118-53465-6 (Adobe PDF) – ISBN 978-1-118-53487-8 (ePub) – ISBN 978-1-118-30798-4 (cloth) 1. Airplanes–Jet propulsion. 2. Rocketry. I. Title.
TL709
629.1′1–dc23

2013027339

A catalogue record for this book is available from the British Library.

ISBN: 978-1-118-30798-4

Set in 10/12 pt Times by Thomson Digital, Noida, India

1 2014

Contents

Series Preface

There are books in the Aerospace Series that deal with propulsion systems for aircraft. They generally treat the engine and its control system as an integral part of the aircraft – as an installed system. The interactions between the propulsion system and the aircraft systems are described.

The power plant of an airborne vehicle is critical to its performance and its safe operation, so it is vital for engineers working in this field to understand the fundamentals of the propulsion system. This book provides a different viewpoint to that of the systems books: it is very much an analytical view of the power plant itself, and it should be read as a complement to the other propulsion books. The author introduces the reader to the principles of thrust and the gas turbine engine before providing a comprehensive mathematical treatment of the major components of the propulsion mechanism and the complex aerodynamic and thermodynamic processes within various engine types – both air-breathing and rocket. This is to provide a basis for developing an understanding of propulsion systems and the modeling tools that can be used to provide a comprehensive and practical knowledge for use in research and industry.

MATLAB® models are provided to reinforce the explanations, and exercises are also set for the diligent student to pursue.

The book covers gas turbine (aeronautical) systems and rocket propulsion (astronautic) systems and is hence of interest to engineers working in the fields of aircraft, missiles and space vehicles. Some novel propulsion systems are also described, that may be pertinent to emerging fields of aerospace transportation systems, setting out to meet environmental objectives.

This is a book for those engineers who wish to understand the fundamental principles of aerospace propulsion systems.

Peter Belobaba, Jonathan Cooper and Allan Seabridge

Preface

Aerospace propulsion devices embody some of the most advanced technologies, ranging from materials, fluid control and heat transfer and combustion. In order to maximize performance, sophisticated testing and computer simulation tools are developed and used. In undergraduate or introductory graduate courses in aerospace propulsion, we only cover the basic elements of fluid mechanics, thermodynamics, heat transfer and combustion science, so that either in industry or in research labs the students/engineers can address some of the modern design and development aspects.

Compressor aerodynamics, for example, is a dynamic process involving rotating blades that see different flows at different radial and axial locations. Cascade and transonic flow behavior can make the analyses more complex and interesting. In turbine flows, the gas temperature is high, and thus various material and heat transfer issues become quite important. Owing to the rotating nature of turbine and compressor fluids, intricate flow control between the axis and the blade section needs to be used, while allowing for cooling flow passage from the compressor to the turbine blades. Combustor flow is even more complex, since liquid-phase fuel needs to be sprayed, atomized, evaporated and burned in a compact volume. High heat release and requirements for downstream dilution and cooling again make the flow design quite difficult and challenging. All of these processes – spray atomization, phase change, combustion, heat transfer (convection and radiation) and mixing – occur in turbulent flows, and no computational tools can accurately reproduce real flows without lengthy modeling and calibration. Any one of the issues mentioned above, such as spray atomization, turbulent flow or combustion, is an unsolved problem in science and engineering, and this is the reason for industry and research labs developing expensive testing and computational analysis methods. This aspect makes aerospace propulsion an important part of engineering curricula, as it provides an interdisciplinary and "tough" training ground for aerospace engineers.

As noted above, owing to the multiple engineering topics involved, we only go into basic elements of aerospace propulsion. After some of the basics are covered, we try to expose the students to projects involving computational fluid dynamic (CFD) software, since this is frequently used in industry and in research labs. There are commercial CFD packages that can be readily made available to the students, using educational licenses. With online documentation and examples, students can learn to operate these codes, individually or in group projects. In addition, the gas-turbine lab at ASU allows the students to use actual testing data for performance analyses. These elements cannot be included in this book without stretching the physical and mental limits, but they are essential components in an aerospace propulsion course, to link the underlying science and engineering to practical applications.

I have included discussions of both gas-turbine and rocket propulsion, for combined or separate aerospace propulsion courses. There are some good interrelations between aeronautical (gas-turbine) and astronautical (rocket) propulsion, based on the same knowledge set. In addition, many students opt to take both aeronautical and astronautical propulsion, unless a combined course is offered, since their final career choices are made many years downstream.

Thank you for reading up to this point, and potentially beyond.

1

Introduction to Propulsion Systems

Propulsion systems include some of the most advanced technologies. The high performance requirements, at low system weight, necessitate advanced thermal-fluid design, materials and system integration. The thrust, generated through a simple-looking principle of conservation of momentum (or Newton's second law), enables many human capabilities, such as high-speed civil transport (approximately 12 hours for trans-Pacific flights), affordable personal aircraft, advanced military aircrafts (e.g. F-22 Raptor, Sukhoi), Earth orbital operations (Space Shuttle) and numerous satellites, planetary probes and possible missions. The propulsion technology can also lead to potentially destructive uses, as in cruise missiles, intercontinental ballistic missiles and many other weapons propelled at high speeds.

A typical gas-engine shown in Figure 1.1 achieves the high exit momentum through a sequence of devices that include compressor, combustor, turbine and nozzle. The ambient air is ingested in gas-turbine engines. The compressor consists of a series of rotating blades, which aerodynamically is a set of airfoils using rotary motion to generate a pressure differential as the air traverses the blade elements. The air pressure is increased in the compressor, and sent into the combustor where the fuel is injected, mixed with the air, and burned. The air energy (enthalpy) increase is now used in the turbines to convert some of the thermal energy (enthalpy) into shaft power. This shaft power is used to power the compressor, by simply having a common axis between the turbine and the compressor in turbojet engines. However, in turbofan engines, the turbine power is used to run both the compressor and the fan. The fan adds enthalpy to the air stream in the fan section. The energy available at the end of the turbine section is converted to air kinetic energy in the nozzle. The high kinetic energy of the exhaust stream also has high momentum, which is useful in generating thrust. Ramjets are a much simpler form of turbojet engines, where "ram compression" of incoming stream at supersonic speeds is sufficient to elevate the pressure of the air. Fuel then needs to be injected into this high-pressure air stream and the resulting flame stabilized in the ramjet combustor, for sustained thrust.

Advances in practically all aspect of engineering, including propulsion technology, can be found in the Lockheed Martin F-22 Raptor (Figure 1.2) that entered service in 2005. New materials such as advanced alloys and composite materials are used in the Raptor

Aerospace Propulsion, First Edition. T.-W. Lee.
© 2014 John Wiley & Sons, Ltd. Published 2014 by John Wiley & Sons, Ltd.
Companion Website: www.wiley.com/go/aerospaceprop

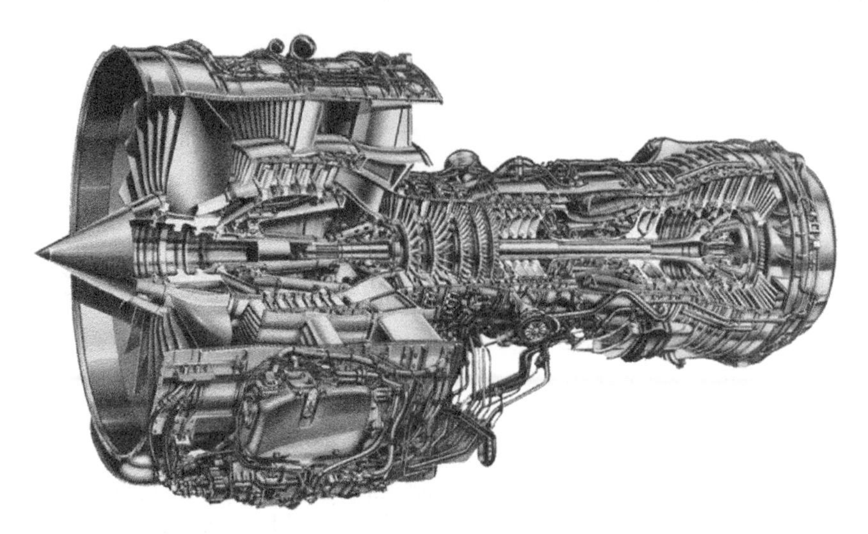

Figure 1.1 A typical gas-turbine engine. Copyright United Technologies Corporation.

airframe, aerodynamic surfaces and engine components. The power plant in the F-22 consists of Pratt-Whitney afterburning turbofans (F119-PW-100) with a high efficiency, which provide supersonic cruise speeds with long range and unmatched agility with pitch-vectoring thrust nozzles. But these technological advances came with a high price tag. Many of the new technologies were researched and developed specifically as part of the F-22 project. If all the development costs are added in, the F-22 carries a price tag of over $300 million per aircraft. Table 1.1 shows some of the main specifications of the F-22, including some of the propulsion characteristics.

The Pratt-Whitney F119-PW-100 engine is another component in the F-22 that is arguably the most advanced in aircraft technology. Each of these engines generates more

Figure 1.2 F-22 Raptor, with advanced embedded technologies, including the power plant (F119-PW-100). Courtesy of US Department of Defence.

Table 1.1 F-22 specifications.

Length	62.1 ft (18.9 m)
Wingspan	44.5 ft (13.56 m)
Height	16.8 ft (5.08 m)
Maximum take-off weight	80 000 lb (36 288 kg)
Power plant	Two Pratt-Whitney F119-PW-100 pitch-vectoring turbofans with afterburners
Total thrust	70 000 lb
Maximum speed	High altitude: Mach 2.42 or 1600 mph (2570 km/h)
	Low altitude: Mach 1.72 or 1140 mph (1826 km/h)
Ceiling	65 000 ft (20 000 m)
Range	2000 miles (5600 km)
Rate of climb	N/A (classified)
Thrust-to-weight ratio	1.26
Maximum g-load	−3/+9.5

thrust without the afterburner than most conventional engines with full afterburner power on, and its supersonic thrust is also about twice that of the other engines in the class. Using two of these engines to develop a total thrust of 70 000 pounds, the F-22 can travel at supersonic speeds without the afterburners for fuel-efficient high-speed cruise to the target area. This level of thrust is more than the aircraft weight, and enables the F-22 to fly vertically upward much like a rocket. The F119 is also unique in fully integrating the vector thrust nozzle into the engine/airframe combination, for a 20-degree up/down redirection of thrust for high-g turn capabilities. The thrust vectoring is designed to enhance the turn rates by up to 50% in comparison to using control surfaces alone. The F119 engine achieves all these functional characteristics with 40% fewer parts than conventional engines to furnish exceptional reliability, and maintenance and repair access. In a design method called integrated product development, inputs from assembly line workers and air force mechanics were incorporated to streamline the entire sequence of engine production, maintenance and repairs. These design innovations are expected to reduce the support equipment, labor and spare parts in demand by approximately half. Similar to the mid-fuselage airframe, the turbine stage, consisting of the disk and blades, is constructed in a single integrated metal piece for high integrity at lower weight, better performance and thermal insulation for the turbine disk cavity. The fan and compressor blade designs went through extensive permutations and modifications using computational fluid dynamic (CFD) simulations, resulting in unprecedented efficiency in both sections. Hardware cut-and-try of different designs would have cost way too much time and money. High-strength and degradation-resistant Alloy C was used in key components such as the compressors, turbines and nozzles to allow the engine to run at higher temperatures, one of the important contributing factors to the increased thrust and durability of F119 engines. The combustor – the hottest component in the engine – uses oxidation-resistant, thermally insulating cobalt coatings. A digital electronic engine control device called FADEC (FADEC is generally meant to signify 'Full Authority Digital Engine Control' the level of redundancy is at the discretion of the engine manufacturer) not only fine tunes the engine operating parameters to deliver the highest performance at the maximum efficiency, but also establishes responsive and precise engine operating parameters with inputs from the pilot control of the throttle and the engine/flight sensors.

As is well known, the F-22 has unique stealth capabilities, in spite of its size. In addition to the external geometry and surfaces, the jet-engine exhaust is a critical component in minimizing infrared signatures that can be detected by forward-looking infrared (FLIR) or IR sensors in heat-seeking missiles. The exhaust of the F-22 is designed to absorb the heat by using ceramic components, rather than conduct heat to the outside surface. Also, the horizontal stabilizers are placed to shield the thermal emission as much as possible.

The F-35 Lightning II Joint Strike Fighter (JSF) Program represents the effort to provide a capable, multi-mission aircraft while containing the budget. The F-35's price tag is about half that of the F-22 Raptor. The argument for wide adoption of this scaled-back aircraft is that the F-22's capabilities are best directed against opponents with similar technological capabilities, and with the changed geo-political environment the United States forces are less likely to be involved in such encounters. A unique variant of the F-35 (Figure 1.3) is the marine STOVL version, F-35B, also planned for adoption by the British Royal Navy to replace the Sea Harrier. The short take-off is facilitated by a number of auxiliary nozzles to divert the thrust. In a normal engine, the jet exhaust is pushed out of the nozzle at the rear of the engine to provide only forward thrust. In engines with thrust reversers, the fan stream is redirected to the forward direction to generate negative thrust. The same concept can be used to redirect the thrust to other directions by using auxiliary nozzles. For the F-35B, there is a lift nozzle that takes the fan exhaust and directs it vertically downward. Also, the pitch nozzle at the main nozzle can be turned to add a vertical component to the thrust. For control of the aircraft during this tricky maneuver, there are four additional nozzles. Two roll nozzles control the roll angle by sending a small fraction of the main exhaust at off-horizontal angles, while two yaw nozzles generate thrust in the forward and backward offset angles.

Rockets, on the other hand, carry all the working fluid (both fuel and oxidizers) on board. The main reason for carrying both the fuel and oxidizer is so that rockets can operate in an

Figure 1.3 F-35 Joint Strike Fighter. Courtesy of US Department of Defence.

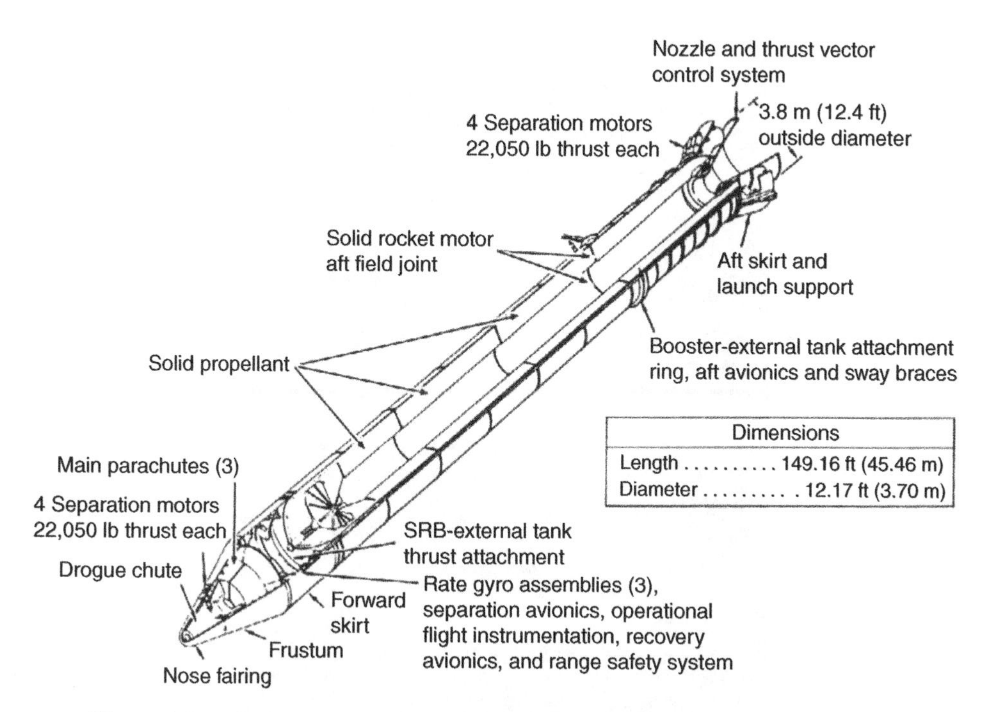

Figure 1.4 Solid-propellant rocket engine. Courtesy of US Department of Defence.

air-less environment (e.g. underwater or in outer space), but this also means zero incoming momentum. In addition, some rocket devices can be quite simple in design. Solid-propellant rockets, for example, only require the propellant and a nozzle (Figure 1.4). The documented use of rocketry dates back to 900 AD in China, where "black powder" was used as crude flame throwers ("fire lance"), grenades, siege weapons and other devices that delivered shock effects against the Mongols in the 10th century. Black powder consists of readily available ingredients – charcoal, sulfur and saltpeter (potassium nitrate), and was probably discovered by accident and perfected through trial-and-error. Combustion of black powder goes roughly as

$$2\,\text{parts saltpeter} + \text{sulfur} + 3\,\text{parts charcoal} \rightarrow \text{combustion products} + \text{nitrogen} + \text{heat}$$

This technology was quickly adopted by the Mongols, and spread to Europe and other parts of the world. Rockets using liquid propellants are, in comparison, relatively new technologies, having been developed in the early 1900s. At the other extreme, modern liquid-propellant rockets contain some of the most advanced technologies (Figure 1.5), due to the high operating pressure and temperatures, in addition to the use of cryogenic propellants such as liquid oxygen and liquid hydrogen. The high operating pressure requires sophisticated pumping devices, while high temperature necessitates advanced combustion control and cooling technologies.

Figure 1.5 Liquid-propellant rocket engine (space shuttle main engine). Courtesy of NASA.

A large altitude change during a rocket flight requires modified designs for each of the stages. At launch, the ambient pressure is roughly equal to the sea level atmospheric pressure, while the pressure decreases with increasing altitude. This results in larger pressure thrust; however, at higher altitudes the nozzle exit pressure becomes greater than the ambient pressure and the nozzle operates in an under-expanded mode. This operation is less than optimum, and the gas expansion continues downstream, creating diamond-shaped shock waves. Upper stages are designed with this aspect in mind, where a larger expansion ratio in the nozzle is used. The first stage of a Delta II launch vehicle, for example, has a nozzle expansion ratio of 12. The propellant is liquid oxygen and RP-1 (a kerosene-based hydrocarbon), which is burned in the combustion chamber at a mixture ratio (O/F) of 2.25 and pressure of 4800 kPa. This combination results in a specific impulse of 255 s. The next stage, on the other hand, has a nozzle expansion ratio of 65. The propellant combination of nitrogen tetroxide and Aerozine 50 (hydrazine/unsymmetrical dimethyl hydrazine) is used at a mixture ratio of 1.90 and chamber pressure of 5700 kPa (830 psia), which provides a specific impulse of 320 s. The space shuttle main engine (SSME) has an even larger nozzle expansion ratio of 77.5.

Liquid oxygen and liquid hydrogen used in the SSME generates a high combustion chamber temperature, and also produces combustion product gases with a low molecular weight. These factors are optimum for producing large exit velocity and thus thrust. For this reason, a liquid hydrogen/oxygen combination is also used in the Atlas Centaur upper stage, the Ariane-4 third stage and the Ariane-5 core stage.

In addition to the boost, rockets are used for various orbit maneuvers, such as station-keeping and attitude adjustments. Various factors can contribute to deviations from the target orbit. Gravitational forces of the sun and moon, for example, can cause the orbital inclination to change by approximately one degree per year. The velocity increment that needs to be expended to compensate for this drift is roughly 50 m/s. Other smaller factors that lead to orbit deviations are the elliptical shape of the Earth's equator and the "solar wind" which is the radiation pressure due to the sun's radiation. Attitude adjustments are performed with a relatively large number of small thrusters, since all three degrees of freedom need to be accessed in addition to start/stop maneuvers. For example, the Ford Aerospace Intelsat V satellite had an array of four 0.44 N (0.1 lbf) thrusters for roll control, ten 2.0 N (0.45 lbf) thrusters for pitch and yaw control and station-keeping, and two 22.2 N (5.0 lbf) thrusters for repositioning and reorientation.

Since the thrust required for orbit maneuvers is small, simpler rocket boosters such as solid propellant or monopropellants can be used. For example, typical satellites in geosynchronous orbits launched during the 1980s were equipped with solid-propellant boosters for apogee maneuver and monopropellant hydrazine thrusters for station-keeping and attitude control. The solid propellant consisted of HTPB (fuel/binder) and ammonium perchlorate (oxidizer). Hydrazine is a monopropellant containing both fuel and oxidizer components in its chemical structure, and only requires a catalytic grid for decomposition. An interesting combination of electric and thermal thrust is the use of electrical heat for the hydrazine monopropellant, which increases the specific thrust.

For more recent satellites, electric or electromagnetic thrusters with high specific thrust are used for low propellant mass requirements and therefore longer mission durations. Arcjets, for example, use an electric arc to superheat propellants such as hydrazine, which nearly doubles the specific impulse to over 500 s with typical thrust levels of 0.20 N. Arcjet thrusters are used on Intelsat VIII and Lockheed Martin A2100 satellites, and Iridium satellites. Another type of electric propulsion system with even higher specific impulse (2000–4000 s) is the ion thruster (Figure 1.6), using xenon as propellant, which produces a typical thrust of less than 0.1 N. Xenon is an inert monatomic gas with a high atomic weight (131 kg/kmol). Xenon atoms are ionized by high-speed electrons, and then these positively charged ions are accelerated to a speed of some 34 000 m/s in an electric field of 750 V in thousands of ion beams. The momentum of these ion beams produces a thrust in the order of 10 mN.

A combination of electric and magnetic fields can also be used in so-called Hall thrusters. Other exotic space propulsion devices include solar sails and nuclear propulsion, still at the experimental stage (Figure 1.7).

1.1 Conservation of Momentum

We can see from the above examples that all propulsion devices generate some high-speed exhaust stream, through a variety of means. Thus, we can say that the objective of

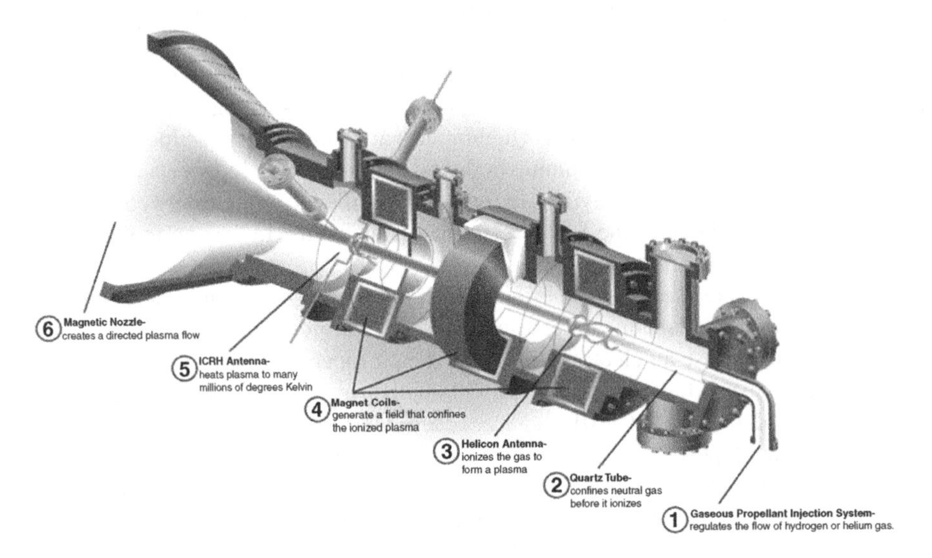

Figure 1.6 Ion propulsion devices. Courtesy of NASA.

propulsion devices, in general, is to obtain excess momentum (higher exit momentum than incoming) by generating high-speed exhaust jets. A simple version (a more precise description is provided in Chapter 2) of the conservation equation of momentum (Newton's second law) can be used to illustrate how this process will work in producing positive thrust.

$$(\text{time rate of change of momentum}) = (\text{force}) \tag{1.1}$$

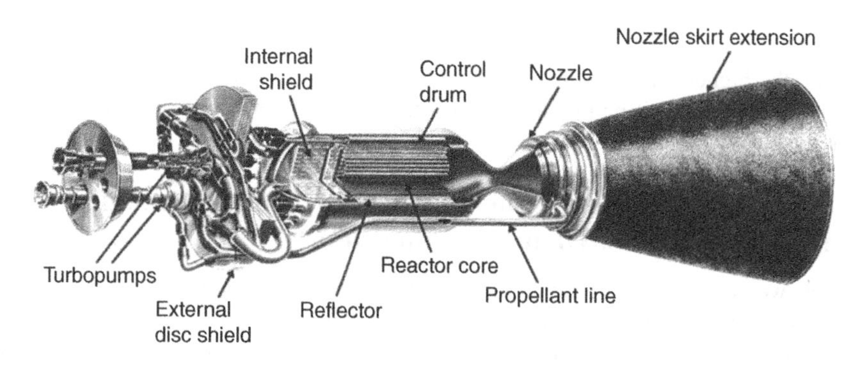

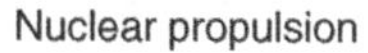

Figure 1.7 Some novel propulsion concepts. Courtesy of NASA.

Or in mathematical form,

$$M\frac{dU}{dt} = F \tag{1.2}$$

M = vehicle mass
U = vehicle velocity
F = thrust force

In a propulsion system, the momentum of some fluid with mass Δm will go from $\Delta m U_{in}$ at the inlet to $\Delta m U_{out}$ at the exit in time Δt. So we can approximate the left-hand side in Eq. (1.2) as $\frac{\Delta m (U_{out} - U_{in})}{\Delta t}$. Here, we can factor out the Δm since we are dealing with the same fluid mass. Moreover, the fluid mass divided by the transit time, $\Delta m / \Delta t$, is the mass flow rate. Thus, we can rewrite Eq. (1.2) as follows.

$$\dot{m}(U_{out} - U_{in}) = F \tag{1.3}$$

Equation (1.3) shows that the higher the exiting momentum with respect to the incoming momentum, the higher the thrust will be, which is the objective of a propulsion device. We may also note that high exiting momentum can be achieved by high exit velocity, large mass flow rate, or both.

1.2 Conservation of Energy (the First Law of Thermodynamics) and Other Thermodynamic Relationships

In this book, we mostly focus on the thermal-fluid aspect of propulsion systems, starting from thermodynamics, fluid dynamics, heat transfer and combustion (chemical reaction). Let us set down some baseline thermodynamic relationships that we will be using in this book. The most important element of thermodynamics is the first law, or the conservation of energy, which simply states that the energy contained in the control volume (E_{cv}) changes at the rate determined by the heat input ($\dot{Q}$) minus the power output ($\dot{W}$) and the net energy input consisting of the enthalpy and the kinetic energy).

$$\frac{dE_{cv}}{dt} = \dot{Q} - \dot{W} + \dot{m}_i\left(h_i + \frac{U_i^2}{2}\right) - \dot{m}_e\left(h_e + \frac{U_e^2}{2}\right) \tag{1.4}$$

For steady state, the left-hand side is set to zero, so that in compressors, turbines, combustors, nozzles and other propulsion components the heat, power and net energy flow rates are all balanced according to Eq. (1.4). And this equation can also be used to define the stagnation enthalpy, which is the total energy of the fluid including the enthalpy and kinetic energy.

$$h^o = h + \frac{U^2}{2} \tag{1.5}$$

Equation (1.4) states that in the absence of heat transfer and power, the stagnation enthalpy will remain the same during a flow process.

For a closed system, there is no energy flux into or out of the volume, and the first law can be written in a differential form.

$$de = dq - dw \tag{1.6}$$

Equation (1.6) states that the internal energy in the system changes as a function of the heat input and work output. Changes in internal energy or enthalpy can be calculated using the specific heats of the fluid.

$$c_v = \left(\frac{de}{dT}\right)_v \rightarrow e_2 - e_1 = \int_{T_1}^{T_2} c_v dT \approx c_v(T_2 - T_1) \tag{1.7}$$

$$c_p = \left(\frac{dh}{dT}\right)_p \rightarrow h_2 - h_1 = \int_{T_1}^{T_2} c_p dT \approx c_p(T_2 - T_1) \tag{1.8}$$

For ideal gases, the pressure (p), density (ρ) and temperature (T) are related by the ideal gas equation of state.

$$p = \rho RT \tag{1.9}$$

Example 1.1 Ideal gas equation of state

For air, the specific gas constant is $$R=\frac{\bar{R}}{M}=\frac{8314\dfrac{\text{J}}{\text{kmoleK}}}{28.97\dfrac{\text{kg}}{\text{kmole}}}=287\frac{\text{kJ}}{\text{kgK}} \quad \text{(E1.1.1)}$$

At the standard atmospheric $p_o=101.325\,\text{kPa}$ and $T_o=288.15\,\text{K}$, the air density is

$$\rho=\frac{p}{RT}=\frac{101.325\dfrac{\text{kN}}{\text{m}^2}}{\left(0.287\dfrac{\text{kN}\cdot\text{m}}{\text{kgK}}\right)(288.15\text{K})}=1.225\frac{\text{kg}}{\text{m}^3} \quad \text{(E1.1.2)}$$

In imperial units, the specific heat for air at 59°F is

$$c_p=0.246\frac{\text{BTU}}{\text{lbm}\cdot{}^\circ\text{R}} \text{ and } R=53.35\frac{\text{ft}\times\text{lbf}}{\text{lbm}\cdot{}^\circ\text{R}}=0.06856\frac{\text{BTU}}{\text{lbm}\cdot{}^\circ\text{R}} \quad \text{(E1.1.3)}$$

At the standard atmospheric $p_o=2116.22\,\text{lbf/ft}^2$ and $T_o=518.67°\text{R}=59°\text{F}$, the air density is

$$\rho=\frac{p}{RT}=\frac{2116.22\dfrac{\text{lbf}}{\text{ft}^2}}{\left(53.35\dfrac{\text{ft}\times\text{lbf}}{\text{lbm}\cdot{}^\circ\text{R}}\right)(518.67°\text{R})}=0.0765\frac{\text{lbm}}{\text{ft}^3} \quad \text{(E1.1.4)}$$

For ideal gases, the relationship between c_p, c_v, and the specific gas constant R follows from the definition of enthalpy.

$$h=e+pv=e+RT\rightarrow dh=de+RdT \quad (1.10)$$

$$\frac{dh}{dT}=\frac{de}{dT}+R\rightarrow c_p=c_v+R \quad (1.11)$$

Using $\gamma=\frac{c_p}{c_v}$,

$$c_p=\frac{\gamma R}{\gamma-1} \quad (1.12)$$

$$c_v=\frac{R}{\gamma-1} \quad (1.13)$$

For isentropic processes, involving ideal gases and constant specific heats, we have the following relationships between the state variables.

$$\frac{p_2}{p_1}=\left(\frac{T_2}{T_1}\right)^{\frac{\gamma}{\gamma-1}} \quad (1.14)$$

$$\frac{p_2}{p_1} = \left(\frac{\rho_2}{\rho_1}\right)^{\gamma} \tag{1.15}$$

Stagnation properties are defined as the condition reached when the flow decelerates to zero speed, isentropically. Using Eq. (1.5), with constant c_p,

$$\frac{1}{2}U^2 + c_p T = c_p T^o \rightarrow \frac{T^o}{T} = 1 + \frac{\gamma - 1}{2}M^2 \tag{1.16}$$

We have used Eq. (1.12) and $a = \sqrt{\gamma RT}$ for ideal gases. Equation (1.15) shows that the stagnation temperature increases as the square of the Mach number, U/a. Using isentropic relationship, we also have

$$\frac{p^o}{p} = \left(\frac{T^o}{T}\right)^{\frac{\gamma}{\gamma-1}} = \left[1 + \frac{\gamma - 1}{2}M^2\right]^{\frac{\gamma}{\gamma-1}} \tag{1.17}$$

Example 1.2 Stagnation temperature

For $T_o = 411.8°R$, the stagnation temperature is

$$T^o = T\left(1 + \frac{\gamma - 1}{2}M^2\right) = 464.5°\text{R for } M_o = 0.8;\ 741.3°\text{R for } M_o$$
$$= 2,2470.8°\text{R for } M_o = 5.$$

Example 1.3 Isentropic work

For an isentropic compression ratio of 10 and $T_2^o = 300$ K, we have

$$\frac{T_3^o}{T_2^o} = \left(\frac{p_3^o}{p_2^o}\right)^{\frac{\gamma-1}{\gamma}} = (10)^{0.3571} = 1.93$$

Here, we use the customary value of $\gamma = 1.4$ for air and $c_p = 1004.76$ J/(kg · K).

The corresponding isentropic work is

$$w_{c,s} = c_p(T_3^o - T_2^o) = c_p T_3^o\left(\frac{T_3^o}{T_2^o} - 1\right) = 280.3\frac{\text{kJ}}{\text{kg}}$$

1.3 One-Dimensional Gas Dynamics

For propulsion systems operating in supersonic flows, some elements of gas dynamics are useful. In adiabatic flows, the stagnation enthalpy is conserved, so that the use of Eq. (1.16) results in a relationship between the static temperatures between two points (1 and 2) in the flow.

$$\frac{T_2}{T_1} = \frac{1 + \frac{\gamma-1}{2} M_1^2}{1 + \frac{\gamma-1}{2} M_2^2} \tag{1.18}$$

Using ideal gas equation of state, and using the fact that for steady-state one-dimensional (area $A = \text{const}$) flows $\rho U = \rho Ma = \rho M \sqrt{\gamma RT} = \text{const}$, we have

$$\frac{T_2}{T_1} = \frac{p_2}{p_1}\frac{\rho_1}{\rho_2} \text{ and } \frac{\rho_1}{\rho_2} = \frac{M_2}{M_1}\sqrt{\frac{T_2}{T_1}} \tag{1.19}$$

Combining Eqs. (1.18) and (1.19) and solving for the pressure ratio, we get

$$\frac{p_2}{p_1} = \frac{M_1}{M_2}\left[\frac{1 + \frac{\gamma-1}{2} M_1^2}{1 + \frac{\gamma-1}{2} M_2^2}\right]^{\frac{1}{2}} \tag{1.20}$$

Conservation of momentum in one-dimensional flows can be written as

$$\rho U dU = -dp \rightarrow p_1 + \rho_1 U_1^2 = p_2 + \rho_2 U_2^2 \tag{1.21}$$

Using $\rho U^2 = \gamma p M^2$, we can rewrite Eq. (1.21) as

$$\frac{p_2}{p_1} = \frac{1 + \gamma M_1^2}{1 + \gamma M_2^2} \tag{1.22}$$

Eliminating pressure from Eqs. (1.20) and (1.22) gives us the relationship between upstream and downstream Mach numbers across, for example, a normal shock.

$$M_2^2 = \frac{(\gamma - 2)M_1^2 + 2}{2\gamma M_1^2 - (\gamma - 1)} \tag{1.23}$$

Then, the ratio of other parameters can also be written as a function of the upstream Mach number.

$$\frac{p_2}{p_1} = \frac{2\gamma M_1^2 - (\gamma - 1)}{\gamma + 1} \tag{1.24}$$

$$\frac{T_2}{T_1} = \frac{[(\gamma - 1)M_1^2 + 2][2\gamma M_1^2 - (\gamma - 1)]}{(\gamma + 1)^2 M_1^2} \tag{1.25}$$

Equations (Equations (1.23)–(1.25) are referred to as the normal shock relationships and are tabulated in the Appendix C.

For isentropic flows in ducts, the local Mach number is a function of the cross-sectional area.

$$\frac{A}{A^*} = \frac{1}{M}\left[\frac{2}{\gamma+1}\left(1+\frac{\gamma-1}{2}M^2\right)\right]^{\frac{\gamma+1}{2(\gamma-1)}} \qquad (1.26)$$

A^* = throat area ($M^* = 1$)

1.4 Heat Transfer

Due to the high temperatures in some components, heat transfer is an important element of propulsion science. There are three modes of heat transfer: conduction, convection and radiation. Each is a subject in itself, but here we briefly state the basic laws of these heat transfer modes. Conduction is due to the molecule-to-molecule transfer of thermal energy, and is described by Fourier's law of conduction.

$$q_{cond} = -kA\frac{dT}{dx} \qquad [\text{W}] \qquad (1.27)$$

k = thermal conductivity [W/(mK)]

The heat is transferred "down" (from hot to cold) the temperature gradient – hence the negative sign. In solids, conduction occurs due to lattice vibration and movement of energy carriers such as electrons in conductors. The latter is the reason why most good electrical conductors are also good thermal conductors. The difference in transfer of heat with electrons is also the basis for devices to measure temperature, thermocouples.

Although heat is also transferred through conduction in fluids, larger amounts of heat can be moved through the motion of the fluid mass itself. Mass times the specific heat is the energy content of the fluid, and if this mass is moved through fluid motion then heat transfer results. This mode of heat transfer is called convection, and is approximated through Newton's law of cooling.

$$q_{conv} = hA(T_\infty - T_s) \qquad [\text{W}] \qquad (1.28)$$

h = heat transfer coefficient [W/(m^2 K)]

The heat transfer coefficient, h, is expressed through correlations of Nusselt numbers (Nu).

$$Nu = \frac{hL}{k} = Nu(Re, \text{Pr}; \text{geometry}) \qquad (1.29)$$

L is the characteristic length of the object. Dimensional arguments show that the Nusselt number is a function of the Reynolds number (flow effects), Prandtl number (fluid properties) and flow geometry. For example, for turbulent flow over a flat plate, the average Nusselt number is given by

$$Nu_L = \frac{\bar{h}L}{k} = 0.037\, Re_L^{4/5}\, \mathrm{Pr}^{1/3} \tag{1.30}$$

Radiation heat transfer occurs due to photon energy, and is determined by the Planck distribution multiplied by the spectral emissivity, ε_λ.

$$q_{rad,\lambda} = \varepsilon_\lambda A \frac{C_1}{\lambda^5 \left[\exp\left(\frac{C_2}{\lambda T}\right) - 1\right]} \qquad [\mathrm{W}/\mu\mathrm{m}] \tag{1.31}$$

λ = wavelength [μm]
$C_1 = 2\pi h c_o^2 = 3.742 \times 10^8 \frac{\mathrm{W}\cdot\mu\mathrm{m}^4}{\mathrm{m}^2}$
$C_2 = hc_o/k = 1.439 \times 10^4\ \mu\mathrm{m}\cdot\mathrm{K}$

The Planck distribution in Eq. (1.31) can be integrated over the wavelength range, to yield the Stafan–Boltzmann law.

$$q_{rad} = \varepsilon A \sigma T^4 \quad [\mathrm{W}] \tag{1.32}$$

$$\sigma = 5.67 \times 10^{-8} \frac{\mathrm{W}}{\mathrm{m}^2\mathrm{K}^4}$$

Since surfaces typically both emit and receive radiation energy, and the emissivity and absorptivity are approximately (exactly at spectral, directional level) equal, the net radiation energy can be written as

$$q = \varepsilon \sigma A (T_\infty^4 - T_s^4) \tag{1.33}$$

1.5 Standard Atmospheric Air Properties

At sea level (zero altitude), the standard atmosphere air properties are as follows.

$$p_{REF} = 101325 \frac{\mathrm{N}}{\mathrm{m}^2} = 2116 \frac{\mathrm{lbf}}{\mathrm{ft}^2} = \text{pressure} \tag{1.34a}$$

$$T_{REF} = 288.15\mathrm{K} = 518.7^\circ\mathrm{R} = \text{temperature} \tag{1.34b}$$

$$\rho_{REF} = 1.225 \frac{\mathrm{kg}}{\mathrm{m}^3} = 0.07647 \frac{\mathrm{lbm}}{\mathrm{ft}^3} = \text{density} \tag{1.34c}$$

$$a_{REF} = 340.294 \frac{\mathrm{m}}{\mathrm{s}} = 1116 \frac{\mathrm{ft}}{\mathrm{s}} = \text{speed of sound} \tag{1.34d}$$

$$R_{air} = 287 \frac{\mathrm{J}}{\mathrm{kg}\cdot\mathrm{K}} = 53.34 \frac{\mathrm{ft}\cdot\mathrm{lbf}}{\mathrm{lbm}\cdot^\circ\mathrm{R}} = \text{specific gas constant} \tag{1.34e}$$

As the altitude increases from sea level, the atmospheric pressure decreases according to hydrostatics, that is, the weight of the air above. The variations of pressure and other air properties are tabulated in Appendix A. A simplified model can also be used to approximate the air pressure.

$$p = p_{REF}\left(1 - 0.0065\frac{h}{T_{REF}}\right)^{5.2561} \quad \text{for h} < \text{h}_{TP} = 11,000\,\text{m} \tag{1.35a}$$

$$p = p_{TP}e^{-\frac{g}{RT_{TP}}(h-h_{TP})} \quad \text{for h} > 11,000\,\text{m} \tag{1.35b}$$

$p_{TP} = 22573Pa$; $T_{TP} = 216.65K$; g = gravitational acceleration = 9.80665 m/s^2

The temperature profile is somewhat more complex, as shown in Figure 1.8. In the troposphere ($h = 1$ to 11 000 m), the temperature decreases linearly as a function of altitude.

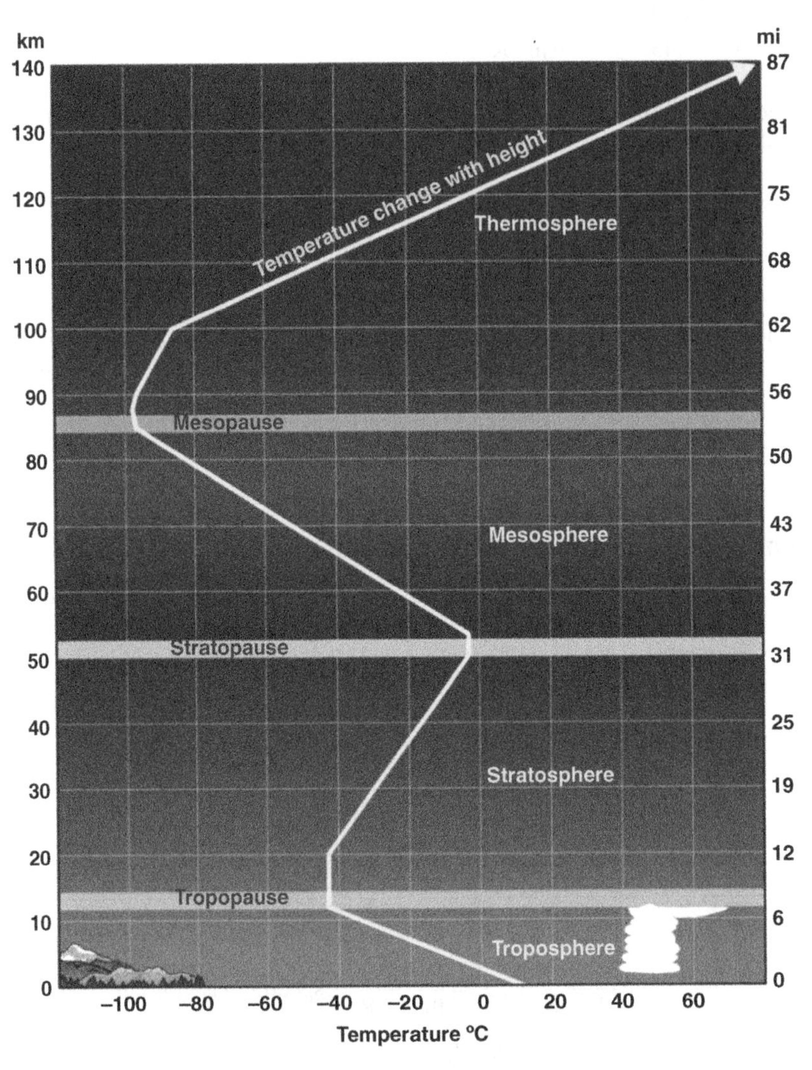

Figure 1.8 The temperature profile in the Earth's atmosphere. Courtesy of NASA.

Then the temperature becomes constant at $T_{TP} = 216.65\,\text{K}$ in the tropopause. This is the typical region for the long-range cruise altitude, so that change in altitude does not result in any appreciable change in temperature. Absence of temperature gradients also means that there is little air movement in the tropopause. Above the tropopause, the temperature begins to increase, due to the absorption of ultraviolet components of solar radiation by the ozone present in the stratosphere. The temperature again becomes constant at an altitude of approximately 50 km, and this region is called the stratopause. Above the stratopause, the temperature decreases again with increasing altitude in the region termed the mesosphere. At high altitudes above 100 km, the temperature rises again due to rarified, but finite amount of oxygen which absorbs ultraviolet radiation. This region is called the thermosphere.

1.6 Unit Conversion

In US industry, imperial units such as lbf, lbm and BTU are used in conjunction with the standard (SI) units. If some conversion relations are remembered, then accurate accounting of the units can be maintained. For example, 1 lbf (pound force) is the force that 1 lbm (pound mass) is subject to under the Earth's gravitational field, through $F = mg$.

$$1\text{lbf} = 1\text{lbm} \times 32.174\frac{\text{ft}}{\text{s}^2} \tag{1.36}$$

This is used to calculate the force due to momentum flow rates, as in Eq. (1.3). For example, for a mass flow rate of 100 lbm/s, and flow velocity difference of 250 ft/s (incident) and 500 ft/s (exit), then the thrust force is in the unit of lbm-ft/s^2, which is not a unit of force in the US unit system.

$$F = \dot{m}(U_{out} - U_{in}) = 100\frac{\text{lbm}}{\text{s}}(500 - 250)\frac{\text{ft}}{\text{s}} \tag{1.37}$$

The unit of force in the US unit system is lbf, which is obtained as follows.

$$1\ \text{lbm} \times \frac{\text{ft}}{\text{s}^2} = \frac{1\ \text{lbf}}{32.174} \tag{1.38}$$

Then, Eq. (1.37) becomes

$$F = \dot{m}(U_{out} - U_{in}) = 100\frac{\text{lbm}}{\text{s}}(500 - 250)\frac{\text{ft}}{\text{s}} = 25000\text{lbm}\frac{\text{ft}}{\text{s}^2} = \frac{25000}{32.174}\text{lbf} = 777\text{lbf} \tag{1.39}$$

Conversion of energy (such as thermal energy) to kinetic energy is somewhat more interesting. A common unit of thermal energy in the US unit system is the BTU (British thermal unit).

$$1\ \text{BTU} = 778.16\ \text{ft-lbf} \tag{1.40}$$

The following example illustrates the unit conversion.

Other commonly used conversion factors are given in Table 1.2.

Example 1.4 Conversion of energy units

For stagnation enthalpy of 850 BTU/lbm, find the flow speed when the static enthalpy is 375 BTU/lbm.

We can start from

$$h^o = h + \frac{U^2}{2} \rightarrow U = \sqrt{2(h^o - h)} \tag{E1.4.1}$$

The enthalpy difference of 850–375 = 475 BTU/lbm is equivalent to

$$475\frac{\text{BTU}}{\text{lbm}} = 475\frac{778.16\text{ft}\cdot\text{lbf}}{\text{lbm}} = 369626\frac{\text{ft}\cdot 32.174\frac{\text{lbm}\cdot\text{ft}}{\text{s}^2}}{\text{lbm}} = 1.19\times10^7\frac{\text{ft}^2}{\text{s}^2} \tag{E1.4.2}$$

Multiplying by 2 and taking the square root in Eq. (E1.4.1) gives us the flow speed of U = 4877 ft/s.

Table 1.2 Conversion factors.

Parameter	Metric	Metric/imperial
Acceleration	1 m/s^2 = 100 cm/s^2	1 m/s^2 = 3.2808 ft/s^2 1 ft/s^2 = 0.3048 m/s^2
Density	1 g/cm^3 = 1 kg/liter = 1000 kg/m^3	1 g/cm^3 = 62.428 lbm/ft^3 = 0.036 127 lbm/in^3 1 lbm/in^3 = 1728 lbm/ft^3 1 kg/m^3 = 0.062 428 lbm/ft^3 = 1.9404 × 10^{-3} slug/ft^3
Energy	1 kJ = 1000 J = 1000 N · m 1 kJ/kg = 1000 m^2/s^2 1 kWh = 3600 kJ 1 cal = 4.184 J 1 Cal = 4.1868 kJ	1 kJ = 0.94 782 BTU 1 BTU = 1.055 056 kJ = 5.40 395 psia · ft^3 = 778.169 lbf · ft 1 BTU/lbm = 25 037 ft^2/s^2 = 2.326* kJ/kg 1 kJ/kg = 0.430 BTU/lbm 1 kWh = 3412.14 BTU
Force	1 N = 1 kg · m/s^2 = 10^5 dyne 1 kgf = 9.80 665 N	1 N = 0.22 481 lbf 1 lbf = 32.174 lbm · ft/s^2 = 4.44 822 N
Length	1 m = 100 cm = 1000 mm = 10^6 μm 1 km = 1000 m 1 astronomical unit (au) = 1.496 × 10^{11} m	1 m = 39.370 in = 3.2808 ft = 1.0926 yd 1 ft = 12 in = 0.3048 m 1 mile = 5280 ft = 1.6093 km 1 in = 2.54 cm 1 nautical mile = 1852.0 m 1 mil = 0.001 in = 0.0 000 254 m
Mass	1 kg = 1000 g 1 metric ton = 1000 kg	1 kg = 2.2 046 226 lbm 1 lbm = 0.45 359 237 kg 1 ounce = 28.3495 g 1 slug = 32.174 lbm = 14.5939 kg 1 short ton = 2000 lbm = 907.1847 kg

Table 1.2 *(Continued)*

Parameter	Metric	Metric/imperial
Power	1 W = 1 J/s 1 kW = 1000 W = 1.341 hp 1 hp = 745.7 W	1 kW = 3412.14 BTU/h = 737.56 lbf ft/s 1 hp = 550 lbf · ft/s = 0.7068 BTU/s = 42.14 BTU/min = 2544.5 BTU/h = 0.74 570 kW 1 BTU/h = 1.055 056 kJ/h 1 ton of refrigeration = 200 BTU/min
Pressure	1 Pa = $1 N/m^2$ 1 kPa = 10^3 Pa = 10^{-3} MPa 1 atm = 101.325 kPa = 1.01 325 bars = 760 mmHg at 0 °C = 1.03 323 kgf/cm^2 1 mmHg = 0.1333 kPa	1 Pa = 1.4504 × 10^{-4} psia = 0.020 886 lbf/ft^2 1 psia = 144 lbf/ft^2 = 6.894 757 kPa 1 atm = 14.696 psia = 2116.2 lbf/ft^2 = 29.92 inHg at 30°F 1 inHg = 3.387 kPa
Specific heat	1 kJ/kg · °C = 1 kJ/kg · K = 1 J/g · °C	1 BTU/lbm · °F = 4.1868 kJ/kg · °C 1 BTU/lbmol · R = 4.1868 kJ/kmol · K 1 kJ/kg · °C = 0.23 885 BTU/lbm · °F = 0.23 885 BTU/lbm · R
Specific volume	1 m^3/kg = 1000 L/kg = 1000 cm^3/g	1 m^3/kg = 16.02 ft^3/lbm 1 ft^3/lbm = 0.062 428 m^3/kg
Temperature	T(K) = T(°C) + 273.15 ΔT(K) = ΔT(°C)	T(R) = T(°F) + 459.67 = 1.8T(K) T(°F) = 1.8T(°C) + 32 ΔT(°F) = ΔT(R) = 1.8ΔT(K)
Velocity	1 m/s = 3.60 km/h	1 m/s = 3.2808 ft/s = 2.237mi/h 1 km/h = 0.2278 m/s = 0.6214 mi/h = 0.9113 ft/s 1 mi/h = 1.46 667 ft/s 1 mi/h = 1.609 km/h 1 knot = 1.15 155 mi/h
Viscosity, dynamic	1 kg/m · s = 1 N · s/m^2 = 1 Pa · s = 10 poise 1 centipoise = 10^{-2} poise = 0.001 Pa · s	1 kg/m · s = 2419.1 lbf/ft · h = 0.020 886 lbf · s/ft^2 = 5.8016 × 10^{-6} lbf · h/ft^2
Viscosity, kinematic	1 m^2/s = 10^4 cm^2/s 1 stoke = 1 cm^2/s = 10^{-4} m^2/s	1 m^2/s = 10.764 ft^2/s = 3.875 × 10^4 ft^2/h 1 m^2/s = 10.764 ft^2/s
Volume	1 m^3 = 1000 liter (L) = 10^6 cm^3 (cc)	1 m^3 = 6.1024 × 10^4 in^3 = 35.315 ft^3 = 264.17 gal (US) 1 ft^3 = 0.028 321 m^3 = 28.32 L 1 US gallon = 231 in^3 = 3.7854 L = 4 qt = 8 pt 1 fl ounce = 29.5735 cm^3 = 0.0 295 735 L 1 US gallon = 128 fl ounces
Consumption (TSFC)	1 lbm/h/lbf = 28.33 mg/N · s	1 mg/N · s = 0.0 352 983 lbm/h/lbf
Specific Thrust	1 lbf/lbm/s = 9.807 N · s/kg	1 N · s/kg = 0.1 019 679 lbf/lbm/s
Gas constant R	1 ft^2/s^2 · R = 0.1672 m^2/s^2 · K	1 m^2/s^2 · K = 5.980 861 ft^2/s^2 · R

1.7 Problems

1.1. What is the operating principle of vertical take-off and landing aircraft? Use sketches as needed.

1.2. What is the operating principle of helicopter thrust? What are the rotor degrees of freedom? Use sketches as needed.

1.3. Describe the components and operating principles of early versions of Goddard's liquid propellant rockets.

1.4. Plot the standard atmospheric pressure and temperature as a function of altitude in US units.

1.5. Plot the standard atmospheric pressure and temperature as a function of altitude in SI units.

1.6. For a turbojet engine with air mass flow rate of $\dot{m} = 100\,\mathrm{lbm/s}$, calculate the thrust in lbf at a flight Mach number of $a_o = 899$ ft/s and jet exhaust velocity of 1500 ft/s?

1.7. In a nozzle with air entering at 112 m/s and static temperature of 650 K, calculate the exit velocity if the exit temperature is 300 K.

1.8. In a nozzle with air entering at 310 ft/s and static enthalpy of 556 BTU/lbm, calculate the exit velocity if the exit static enthalpy is 442 BTU.

Bibliography

Hill, Philip and Petersen, Carl (1992) *Mechanics and Thermodynamics of Propulsion*, 2nd edn, Addison-Wesley Publishing Company.

2

Principle of Thrust

2.1 Thrust Configurations

The role of a propulsion device, whether it is a propeller, a gas-turbine or a rocket, is to generate a thrust force in order to change the momentum of the attached vehicle. The momentum or the velocity of the vehicle will be altered following the conservation of momentum (Newton's second law).

$$M\frac{dU}{dt} = F \tag{2.1}$$

M = vehicle mass
U = vehicle velocity
F = thrust force

In the above equation, we assume that the vehicle mass stays constant. Therefore, the larger the thrust, F, the higher the vehicle acceleration will be. Conversely, the same principle of conservation of momentum states that force will be generated if the momentum change occurs through a propulsion device. To visualize this effect, we draw a generic propulsion device that encloses the fluid going through the device, as in Figure 2.1. For the time being, we need not concern ourselves with all the internal components of the device, only the fluid momentum change through the device. A gas-turbine engine works in the way shown schematically in Figure 2.1. Air is ingested at the inlet and exits at higher speed through the nozzle exit, thus creating a momentum increase.

For a device like a propeller, we can consider a stream-tube of the air passing across the propeller. Although this air stream is not contained in any enclosure, the air speed increases across this stream-tube, again creating higher momentum at the exit in excess of the momentum that entered the control volume. Schematics of propeller-driven and other propulsion configurations are shown in Figure 2.2. For a turbofan, typically there are two streams: one through the core section and the other through the fan section, as shown in Figure 2.2. Both sections are designed to generate excess momentum on the exiting sides.

Aerospace Propulsion, First Edition. T.-W. Lee.
© 2014 John Wiley & Sons, Ltd. Published 2014 by John Wiley & Sons, Ltd.
Companion Website: www.wiley.com/go/aerospaceprop

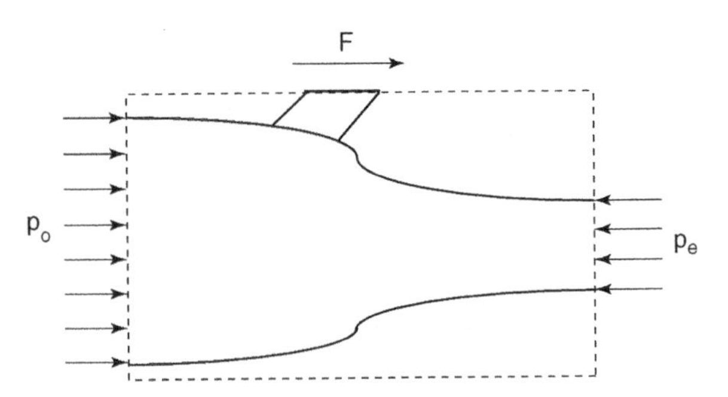

Figure 2.1 A generic propulsion device.

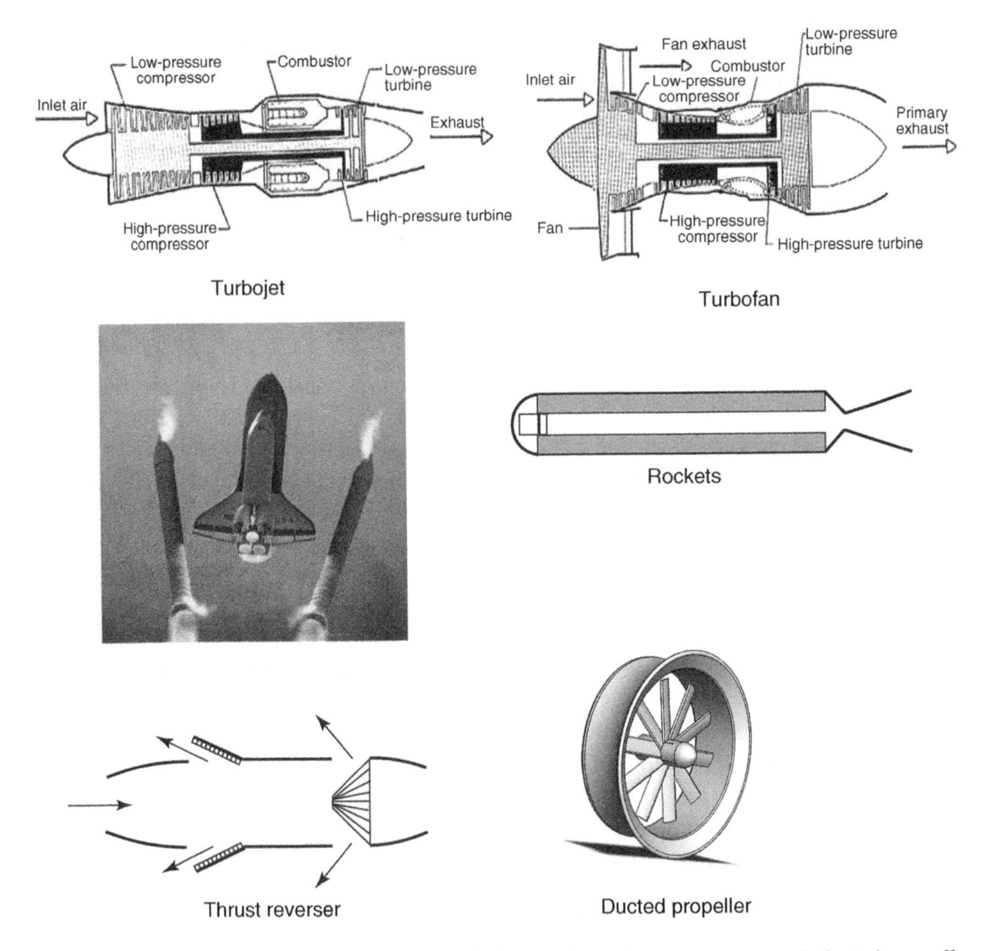

Figure 2.2 Various thrust configurations: turbofan; rocket; thrust reverser; and ducted propeller. Courtesy of NASA.

A gas-turbine engine with thrust reverse is also shown in Figure 2.2. A portion of the gas is sent back toward the front, and as the term "thrust reverse" suggests there will be a great decrease or reversal in the thrust force. In some cases, a vectored nozzle is used to direct the exhaust, to gain thrust components in offline directions. For a rocket, there is no inlet; just the exit. High enthalpy created in the combustion chamber is converted into kinetic energy through the nozzle, again resulting in excess momentum exiting the system.

2.2 Thrust Equation

Now, let us consider a fluid of mass, Δm, that enters the propulsion device as shown by the dotted box in Figure 2.1. If the inlet velocity of this fluid mass is U_{in}, then the momentum associated with the fluid mass would simply be the mass times its velocity: $\Delta m U_{in}$. This mass of fluid would take some time to transit through the device – let us call this transit time Δt. After Δt, if the propulsion device functions as intended, then it would have added some momentum to the fluid. By the time the fluid mass is exhausted through the nozzle, its velocity is U_{out}, and corresponding momentum is $\Delta m U_{out}$. The conservation of momentum can be verbally stated as

$$\text{(time rate of change of momentum)} = \text{(force)} \tag{2.2}$$

Since the momentum went from $\Delta m U_{in}$ to $\Delta m U_{out}$ in time Δt, we can write the left-hand side in Eq. (2.2) as $\frac{\Delta m(U_{out}-U_{in})}{\Delta t}$. Here, we can factor out the Δm since we are dealing with the same fluid mass. Moreover, the fluid mass divided by the transit time, $\Delta m/\Delta t$, is the mass flow rate. Thus, we can write Eq. (2.2) mathematically as follows.

$$\dot{m}(U_{out} - U_{in}) = F \tag{2.3}$$

$\dot{m}$ = mass flow rate of the fluid
F = thrust force

Example 2.1 For a turbojet engine with air mass flow rate of $\dot{m}$ = 164 lbm/s, flight speed of 500 ft/s and jet exhaust velocity of 1973 ft/s, what is the thrust in lbf?

U_{in} = flight speed and U_{out} = jet exhaust velocity, so that using Eq. (2.3),

$$\dot{m}(U_{out} - U_{in}) = F = 164\frac{\text{lbm}}{\text{s}}\left(1973\frac{\text{ft}}{\text{s}} - 500\frac{\text{ft}}{\text{s}}\right) = 241506\,\text{lbm}\frac{\text{ft}}{\text{s}^2} = 7506\,\text{lbf} \tag{E2.1.1}$$

Here, a conversion factor is used.

$$1\text{lbf} = 1\frac{\text{lbm}}{\text{s}}\left(32.174\frac{\text{ft}}{\text{s}^2}\right) = 32.174\,\text{lbm}\frac{\text{ft}}{\text{s}^2} \tag{E2.1.2}$$

Equation (2.3) states that if the exiting velocity is greater than the incoming, then it will generate a positive force. The greater the difference in the exiting and incoming velocity, the larger the thrust will be. The mass flow rate has the same effect, in that for a given velocity difference, a larger mass flow rate would enhance the thrust. The fluid can be air, water, combustion products; in fact, it does not even have to be fluid as long as it possesses momentum. For example, photons are massless particles but they do possess momentum so that photon momentum can be used to propel a spacecraft through very large "solar sails." Equation (2.3) is illustrative but omits the pressure effects at the inlet and the nozzle. Thus, in order to calculate the thrust force more accurately, we need to take a closer look at the conservation of momentum.

The conservation of momentum can first be stated verbally:

$$\begin{matrix}\text{time rate of change}\\ \text{of momentum inside a}\\ \text{control volume}\end{matrix} = \begin{matrix}\text{net rate of influx of}\\ \text{momentum into the}\\ \text{control volume}\end{matrix} + \begin{matrix}\text{net forces}\\ \text{acting on the}\\ \text{control volume}\end{matrix} \tag{2.4}$$

Equation (2.3) is an abbreviated version of the above conservation principle, with steady-state assumption and without inclusion of the pressure forces. We can visualize the terms in Eq. (2.4) more closely in Figure 2.2. Also, we can start expressing the terms in Eq. (2.4) in more precise, mathematical ways. For example, the momentum is mass times the fluid velocity, and in turn the mass can be written as density times volume. For a control volume, we can write the fluid momentum as an integral, and take the time derivative of this quantity to represent the left-hand side of Eq. (2.4).

$$\begin{matrix}\text{time rate of change}\\ \text{of momentum inside a}\\ \text{control volume}\end{matrix} = \frac{d(m\vec{U})_{CV}}{dt} = \frac{d}{dt}\int_{CV} \rho\vec{U}\,dV \tag{2.5}$$

As for the first term on the right-hand side, the net rate of momentum flux is the amount of momentum entering minus the amount exiting the control volume per unit time, and has the same unit as the mass flow rate times the velocity $\left[\frac{\text{kg}\cdot\text{m}}{\text{s}^2}\right]$. In an integral form, this takes on a "surface integral" since the fluxes can only happen across the control surfaces (CS).

$$\begin{matrix}\text{net rate of influx}\\ \text{of momentum into the}\\ \text{control volume}\end{matrix} = -\int_{CS} \rho\vec{U}(\vec{U}\cdot\vec{n})dA \tag{2.6}$$

The velocities are expressed as vectors and $\vec{n}$ is the normal vector, pointing outward from the surface elements. If we consider the x-component of the momentum, for example, it can easily be shown that the integral in Eq. (2.6) gives us the incoming and outgoing momentum flux terms with correct signs.

$$\begin{matrix}\text{net rate of influx}\\ \text{of momentum into the}\\ \text{control volume}\end{matrix} = \dot{m}_{in}U_{in} - \dot{m}_{out}U_{out} \tag{2.7}$$

Example 2.2 Show that the integral in Eq. (2.6) gives the net rate of momentum flux into the control volume.

$$
\begin{aligned}
-\int_{CS} \rho\vec{U}(\vec{U}\cdot\vec{n})dA &= -\left(\int_{in} \rho_{in}U_{in}(\vec{U}_{in}\cdot\vec{n}_{in})dA + \int_{out} \rho_{out}U_{out}(\vec{U}_{out}\cdot\vec{n}_{out})dA\right) \\
&= -\int_{in} \rho_{in}U_{in}(-U_{in})dA - \int_{out} \rho_{out}U_{out}(+U_{out})dA \\
&= \dot{m}_{in}U_{in} - \dot{m}_{out}U_{out}
\end{aligned}
$$

Note that the dot products give the inlet velocity with a negative sign, since the inlet velocity is at 180° from the normal vector at the inlet surface, and a positive sign for the exit velocity since the vectors are aligned in the same direction at the exit plane. Also, the density, velocity and area integrate to the mass flow rate.

Now, we just need to sum the forces in the third term in Eq. (2.4), and Figure 2.2 illustrates the manner in which external forces act on the control volume. First, the thrust force, F, will propel the vehicle forward and thus is drawn in the "forward" direction. A free-body diagram at the junction between the vehicle and the propulsion device will then show that an equal and opposite force acts on the control volume, as shown in Figure 2.2. The pressure force is isotropic, and acts on the surfaces in the normal "inward" direction. However, the pressure is due to the ambient pressure at all surfaces except obviously at surfaces that are at a different pressure. For a propulsion device, the pressure at the inlet plane is also assumed to be at ambient pressure, and only the exit plane may be at a different pressure due to nozzle expansion or other effects. Therefore, the pressure forces in the horizontal direction, as illustrated in Figure 2.2, will cancel one another except along the lines extended from the exit surface. The net pressure force is then $(p_a - p_e)A_e$ and acts in the positive x-direction (left to right). Here, p_a, p_e and A_e are the ambient pressure, exit pressure and exit area, respectively. Thus, the net forces in the x-direction can be expressed as follows:

$$
\begin{matrix}\text{net forces}\\ \text{acting on the}\\ \text{control volume}\end{matrix} = F + (p_a - p_e)A_e \tag{2.8}
$$

Now, we can summarize the conservation of momentum in the thrust direction (x-direction) by combining Eqs. (2.5), (2.7) and (2.8).

$$
\frac{d(m\vec{U})_{CV}}{dt} = \dot{m}_{in}U_{in} - \dot{m}_{out}U_{out} + F + (p_a - p_e)A_e \tag{2.9}
$$

Most of the time, we are interested in the steady-state thrust of the propulsion system, so we can set the left-hand side of the above equation to zero and solve for the thrust, F. In so doing, we also write $\dot{m}_{in} = \dot{m}_o$ and $\dot{m}_{out} = \dot{m}_o + \dot{m}_f = \dot{m}_o(1+f)$, where $\dot{m}_o$ is the inlet mass flow rate and f is the fuel–air ratio, that is, $f = \dot{m}_f/\dot{m}_o$.

$$
F = \dot{m}_o[(1+f)U_e - U_o] + (p_e - p_o)A_e \tag{2.10}
$$

Again, for incoming and exiting velocities, the subscripts have been replaced with those representing the inlet "o" and exit "e" conditions. We will call Eq. (2.10) the thrust equation, and we can use it for any jet-propulsion devices such as turbojet, turbofan and rocket engines.

If the fuel mass flow rate is small in comparison to the air mass flow rate ($f \ll 1$), and also if the pressure term in Eq. (2.10) is relatively small, then Eq. (2.10) boils down to the simple form initially obtained in Eq. (2.3). For a typical turbojet, Eq. (2.3) gives a good estimate of the thrust. For propulsion devices with multiple streams, such as turbofans, then the momentum differences need to be considered for different streams. The subscript "C" refers to the core stream (the part of the air that goes through the main gas-turbine engine) and "f" refers to the fan stream. Here, the ratio f is defined with respect to the core mass flow rate: $f = \dot{m}_f/\dot{m}_C$.

$$F = \dot{m}_C[(1+f)U_{eC} - U_o] + \dot{m}_{fan}(U_{e,fan} - U_o) + (p_{eC} - p_o)A_{eC} + (p_{e,fan} - p_o)A_{e,fan} \tag{2.11}$$

Example 2.3 GE90-85B turbofan engine used in the Boeing 777-200 has the following specifications.

M_o = flight Mach number = 0.83 at altitude of 35 kft.

$BPR = a$ = bypass ratio = 8.4

$$\dot{m}_o = 3037 \text{ lbm/s}$$

The core and fan exhaust velocities are estimated as U_{eC} = 1417 ft/s and $U_{e,fan}$ = 934 ft/s.

The thrust can be estimated using Eq. (2.11). f is usually small in turbofan engines and also the pressure forces can be neglected.

We first need the flight speed from the flight Mach number. At 35 kft, from the standard atmospheric air data, we have

$$\frac{T}{T_{ref}} = 0.75975 \rightarrow T_o = 394.1°\text{R} = -65.9°\text{F, from } \text{T}_{\text{ref}} = 518.7°\text{R}.$$

$$\frac{a}{a_{ref}} = 0.8795 \rightarrow a_o = 981.5 \text{ ft/s, from } a_{ref} = 1116 \text{ ft/s}.$$

So that $U_o = M_o a_o$ = 814.6 ft/s.

$$\text{Also, } \dot{m}_o = \dot{m}_C + \dot{m}_{fan}; \alpha = \frac{\dot{m}_{fan}}{\dot{m}_C} \rightarrow \dot{m}_C = \frac{1}{1+\alpha}\dot{m}_o; \dot{m}_{fan} = \frac{\alpha}{1+\alpha}\dot{m}_o \tag{E2.3.1}$$

$$F \approx \dot{m}_C(U_{eC} - U_o) + \dot{m}_{fan}(U_{e,fan} - U_o)$$
$$= 323\frac{\text{lbm}}{\text{s}}(1417 - 814.6)\frac{\text{ft}}{\text{s}} + 2714\frac{\text{lbm}}{\text{s}}(934 - 814.6)\frac{\text{ft}}{\text{s}} = 16118 \text{ lbf} \tag{E2.3.2}$$

$\dot{m}_C$ = mass flow rate of air through the core part of the engine
U_{eC} = core exhaust velocity
p_{eC} = pressure at the core exit
$\dot{m}_{fan}$ = mass flow rate of air through the fan section
$U_{e,fan}$ = fan exhaust velocity
$p_{e,fan}$ = pressure at the fan exit

For rocket engines, there is no incident momentum. Thus, the thrust equation becomes

$$F = \dot{m}_p U_e + (p_e - p_o)A_e \tag{2.12}$$

$\dot{m}_p$ = mass flow rate of the propellant

Thrust can be modified in some devices, such as the thrust reverser and vectored nozzles. In those instances, we need a vector form of the thrust equation. If n_e denotes the unit normal vector at the exit (Figure 2.2), then the vector form of the thrust equation is:

$$\vec{F} = \dot{m}_o\left[(1+f)\vec{U}_e - \vec{U}_o\right] + (p_e - p_o)A_e\vec{n}_e \tag{2.13}$$

Example 2.4 Let us consider the thrust of the space shuttle main engine (SSME), at burn rate of 1083 gallons/s of liquid oxygen (LOX) and hydrogen (LH_2). The mixture ratio LOX: LH_2 is 4: 1, with the mixture density of 0.28 g/cc.

For rockets, a convenient parameter called the equivalent exhaust velocity is often used, which combines the effect of jet momentum and pressure thrust in Eq. (2.12).

$$U_{eq} = U_e + \frac{(p_e - p_o)A_e}{\dot{m}_p} \tag{E2.4.1}$$

For SSME, U_{eq} = 14 520 ft/s.

Also, 1 gallon = 4.546 liter = 4.546×10^{-3} m^3 and 0.28 g/cc = 280 kg/m^3.

So the mass flow rate is

$$\dot{m}_p = \left(280\frac{\text{kg}}{\text{m}^3}\right)(1\ 083\ \text{gallons})\left(4.546 \times 10^{-3}\frac{\text{m}^3}{\text{gallon}}\right) = 1378.5\frac{\text{kg}}{\text{s}} \tag{E2.4.2}$$

$$F = \dot{m}_p U_{eq} = 6.1 \times 10^6\text{N} \tag{E2.4.3}$$

Example 2.5 Vectored thrust

The F119-PW100 turbofan engine used in F-22 Raptors has the following nominal specifications, along with ±20° thrust vectoring in the vertical plane.

$$\mathrm{F} = 35\ 000\ \mathrm{lbf\ at\ U_o} = 558\ \mathrm{ft/s\ and}\ \dot{m}_o = 464\frac{\mathrm{lbm}}{\mathrm{s}}$$

From the above, we can estimate the nominal exhaust.

$$U_e = \frac{F}{\dot{m}_o} + U_o = \frac{35\ 000\ \mathrm{lbf}\dfrac{32.174\ \mathrm{lbm}\frac{\mathrm{ft}}{\mathrm{s}^2}}{1\ \mathrm{lbf}}}{464\dfrac{\mathrm{lbm}}{\mathrm{s}}} + 558\frac{\mathrm{ft}}{\mathrm{s}} = 2985\frac{\mathrm{ft}}{\mathrm{s}}$$

$$F_x = \dot{m}_o(U_e \cos = \frac{35\ 000\ \mathrm{lbf}\dfrac{32.174\ \mathrm{lbm}\frac{\mathrm{ft}}{\mathrm{s}^2}}{1\ \mathrm{lbf}}}{464\dfrac{\mathrm{lbm}}{\mathrm{s}}} + 558\frac{\mathrm{ft}}{\mathrm{s}} = 2985\frac{\mathrm{ft}}{\mathrm{s}} \tag{E2.5.1}$$

If this exhaust is vectored to $\theta = +20°$, then the resulting thrust components are

$$F_x = \dot{m}_o(U_e \cos\theta - U_o) = 32\ 379\ \mathrm{lbf} \tag{E2.5.2}$$

$$F_y = \dot{m}_o(U_e \sin\theta - U_o) = 14\ 711\ \mathrm{lbf} \tag{E2.5.3}$$

2.3 Basic Engine Performance Parameters

The thrust equation shows that we can obtain larger thrust, either by increasing the exhaust velocity with respect to the incoming velocity or by increasing the mass flow rate. The latter can be achieved simply by making the engine bigger, so in some sense it is not a true improvement of the engine performance. In general, it is of interest to determine the thrust for a given mass flow rate entering the engine. To this end, specific thrust is used:

$$F_s = \text{specific thrust} = F/\dot{m}_o \tag{2.14a}$$

For dual-stream engines such as turbofans, it is customary to use the total mass flow rate ($\dot{m}_o = \dot{m}_C + \dot{m}_{fan}$) in Eq. (2.14a). Using Eq. (2.11), the specific thrust for turbofans (for $f \ll 1$ and negligible pressure thrusts) is

$$F_s = \frac{\dot{m}_C}{\dot{m}_o}(U_{eC} - U_o) + \frac{\dot{m}_{fan}}{\dot{m}_o}(U_{e,fan} - U_o) = \frac{1}{1+\alpha}(U_{eC} - U_o) + \frac{\alpha}{1+\alpha}(U_{e,fan} - U_o). \tag{2.14b}$$

$$\alpha = \text{bypass ratio} = \dot{m}_{fan}/\dot{m}_{fan}$$

For turbofans (again with $f \ll 1$ and negligible pressure thrusts), we can define a "mass-weighted average" exhaust velocity:

$$\bar{U}_e = \frac{\dot{m}_C U_{eC} + \dot{m}_{fan} U_{e,fan}}{\dot{m}_C + \dot{m}_{fan}} \tag{2.15}$$

Using the averaged velocity, a general expression for specific thrust is then

$$F_s = \bar{U}_e - U_o = \bar{U}_e - a_o M_o \tag{2.14c}$$

a_o = speed of sound at the ambient temperature
M_o = flight Mach number

Equation (2.14c) is applicable for turbojets and turbofans for $f \ll 1$ and negligible pressure thrusts.

Another important parameter is the fuel consumption normalized by the thrust that is be produced. This parameter is called thrust-specific fuel consumption (TSFC).

$$\text{TSFC} = \dot{m}_f / F \tag{2.16}$$

We can imagine that the fuel consumption would be related to the thermal efficiency of the propulsion engine. In thermodynamics, the thermal efficiency of power cycles is defined as the power output divided by the heat input. For gas-turbine engines used for generating shaft power, as in turboprops, the power output would be the product of the torque and the rotational speed. However, what is the power output of a propulsion engine that develops thrust through a high-momentum exhaust jet? One way to rate this power output is to determine the rate of kinetic energy that is being generated by the engine, while taking into account the fact that incident air carries some kinetic energy into the engine. Then, the heat input to the engine is the heat of combustion produced when fuel is consumed in the combustion chambers at the rate of $\dot{m}_f$. Therefore, the thermal efficiency of propulsion engine can be written as:

$$\eta_T = \frac{\text{engine power output}}{\text{heat input rate}} = \frac{(1+f)\dot{m}_o \frac{U_e^2}{2} - \dot{m}_o \frac{U_o^2}{2}}{\dot{m}_f h_{PR}} \tag{2.17}$$

h_{PR} = heat of combustion [kJ/kg or BTU/lbm]

The thermal efficiency refers to the effectiveness of the engine in developing the excess kinetic energy in the exhaust stream. However, the final desired outcome is motion of the aircraft at some flight speed, U_o, due to the developed thrust, F. Force times the velocity has the unit of power, and the "propulsion power" is product of the thrust and flight speed.

$$P_T = \text{propulsion power} = F U_o \tag{2.18}$$

Propulsion efficiency defines how much of the engine power output in Eq. (2.17) is converted to the propulsion power.

$$\eta_P = \frac{\text{propulsion power}}{\text{engine power output}} = \frac{F U_o}{(1+f)\dot{m}_o \frac{U_e^2}{2} - \dot{m}_o \frac{U_o^2}{2}} \tag{2.19}$$

We can also combine the two efficiencies above, as a measure of how much of the heat input is being converted to the final propulsion power.

$$\eta_o = \text{overall efficiency} = \eta_T \eta_P = \frac{FU_o}{\dot{m}_f h_{PR}} \tag{2.20}$$

For typical turbojets, for which the pressure term in the thrust can be neglected and $f \ll 1$, the propulsion efficiency can be shown to be a function only of the ratio of the flight speed and the exhaust speed.

$$\eta_P = \frac{2\dfrac{U_o}{U_e}}{1 + \dfrac{U_o}{U_e}} \quad \text{for turbojets} \tag{2.21a}$$

We can see from Eq. (2.20) that the propulsion efficiency will approach the maximum of one, if the flight speed is close to the exhaust speed. Obviously, the flight speed cannot exceed the exhaust speed, since the thrust will become negative. Conversely, if the flight speed is very

Example 2.6 For a turbojet engine with $\eta_T = 0.4$, $h_{PR} = 18\,400$ BTU/lbm, $U_o = 500$ ft/s, and $F_s = 120$ lbf/(lbm/s), determine U_e, TSFC and the propulsion and overall efficiencies.

Using $F = \dot{m}_o(U_e - U_o)$, we have

$$U_e = \frac{F}{\dot{m}_o} - U_o = F_s - U_o = 120\frac{\dfrac{\text{lbf}}{\text{lbm}}}{\text{s}}\frac{32.174\ \text{lbm}\dfrac{\text{ft}}{\text{s}^2}}{1\ \text{lbf}} - 500\frac{\text{ft}}{\text{s}} = 4360\frac{\text{ft}}{\text{s}} \tag{E2.6.1}$$

TSFC can be calculated using Eq. (2.22).

$$\begin{aligned} \text{TSFC} &= \frac{\dfrac{F}{\dot{m}_o} + 2U_o}{2\eta_T h_{PR}} = \frac{F_s + 2U_o}{2\eta_T h_{PR}} \\ &= \frac{120\dfrac{\dfrac{\text{lbf}}{\text{lbm}}}{\text{s}}\dfrac{32.174\ \text{lbm}\dfrac{\text{ft}}{\text{s}^2}}{1\ \text{lbf}} - 2\times 500\dfrac{\text{ft}}{\text{s}}}{2(0.4)18400\dfrac{\text{BTU}}{\text{lbm}}\dfrac{778.16\ \text{ft}\cdot\text{lbf}}{1\ \text{BTU}}} \times \frac{3,600\ \text{s}}{1\ \text{hr}} = 1.528\frac{\text{lbm/hr}}{\text{lbf}} \end{aligned} \tag{E2.6.2}$$

With $U_o/U_e = 0.115$ in Eq. (2.21a),

$$\eta_P = \frac{2\dfrac{U_e}{U_o}}{1 + \dfrac{U_e}{U_o}} = 0.2057 \tag{E2.6.3}$$

$$\text{So, } \eta_o = \eta_P \eta_T = 0.082. \tag{E.2.6.4}$$

much lower than the exhaust speed, then the conversion of the engine power output to propulsion power is not efficient. This is an important observation, in that the choice of the exhaust speed (or the engine type) will determine the optimum flight speed regime of the aircraft, that is, if the desired cruise speed is low, then an engine with lower exhaust speed would in general be a better choice.

For rockets with negligible pressure thrust, we can also obtain a simple expression for the propulsion efficiency.

$$\eta_P = \frac{2\dfrac{U_o}{U_e}}{1+\left(\dfrac{U_o}{U_e}\right)^2} \quad \text{for rockets} \tag{2.21b}$$

Example 2.7 A similar set of calculations in SI units is more straightforward.

Inputs: $M_o = 0.8$ at $H =$ altitude $= 10\,\text{km}$; $h_{PR} =$ fuel heating value $= 42\ 800\,\text{kJ/kg}$; $F = 50$ kN, air mass flow rate $= 45\,\text{kg/s}$; and fuel mass flow rate $= 2.65\,\text{kg/s}$.

Outputs: f, F_s, U_e, $TSFC$, $P_e =$ engine power output, η_P, η_T, and η_o.

$$f = \frac{\dot{m}_f}{\dot{m}_o} = \frac{2.65}{45} = 0.059 \tag{E2.7.1}$$

$$F_s = \frac{F}{\dot{m}_o} = 1111.1\frac{\text{N}}{\dfrac{\text{kg}}{\text{s}}} \tag{E2.7.2}$$

At the altitude of 10 000 m, $U_o = M_o a_o = 239.6\,\text{m/s}$.

$$U_e = \frac{F + \dot{m}_o U_o}{\dot{m}_o + \dot{m}_f} = \frac{50\ 000\ \text{kg}\dfrac{\text{m}}{\text{s}^2} + 45\dfrac{\text{kg}}{\text{s}}239.6\dfrac{\text{m}}{\text{s}}}{(45+2.65)\dfrac{\text{kg}}{\text{s}}} = 1275.6\frac{\text{m}}{\text{s}} \tag{E2.7.3}$$

Engine power output is given by

$$P_e = FU_o + \frac{\dot{m}_o}{2}(1+f)(U_e - U_o)^2 = \frac{(\dot{m}_o + \dot{m}_f)U_e^2}{2} - \frac{\dot{m}_o U_e^2}{2} = 37.55\,\text{MW} \tag{E2.7.4}$$

The contribution of fuel kinetic energy is small, can be neglected.

$$\frac{\dot{m}_f U_o^2}{2} = 76.1\ \text{kW} \tag{E2.7.5}$$

$$\eta_T = \frac{P_e}{\dot{m}_f h_{PR}} = \frac{37.55\ \text{MW}}{113.42\ \text{MW}} = 0.33 \tag{E2.7.6}$$

$$\eta_P = \frac{FU_o}{P_e} = \frac{50000 \times 239.6\ \text{W}}{37.55 \times 10^6\ \text{W}} = 0.32 \tag{E2.7.7}$$

$$\eta_o = \eta_P \eta_T = 0.106 \tag{E.2.7.8}$$

Example 2.8 Optimizing the overall efficiency in turbojets

We can see from Eq. (2.21a) that the propulsion efficiency would approach the maximum of one ($\eta_P \to 1$) as the flight speed is close to the exhaust speed ($U_o \to U_e$), for turbojets. Let us see where the maximum in the overall efficiency occurs.

From Eq. (2.20), $\eta_o = \eta_T \eta_P = \frac{\dot{m}_o\left(U_e^2 - U_o^2\right)}{2\dot{m}_f h_{PR}} \eta_P$, by using Eq. (2.17) for the thermal efficiency ($f \ll 1$). We already have the expression for the propulsion efficiency for turbojets (Eq. (2.20a)). Combining these expressions, we obtain

$$\eta_o = \frac{\dot{m}_o\left(U_e^2 - U_o^2\right)}{2\dot{m}_f h_{PR}} \times \frac{2\dfrac{U_o}{U_e}}{1 + \dfrac{U_o}{U_e}} = \frac{\dot{m}_o U_e^2}{\dot{m}_f h_{PR}} \frac{U_o}{U_e}\left(1 - \frac{U_o}{U_e}\right) \tag{E.2.8.1}$$

For a given engine operating condition, the mass flow rate, the exhaust velocity and the heat input can be considered constant. Then, from the above expression we find that the maximum in the overall efficiency occurs when $U_o = U_e/2$. The same analysis can be applied for turbofans, simply by using the mass-weighted average exhaust velocity, $\bar{U}_e$, with the identical result that the maximum in the overall efficiency for turbofans occurs at $U_o = \bar{U}_e/2$.

Using the averaged velocity, $\bar{U}_e$, we can examine the trends in the basic engine performance parameters for turbojets, low bypass-ratio (BPR) turbofans, high BPR turbofans and turboprops. Straight turbojets convert all of the available thermal energy into kinetic energy of the exhaust, so they have the largest exhaust velocity. Turbofans use small (low BPR) to large (high BPR) portions of the turbine power to turn the fans. The fans have a relatively low exhaust speed, and also removal of turbine power toward the fans leaves lower thermal energy to be converted to exhaust kinetic energy in the engine core. For that reason, low BPR turbofans have lower averaged exhaust velocity, $\bar{U}_e$, than turbojets, and high BPR yet lower. Turboprops convert nearly all of the turbine power into propeller power, so they have the lowest exit stream velocity. Propeller propulsion is addressed in Section 2.5. For the above reason, the specific thrust (Eq. (2.14c): $F_s = \bar{U}_e - U_o = \bar{U}_e - a_o M_o$) would be the highest for turbojets, and then decrease for low BPR turbofans, high BPR turbofans and turboprops, in that order. From Eq. (2.14c), we can also see that higher flight speed or the Mach number increases the momentum entering the engine, and therefore reduces the specific thrust. For typical gas-turbine engines, these effects can be plotted, as in Figure 2.3.

For TSFC, we can combine Eqs. (2.15), (2.16), (2.19) and (2.20) to find that

$$\text{TSFC} \approx \frac{F_s + 2a_o M_o}{2\eta_T h_{PR}}. \tag{2.22}$$

Equation (2.22) shows that the higher specific thrust, for turbojets for example, comes at the price of higher fuel consumption. Also, TSFC increases to overcome higher incident momentum at high flight Mach numbers. On the other hand, lower speed of sound at higher altitudes would have the effect of reducing TSFC at a given Mach number. More efficient

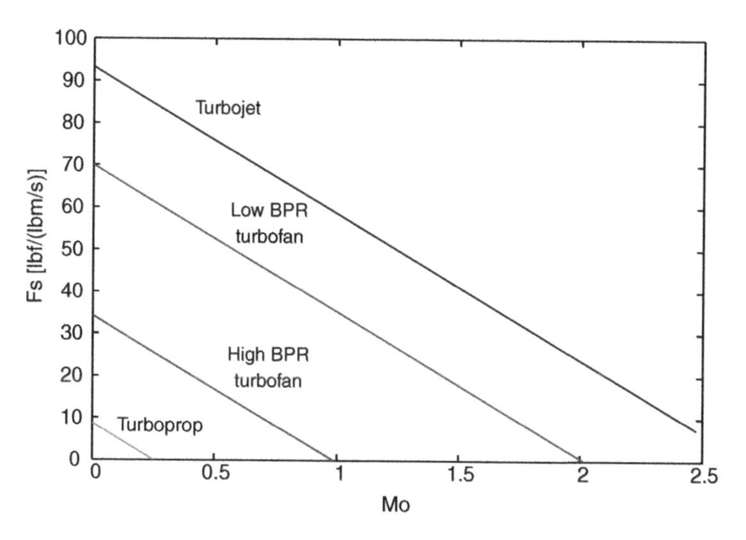

Figure 2.3 Specific thrust for typical gas-turbine engines.

Example 2.9 MATLAB® program for basic engine operating parameters

The MATLAB® program (MCE29) to compute the specific thrust and TSFC as a function of the flight Mach number is given at the end of this chapter. The results are shown in Figures E2.9.1 and Figure E2.9.2.

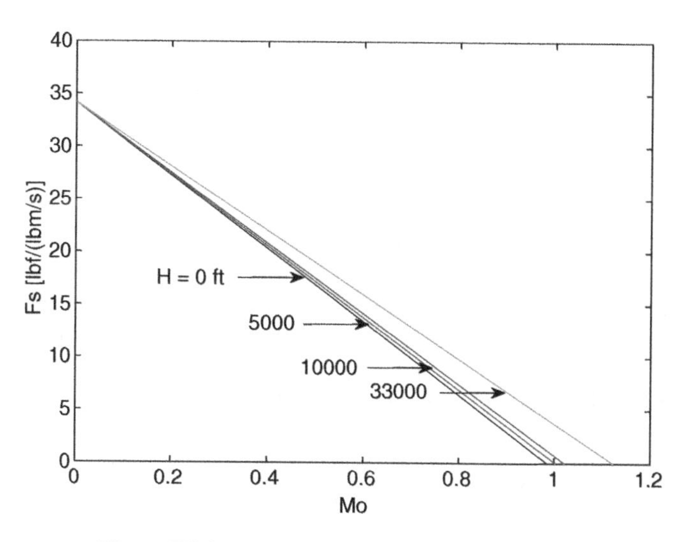

Figure E2.9.1 Specific thrust at various altitudes.

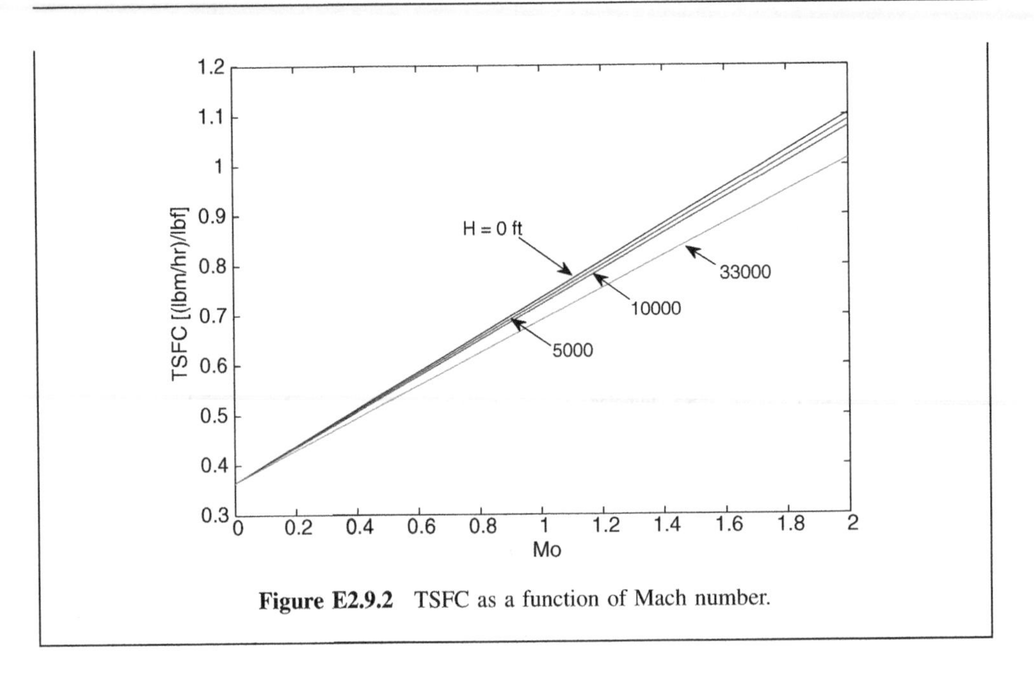

Figure E2.9.2 TSFC as a function of Mach number.

engines, with higher thermal efficiency, would decrease TSFC, as would high-yield fuels with large h_{PR}. However, for most gas-turbine engines for aircraft, different types of jet fuels have more or less the same heating value (h_{PR}).

2.4 Propulsion and Aircraft Performance

Let us look at how some of the propulsion parameters affect aircraft performance. The energy balance for an aircraft of mass m can be written as

$$(F - D)U_o = mg\frac{dH}{dt} + m\frac{d}{dt}\left(\frac{U_o^2}{2}\right) \tag{2.23}$$

D = drag force
H = altitude
g = gravitational acceleration (9.81 m/s^2)

$(F - D)$ is the net force in the direction of flight, so that the left-hand side of Eq. (2.23) represents the power input to the aircraft. The power is used to increase the potential energy (the first term on the right-hand side) and/or the kinetic energy of the aircraft. Under steady climb, the flight speed is constant. The rate of climb is then

$$\text{Rate of climb} = \frac{dH}{dt} = \frac{(F - D)U_o}{mg} \tag{2.24}$$

Thus, the net thrust of the aircraft has a direct effect on the rate of climb.

Example 2.10 Rate of climb

The Airbus A380 has four turbofan engines, developing a nominal thrust of 1 061 800 N at sea level, for take-off mass of 540 000 kg. The aircraft speed is 110 m/s for an air density of 1.16 kg/m^3. Wing area is 858 m^2, and the drag polar is given by

$$C_D = 0.02 + 0.042C_L^2 \quad \text{(E2.10.1)}$$

We can estimate the rate of climb based on these numbers.

First, the required lift coefficient is

$$C_L = \frac{2mg}{\rho U_o^2 A} = 0.876 \quad \text{(E2.10.2)}$$

From Eq. (2.10.1), $C_D = 0.0523$.

$$\text{Thus, the drag is } D = C_D \frac{1}{2}\rho U_o^2 A = 316\ 000\ \text{N} \quad \text{(E2.10.3)}$$

Now we can use Eq. (2.24).

$$\frac{dH}{dt} = \frac{(F-D)U_o}{mg} = \frac{(1{,}061{,}800 - 316{,}340)(110)}{(540{,}000)(9.81)} = 15.5\frac{\text{m}}{\text{s}} \quad \text{(E.2.10.4)}$$

This is equal to 3058 ft/min of climb rate.

We can also see from Eq. (2.24) that a decrease in thrust, at higher altitude or due to higher flight speed, will result in a lower climb rate. In addition, higher flight speed also increases the drag force.

We can see from Eq. (2.21a) that the propulsion efficiency will approach the maximum of one ($\eta_P \rightarrow 1$) as the flight speed is close to the exhaust speed ($U_o \rightarrow U_e$), for turbojets. Let us see where the maximum in the overall efficiency occurs.

From Eq. (2.20), by using Eq. (2.17) for the thermal efficiency ($f \ll 1$). We already have the expression for the propulsion efficiency for turbojets (Eq. (2.20a)). Combining these expressions, we obtain

$$\eta_o = \frac{\dot{m}_o\left(U_e^2 - U_o^2\right)}{2\dot{m}_f h_{PR}} \times \frac{2\frac{U_o}{U_e}}{1 + \frac{U_o}{U_e}} = \frac{\dot{m}_o U_e^2}{\dot{m}_f h_{PR}} \frac{U_o}{U_e}\left(1 - \frac{U_o}{U_e}\right) \quad \text{(E.2.7.1)}$$

For a given engine operating condition, the mass flow rate, the exhaust velocity and the heat input can be considered constant. Then, from the above expression we find that the maximum in the overall efficiency occurs when $U_o = U_e/2$. The same analysis can be applied for turbofans, simply by using the mass-weighted average exhaust velocity, $\bar{U}_e$, with the identical result that the maximum in the overall efficiency for turbofans occurs at $U_o = \bar{U}_e/2$.

For take-off considerations, the thrust at static conditions ($U_o \approx 0$) is useful.

$$\text{static thrust} = F_o = \dot{m}_o U_e \tag{2.25}$$

For an aircraft of mass m, the vehicle acceleration is

$$m\frac{dU_o}{dt} = F - D - \mu(mg - L) \tag{2.26a}$$

L = lift force
μ = coefficient of rolling friction

The right-hand side is the sum of all the forces in the direction of take-off motion. Using the lift and drag coefficients, the vehicle acceleration can be rewritten:

$$\frac{dU_o}{dt} = \left(\frac{F}{m} - \mu g\right) - \frac{1}{m}\left[\frac{1}{2}\rho A(C_D - \mu C_L)\right]U_o^2 \tag{2.26b}$$

ρ = air density
A = reference area for the lift and drag
C_L = lift coefficient
C_D = drag coefficient

But thrust, F, is also a function of the aircraft speed, U_o.

$$F = \dot{m}_o(U_e - U_o) = F_o - \rho A_o U_o^2 \tag{2.27}$$

A_o = inlet area

So Eq. (2.26b) becomes

$$\frac{dU_o}{dt} = \left(\frac{F_o}{m} - \mu g\right) - \frac{1}{m}\left[\frac{1}{2}\rho A(C_D - \mu C_L) + \rho A_o\right]U_o^2 = A - BU_o^2 \tag{2.26c}$$

$$A = \left(\frac{F_o}{m} - \mu g\right)$$

$$B = \frac{1}{m}\left[\frac{1}{2}\rho A(C_D - \mu C_L) + \rho A_o\right]$$

Equation (2.26c) shows that under take-off conditions, the static thrust would contribute directly to the vehicle acceleration, while large drag coefficient and inlet area would deduct from the vehicle acceleration. The take-off speed is typically set at 1.2 times the stall speed.

Using $U_{TO} = 1.2U_{stall}$, and integrating Eq. (2.26c) from $t = 0$, $U_o = 0$, we obtain the time to reach the take-off speed is

$$\text{time to reach take-off speed} = t_{TO} = \frac{1}{\sqrt{AB}} \frac{1}{\tanh\left(\sqrt{\frac{A}{B}} U_{TO}\right)} \tag{2.28}$$

A more important consideration than the time is the take-off distance. By rewriting $dU_o/dt = U_o(dU_o/ds)$, we can integrate Eq. (2.26c) for the take-off distance.

$$\text{take-off distance} = s_{TO} = \frac{1}{2B} \ln \frac{A}{A - BU_{TO}} \tag{2.29}$$

We see that the static thrust and the engine dimensions affect the take-off thrust through the parameters A and B.

During cruise (level, constant speed), the thrust and drag forces are equal, as are the aircraft weight and lift.

$$F = D = \frac{L}{L/D} = \frac{Mg}{L/D} \tag{2.30}$$

The thrust power is then

$$FU_o = \frac{MgU_o}{L/D} = \eta_o \dot{m}_f h_{PR} \tag{2.31}$$

The latter part of Eq. (2.31) is obtained by using the definition of the overall efficiency (Eq. (2.20)). The vehicle mass is not constant, and decreases as the fuel is consumed.

$$\frac{dM}{dt} = \frac{dM}{ds}\frac{ds}{dt} = U_o \frac{dM}{ds} = -\dot{m}_f \tag{2.32}$$

Solving for the fuel mass flow rate in Eq. (2.31) and using it in Eq. (2.32), we get an expression for the rate of change of vehicle mass as a function of the distance.

$$\frac{dM}{ds} = -\frac{Mg}{\eta_o h_{PR}(L/D)} \tag{2.33}$$

The above equation can be integrated from the initial to the final vehicle mass to give the aircraft range.

$$\text{aircraft range} = R = \eta_o \frac{C_L}{C_D} \ln \frac{M_i}{M_f} \frac{h_{PR}}{g} \tag{2.34a}$$

M_i = initial mass
M_f = final mass

Using Eq. (2.20) back in the above expression gives us

$$R = \frac{C_L}{C_D} \ln \frac{M_i}{M_f} \text{TSFC} \frac{U_o}{g} \tag{2.34b}$$

Thus, optimizing the overall efficiency (e.g. $U_o = \frac{U_e}{2}$) has direct influence on maximizing the range, as would the amount of fuel. Jettisoning non-critical mass from the aircraft also increases the range, by reducing the final vehicle mass, M_f. TSFC, the thrust-specific fuel consumption, is the key parameter in determining the range.

2.5 Propeller Propulsion

For conventional (unducted) propellers, we need a slightly different method to determine their thrust. Look at the control volume around a stream-tube caused by the propeller, in Figure 2.4. Due to the acceleration of the flow, the streamlines become narrow at the exit, and some air is entrained into the control volume. If we call the volumetric flow rate of this entrainment, ΔQ, then the mass balance for constant air density ($\rho = \text{const}$) is:

$$A_o U_o + \Delta Q = A_e U_e + (A_o - A_e) U_o \quad (\rho = \text{const}) \tag{2.35a}$$

So, at the exit plane of the control volume in Figure 2.4, the flow speed is U_o, except in the stream-tube, where the speed is U_e. Other notations are as shown in Figure 2.4.

Solving for ΔQ, we obtain

$$\Delta Q = A_e (U_e - U_o). \tag{2.35b}$$

Now we apply the momentum equation (Eq. (2.9)) for steady-state thrust, for the control volume in Figure 2.4.

$$\begin{aligned} F &= \dot{m}_{out} U_{out} - \dot{m}_{in} U_{in} \\ &= \rho [A_e U_e^2 + (A_o - A_e) U_o^2] - \rho A_o U_o^2 - \rho \Delta Q U_o \end{aligned} \tag{2.36a}$$

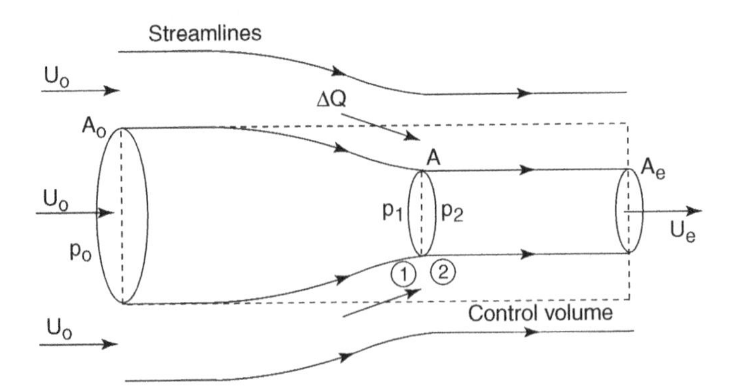

Figure 2.4 Control volume analysis for propeller thrust.

After using the result for ΔQ (Eq. (2.35b)) in Eq. (2.36a) and simplifying, we get an expression for the thrust.

$$F = \rho A_e U_e (U_e - U_o) \tag{2.36b}$$

2.6 MATLAB® Program

MCE29

```
%MCE29: Basic Aircraft Propulsion Parameters
% Specific thrust, TSFC as a function of flight Mach no and altitude
%
% Fs=specific thrust[ lbf/(lbm/s)] (Eq.2.14c)
% TSFC=TSFC[ (lbm/hr)/lbf] (Eq.2.22)
% H= altitude[ ft]
% Temperature model for standard atmosphere
% T=Tref-0.0036H; Tref=518.67 oR
Tref=518.67;
% a= speed of sound =sqrt(gamma* R* T); R=gas constant=1716.2[ ft^2/(s^2oR)]
% gamma=cp/cv=1.4
gamma=1.4;
R=1716.2;
% Start of calculations
H=0;
T=Tref-0.0036* H;
a=sqrt(gamma* R* T);
%1=turbojet (Ue=3000); 2=low bypass ratio turbofan (Ue=2250)
%3=high bypass ratio turbofan (Ue =1500); 4=turboprop (Ue=1000)
Ue1=3000; Ue2=2250; Ue3=1100; Ue4=280;
%hT=thermal efficiency; hPR=fuel heating value[ kJ/kg]
hT=0.38; hPR=18400;
for i=1:100;
  % Mo from 0 to 2.5;
  Mo(i)=(i-1)* (2.5/100);


  Fs1(i)=(1/32.174)* (Ue1-a* Mo(i));
  Fs2(i)=(1/32.174)* (Ue2-a* Mo(i));
  Fs3(i)=(1/32.174)* (Ue3-a* Mo(i));
  Fs4(i)=(1/32.174)* (Ue4-a* Mo(i));

  TSFC1(i)=(Fs1(i)+(1/32.174)*2* a* Mo(i))/(2* hT* hPR* 778.16)* 3600* 32.174;
  TSFC2(i)=(Fs2(i)+(1/32.174)*2* a* Mo(i))/(2* hT* hPR* 778.16)* 3600* 32.174;
  TSFC3(i)=(Fs3(i)+(1/32.174)*2* a* Mo(i))/(2* hT* hPR* 778.16)* 3600* 32.174;
  TSFC4(i)=(Fs4(i)+(1/32.174)*2* a* Mo(i))/(2* hT* hPR* 778.16)* 3600* 32.174;

end
%Altitude effect
H1=0; H2=5000; H3=10000; H4=33000;
T1=Tref-0.0036* H1
```

```
T2=Tref-0.0036*H2
T3=Tref-0.0036*H3
T4=Tref-0.0036*H4
a1=sqrt(gamma*R*T1);
a2=sqrt(gamma*R*T2);
a3=sqrt(gamma*R*T3);
a4=sqrt(gamma*R*T4);
for i=1:100;
    % Mo from 0 to 2.5;
    Mo(i)=(i-1)*(2.5/100);


     Fs11(i)=(1/32.174)*(Ue3-a1*Mo(i));
     Fs12(i)=(1/32.174)*(Ue3-a2*Mo(i));
     Fs13(i)=(1/32.174)*(Ue3-a3*Mo(i));
     Fs14(i)=(1/32.174)*(Ue3-a4*Mo(i));

  TSFC11(i)=(Fs11(i)+(1/32.174)*2*a1*Mo(i))/(2*hT*hPR*778.16)*3600*32.174;
  TSFC12(i)=(Fs12(i)+(1/32.174)*2*a2*Mo(i))/(2*hT*hPR*778.16)*3600*32.174;
  TSFC13(i)=(Fs13(i)+(1/32.174)*2*a3*Mo(i))/(2*hT*hPR*778.16)*3600*32.174;
  TSFC14(i)=(Fs14(i)+(1/32.174)*2*a4*Mo(i))/(2*hT*hPR*778.16)*3600*32.174;

end
plot(Mo,Fs1,Mo,Fs2,Mo,Fs3,Mo,Fs4)
plot(Mo,TSFC1,Mo,TSFC2,Mo,TSFC3,Mo,TSFC4)
plot(Mo,Fs11,Mo,Fs12,Mo,Fs13,Mo,Fs14)
plot(Mo,TSFC11,Mo,TSFC12,Mo,TSFC13,Mo,TSFC14)
```

2.7 Problems

2.1. For a turbojet engine with air mass flow rate of $\dot{m} = 164$ lbm/s, plot the thrust in lbf as a function of the flight Mach number for $a_o = 899$ ft/s and jet exhaust velocity of 1973 ft/s?

2.2. The F119-PW-100 turbofan engine generates a nominal 35 000 lbf, at mass flow rate = 170 lbm/s and $U_o = 558$ ft/s.

a. Estimate the exhaust velocity, neglecting pressure forces.
b. If the exhaust is vectored at 20° from horizontal, what is the resultant thrust in the x- and y-directions?

2.3. For GE J-57 turbojet, the following parameters are given:

$p_O = 14.696$ psia
$T_O = 518.7$ oR
static thrust = F (@$U_o = 0.0$) = 10 200 lbf
air mass flow rate = 164 lbm/s
fuel mass flow rate = 8520 lbm/hr
$p_e = p_O$

a. Determine U_e.
b. For $M_o = 0.5$, what is the thrust if U_e remains the same as above?
c. What is the propulsion efficiency?
d. If $h_{PR} = 18\,400$ BTU/lbm, what are the thermal and overall efficiencies?

2.4. A turbofan engine has the following specifications.
M_o = flight Mach number = 0.83 at altitude of 35 kft

$$\dot{m}_o = 2037\ \text{lbm/s}$$

The core and fan exhaust velocities are estimated as $U_{eC} = 1417$ ft/s and $U_{e,fan} = 934$ ft/s. Plot the thrust as a function of the bypass ratio.

2.5. Calculate the propulsive efficiency of a turbojet engine under the following two flight conditions: (a) $U_o = 100$ m/s and $U_e = 2000$ m/s; and (b) $U_o = 750$ m/s and $U_e = 2000$ m/s.

2.6. Determine the thrust of a turbofan engine (see the figure below), for the following conditions. Include the pressure terms.

	1	7	9
P [kPa]	101.3	75.2	101.3
T [K]	225	1020	405
flow area [m^2]	5.75	3.25	1.25
mass flow rate [kg/s]	927	627	300

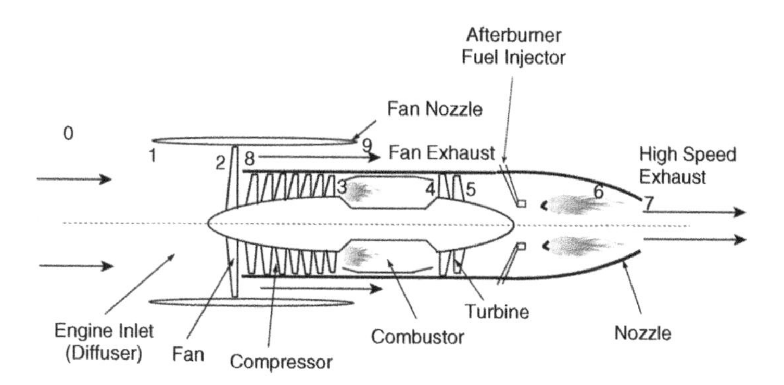

2.7. Consider the use of a turbojet engine for reverse thrust. If the aircraft equipped with a turbojet engine is rolling on the runway at 160 km/h, and the engine mass flow rate is 55 kg/s with a jet exhaust velocity of 175 m/s. What is the forward thrust of this engine? What are the magnitude and direction of the thrust (forward or reverse) if the jet exhaust is deflected by 90°?

2.8. One method of reducing an aircraft landing distance is through the use of thrust reversers. Consider the following thrust reverser configuration and specs, where the all of the fan bypass air is turned at a reverse 60° from the horizontal. Determine the magnitude and direction of the net horizontal force on the strut.

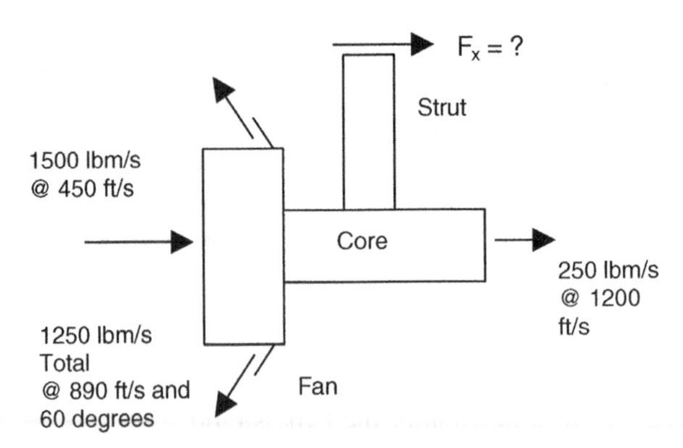

2.9. For a turbojet engine with $\eta_T = 0.35$, $h_{PR} = 18\ 400$ BTU/lbm, $U_o = 650$ ft/s, and $F_s = 120$ lbf/(lbm/s), determine U_e, TSFC and the propulsion and overall efficiencies.

2.10. A turbojet engine has a compression ratio of 25 and turbine inlet temperature of 1600 K. For $M_o = 0.75$, find (a) specific thrust; (b) TSFC; (c) thermal efficiency; and (d) propulsion efficiency.

2.11. For a GE CF-6-50C2 engine, write a MATLAB® (or equivalent) program to calculate and plot the thrust, propulsion and overall efficiency as a function of the Mach number at the sea level, by using a constant air mass flow rate of 1476 lbm/s and TSFC of 0.63 (lbm/hr)/lbf. It is assumed that the ratio of the fan and the core exhaust velocity is 0.35. Then, the take-off thrust (52 500 lbf) can be used to determine the exhaust velocities for $U_o = 0$. For a constant flight Mach number of 0.8, plot the above parameters as a function of the altitude from 0 to 50 000 ft. Assume that the air mass flow rate scales inversely proportional to the altitude, while exhaust velocities remain constant. The bypass ratio is 4.31.

Bibliography

Hill, Philip and Petersen, Carl (1992) *Mechanics and Thermodynamics of Propulsion*, 2nd edn, Addison-Wesley Publishing Company.

3

Basic Analyses of Gas-Turbine Engines

3.1 Introduction

Tables 3.1 and 3.2 list some of the basic engine operating parameters for civil (Table 3.1) and military (Table 3.2) gas-turbine power plants. We saw in Chapter 2 that we need parameters such as the mass flow rates of air and fuel, flight speed and exhaust velocity of gas-turbine engines to find their thrust and specific fuel consumptions. In addition, we would like to determine propulsion and thermal efficiencies. A brief review of gas-turbine engines treated as a thermodynamic power cycle, or the so-called Brayton cycle, will show that the thermal efficiency of gas-turbine engines depends on one of its operating parameters, the compression ratio. In this chapter, we will apply the conservation equations of mass and energy, to perform basic analyses of various types of gas-turbine engines.

3.2 Gas-Turbine Engine as a Power Cycle (Brayton Cycle)

The gas-turbine engine can be viewed as a power cycle, called the Brayton cycle from classical thermodynamics, where the heat input is converted to a net power output. A simplified model of the gas-turbine power cycle is shown in Figure 3.1. It consists of the following processes.

1. Isentropic compression of air through the compressor
2. Constant-pressure heat addition in the combustor
3. Isentropic expansion through the turbine, that generates shaft power
4. Isentropic acceleration of the exhaust gas through the nozzle, into the ambient

Sometimes a link between steps 4 and 1 is included to make this cycle a "closed" cycle, but for our purposes the "open" cycle shown in Figure 3.1 is more realistic. The isentropic compression, in Figure 3.1, goes from state 0 to 3°, and the nozzle expansion from state 4° to 7. The superscript, "°", refers to the stagnation condition, and the reason the numbering of the states is not sequential is to make the number notation consistent with that of the full gas-turbine

Aerospace Propulsion, First Edition. T.-W. Lee.
© 2014 John Wiley & Sons, Ltd. Published 2014 by John Wiley & Sons, Ltd.
Companion Website: www.wiley.com/go/aerospaceprop

Table 3.1a Basic engine operating parameters for civil gas-turbine engines.

Model no.	Manufacturer	Takeoff				Cruise				Aircraft
		Thrust [lbf]	BR	OPR	Airflow [lbm/s]	Alt. (kft)	M_0	Thrust [lbf]	TSFC [(lbm/hr)/lbf]	
CF6-50C2	General Electric	52 500	4.31	30.4	1476	35	0.80	11 555	0.630	DC10-10, A300B, 747-200
CF6-80C2	General Electric	52 500	5.31	27.4	1.650	35	0.80	12 000	0.576	767-200,-300,-200ER
GE90-B4	General Electric	87 400	8.40	39.3	3.037	35	0.80	17 500		777
JT8D-15A	Pratt and Whitney	15 500	1.04	16.6	327	30	0.80	4920	0.779	727, 737, DC9
JT9D-59A	Pratt and Whitney	53 000	4.90	24.5	1.639	35	0.85	11 950	0.646	DC10-40, A300B, 747-200
PW2037	Pratt and Whitney	38 250	6.00	27.6	1.210	35	0.85	6500	0.582	757-200
PW4052	Pratt and Whitney	52 000	5.00	27.5	1700					767, A310-300
PW4084	Pratt and Whitney	87 900	6.41	34.4	2550	35	0.83			777
CFM56-3	CFM International	23 500	5.00	22.6	655	35	0.85	4890	0.667	737-300, -400, -500
CFM56-5C	CFM International	31 200	6.60	31.5	1.027	35	0.80	6600	0.545	A340
RB211-524B	Rolls Royce	50 000	4.50	28.4	1.513	35	0.85	11 000	0.643	L1011-200, 747-200
RB211-535E	Rolls Royce	40 100	4.30	25.8	1.151	35	0.80	8495	0.607	757-200
RB211-882	Rolls Royce	84 700	6.01	39.0	2.640	35	0.83	16 200	0.557	777
V2528-D5	International Aero Engines	28 000	4.70	30.5	825	35	0.80	5773	0.574	MD-90
ALF502R-5	Textron Lycoming	6970	5.70	12.2		25	0.70	2250	0.720	BAe 146-100, -200
TFE731-5	Garrett	4500	3.34	14.4	140	40	0.80	986	0.771	BAe 125-800
PW300	Pratt and Whitney Canada	4750	4.50	23.0	180	40	0.80	1113	0.675	BAe 1000
FJ44	William Rolls	1900	3.28	12.8	63.3	30	0.70	600	0.750	
OLYMPUS 593	Rolls Royce/SNECMA	38 000	0	*11.3	410	53	2.00	10 030	1.190	Concorde

OPR = overall pressure ratio; TSFC = thrust specific fuel consumption; BR = bypass ratio.

Table 3.1b Basic engine operating parameters for military gas-turbine engines.

Data for some military turbofan engines											
Model no.	Thrust [lbf]	TSFC [(lbm/hr)/lbf]	Airflow [lbm/s]	OPR	D [in]	L [in]	Weight [lb]	Turbine Inlet Temp. [°F]	FPR	BR	Aircraft
F100-PW-229	29 000/17 800	2.05/0.74	248	23.0	47	191	3036	2700	3.8	0.4	F-15, F-16
F101-GE-102	30 780/17 390	2.460/0.562	356	26.8	55.2	180.7	4448	2550	2.31	1.91	B-1B
F103-GE-101	51 711	0.399	1476	30.2	86.4	173	8768	2490		4.31	KC-10A
F107-WR-101	635	0.685	13.6	13.8	12	48.5	141		2.1	1.0	air launch cruise missile
F108-CF-100	21 364	0.363	785	23.7	72	115.4	4610	2228	1.5	6.0	KC-135R
F110-GE-100	28 620/18 330	2.08/1.47	254	30.4	46.5	182	3895		2.98	0.80	F-16
F117-PW-100	41 700	0.33		31.8	84.5	146.8	7100			5.8	(PW2040) C-17A
F118-GE-100	19 000										B-2
F404-GE-FID	10 000			25	34.5	87	1730				F-1117A
F404-GE-400	16 000			25	35	159				0.34	F-18, F-5G
JT3D-3B	18 000	0.535	458	13.6	53	136.4	4300	1600	1.74	1.37	(TF33-102) EC/RC-135
JT8D-7B	14 500	0.585	318	16.9	45	123.7	3252	1076		1.03	C-22, C-9, T-43A
TF30-P-111	25 100/14 560	2.450/0.686	260	21.8	49	241.7	3999	2055	2.43	0.73	F-111F
TF33-P-3	17 000	0.52	450	13.0	53	136	3900	1600	1.7	1.55	B-52H
TF33-P-7	21 000	0.56	498	16.0	54	142	4650	1750	1.9	1.21	C-141
TF34-GE-100	9065	0.37	333	20.0	50	100	1421	2234	1.5	6.42	A-10
TF39-GE-1	40 805	0.315	1549	26.0	100	203	7186	2350	1.56	8.0	C-5A
TF41-A-1B	14 500	0.647	260	20.0	40	114.5	3511	2165	2.45	0.76	A-7D, K
TFE731-2	3500	0.504	113	17.7	40	50	625		1.54	2.67	C-21A

Table 3.2 Properties of jet fuels.

	Jet-A (commercial)	JP-4 (USAF)	JP-5 (Navy)
h_{PR}	18 400 BTU/lbm, 43 000 kJ/kg	18 314 BTU/lbm, 42 800 kJ/kg	18 314 BTU/lbm, 42 800 kJ/kg
Specific gravity	0.710	0.758	0.828
Vapor pressure [atm] at 100°F	0.007	0.18	0.003
Initial boiling point [°C]	169	60	182
Flashpoint [°C]	52	−25	65

cycle analyses, where we will add other intermediate states. We use the stagnation quantities inside the gas-turbine engines because the air flow has appreciable velocity, and we would like to keep track of both the enthalpy and this kinetic energy associated with the air velocities.

From thermodynamics, isentropic compression will involve increase of both temperature and pressure, and for gas-turbine engines the compression is achieved by a compressor, which requires power input. This power input is drawn from the turbine. Of course, the turbine needs to generate more power than is needed for the compressor, to produce useful power output. In fact, the net power output is defined as the turbine power output minus the compressor power input. The thermal efficiency of a power cycle is defined as the ratio of this net power output to the heat input.

$$\text{thermal efficiency} = \eta_T = \frac{\text{net power output}}{\text{heat input rate}} = \frac{\dot{W}_{out} - \dot{W}_{in}}{\dot{Q}_{in}} \tag{3.1}$$

From basic thermodynamics, the steady-state energy balance equation (first law of thermodynamics) relates the energy flow rates, power and heat rates.

$$\begin{aligned}\frac{dE_{CV}}{dt} &= 0 = \dot{Q}_{CV} - \dot{W}_{CV} + \dot{m}_{in}\left(h_{in} + \frac{U_{in}^2}{2}\right) - \dot{m}_{out}\left(h_{out} + \frac{U_{out}^2}{2}\right) \\ &= \dot{Q}_{CV} - \dot{W}_{CV} + \dot{m}_{in}h_{in}^o - \dot{m}_{out}h_{out}^o\end{aligned} \tag{3.2}$$

The left-hand side of Eq. (3.2) is set to zero, since we are considering steady-state processes. Also, the mass flow rates are the same under steady state: $\dot{m}_{in} = \dot{m}_{out} = \dot{m}$. The stagnation enthalpy is the sum of the static enthalpy and the kinetic energy. Now, if we assume constant specific heat, c_p, then the enthalpy differences can be converted into the temperature differences in Eq. (3.3).

$$\dot{Q}_{CV} - \dot{W}_{CV} + \dot{m}c_p(T_{in}^o - T_{out}^o) = 0 \tag{3.3}$$

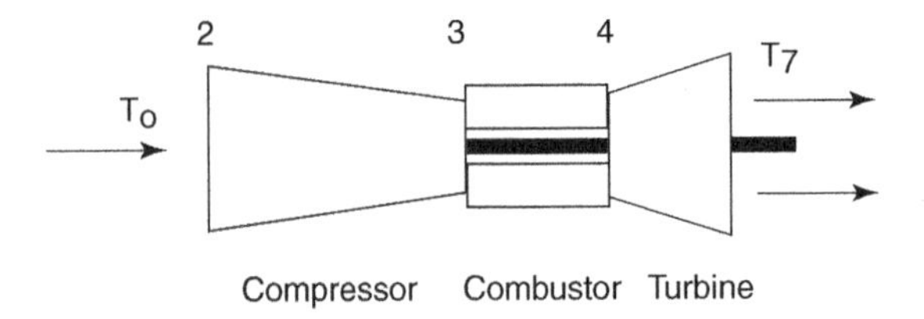

Figure 3.1 Idealized model of the gas-turbine engine: a Brayton cycle.

We apply the above Eq. (3.3) sequentially to the gas-turbine processes, in order to estimate the power and heat input rate. First, for the compression we consider the compression through the inlet and compressor of the gas-turbine engine from the ambient static temperature, T_o, to the stagnation temperature T_3^o. This process is isentropic, and therefore adiabatic (heat rate in Eq. (3.3) should be set equal to zero).

$$-\dot{W}_{CV} = \dot{W}_{in} = \dot{m}c_p(T_3^o - T_o) \tag{3.4}$$

So the power input is the compressor power. We are making a sign change in Eq. (3.4) to reflect the fact that power is defined positive for power output (i.e. $\dot{W}_{CV} < 0, \dot{W}_{in} > 0$ in Eq. (3.4)). Next, we consider the heat input in the combustor, where the temperature increases from T_3^o to T_4^o in a constant-pressure mode. The reason this heat addition occurs at a constant pressure is because gas-turbine combustors are "throughput" devices where the flow enters the combustor and then is allowed to expand out without flow restrictions. A contrary example would be the combustion chamber in a spark-ignition engine, where the piston impedes the expansion of the gas, so that the combustion is best approximated as a constant-volume process. In any event, the heat input rate can be related as follows.

$$\dot{Q}_{CV} = \dot{Q}_{in} = \dot{m}c_p(T_4^o - T_3^o) \tag{3.5}$$

For the turbine expansion, we consider the full conversion of the energy from the stagnation enthalpy at "4" to the static enthalpy at "7".

$$\dot{W}_{CV} = \dot{W}_{out} = \dot{m}c_p(T_4^o - T_7) \tag{3.6}$$

The turbine power is positive since it is a power output. Substituting these terms in Eq. (3.1), after canceling constant mass flow rate and specific heats, gives us the thermal efficiency purely in terms of the state temperatures.

$$\eta_T = \frac{\dot{W}_{out} - \dot{W}_{in}}{\dot{Q}_{in}} = \frac{(T_4^o - T_7) - (T_3^o - T_o)}{T_4^o - T_3^o} = 1 - \frac{T_7 - T_o}{T_4^o - T_3^o} = 1 - \frac{T_o\dfrac{T_7}{T_o} - 1}{T_3^o\dfrac{T_4^o}{T_3^o} - 1} \tag{3.7}$$

We can obtain a simpler and more insightful result by considering the temperature ratios in Eq. (3.7). We start by noting that the compression and expansion processes are isentropic in the idealized gas-turbine cycle, so that we may apply the isentropic relationships. For the compression, the isentropic relationship between the temperature and pressure is

$$\frac{T_3^o}{T_o} = \left(\frac{p_3^o}{p_o}\right)^{\frac{\gamma-1}{\gamma}} \tag{3.8}$$

The pressure ratio inside the parentheses can be considered as the "overall compression ratio", and γ is the ratio of the specific heats, $\gamma = c_p/c_v$. We can write another isentropic relationship for the expansion process.

$$\frac{T_4^o}{T_7} = \left(\frac{p_4^o}{p_7}\right)^{\frac{\gamma-1}{\gamma}} = \left(\frac{p_3^o}{p_o}\right)^{\frac{\gamma-1}{\gamma}} \tag{3.9}$$

In the second part of the equality in Eq. (3.8), we make use of the facts that combustion occurs at constant pressure ($p_3^o = p_4^o$) and that the expansion continues on to the ambient pressure ($p_7 = p_o$). Since the pressure ratios on the right-hand sides of Eqs. (3.8) and (3.9) are identical, the temperature ratios must also be equal.

$$\frac{T_4^o}{T_7} = \frac{T_3^o}{T_o} \quad \text{or} \quad \frac{T_4^o}{T_3^o} = \frac{T_7}{T_o} \tag{3.10}$$

We return to Eq. (3.7) with this result, which now becomes

$$\eta_T = 1 - \frac{T_o}{T_3^o} = 1 - \frac{1}{\dfrac{T_3^o}{T_o}} = 1 - \frac{1}{\left(\dfrac{p_3^o}{p_o}\right)^{\frac{\gamma-1}{\gamma}}} \tag{3.11}$$

Again, if we call the pressure ratio in the denominator of Eq. (3.11) the "overall compression ratio", then the thermal efficiency of gas-turbine engines depends on, and only on, this overall compression ratio. Eq. (3.11) also tells us that without compression, a gas-turbine engine cannot produce power. For example, if there is no compression and the pressure ratio is 1 in Eq. (3.11), then the thermal efficiency is zero. Figure 3.2 shows the plot of the thermal efficiency as a function of the overall compression ratio, using Eq. (3.11). We can see that the initial increase in the thermal efficiency is rapid for compression ratio up to about 10. After that, the gain in thermal efficiency is slow. If we consider that higher compression ratio is achieved with larger and more costly compressors, the gain in thermal efficiency comes at some cost beyond compression ratio of 20.

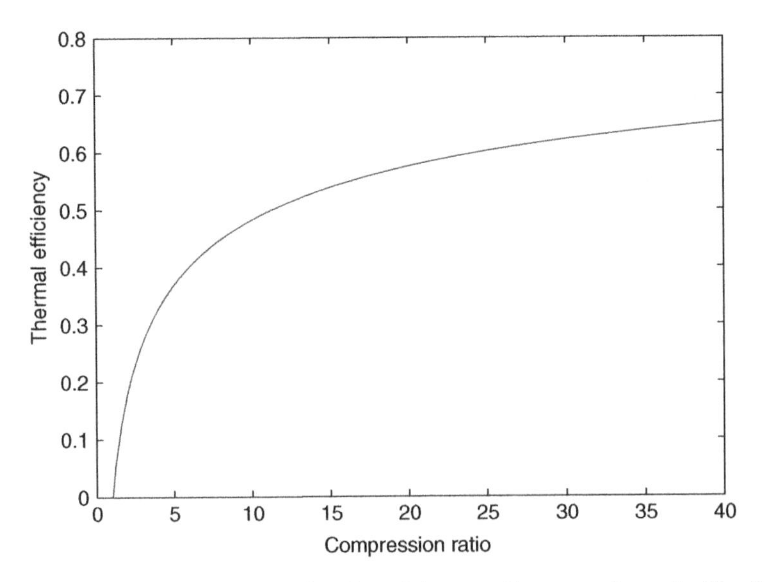

Figure 3.2 Thermal efficiency as a function of the overall compression ratio (Eq. (3.11)).

Example 3.1

From Tables 3.1a and 3.1b, the overall compression ratio (OPR) of some of the civil and military gas-turbine engines can be found. The JT-8D-15A used in Boeing 727, for example, has an OPR of 16.6, while modern high-technology engines such as the GE90-B4 used in the Boeing 777 has an OPR of 39.3. Using Eq. (3.22), with $\gamma = 1.4$, these OPRs translate to thermal efficiencies of

$$\eta_T = 1 - \frac{1}{\left(\frac{p_3^o}{p_o}\right)^{\frac{\gamma-1}{\gamma}}} = 1 - \frac{1}{(\text{OPR})^{\frac{\gamma-1}{\gamma}}} = 0.552\ (\text{OPR} = 16.6)\ \text{or}\ 0.650\ (\text{OPR} = 39.3) \tag{E3.1.1}$$

Since exponent for OPR is $(\gamma - 1)/\gamma = 0.2857$, the rate of increase of thermal efficiency with OPR is very low.

Example 3.2

Air enters the compressor of a gas-turbine engine at $p_2 = 0.95$ bar and $T_2 = 22°C$. For the turbine, the inlet pressure and temperature are $p_4 = 5.7$ bar and $T_4 = 826.85°C$, while the exit pressure is $p_5 = 0.95$ bar. What is the net power output per unit mass of air in kJ/kg?

The net power is given by

$$\dot{W}_{net}\dot{m} = (h_4 - h_5) - (h_3 - h_2) = c_p[(T_4 - T_5) - (T_3 - T_2)] \tag{E3.2.1}$$

Since the combustor is at constant pressure, $p_2 = p_3$. Assuming that the compressor and turbine processes are isentropic, we can use

$$\frac{T_3}{T_2} = \left(\frac{p_3}{p_2}\right)^{\frac{\gamma-1}{\gamma}}; \frac{T_5}{T_4} = \left(\frac{p_5}{p_4}\right)^{\frac{\gamma-1}{\gamma}} \rightarrow T_3 = 492.5\ \text{K}, T_5 = 659.3\ \text{K} \tag{E3.2.2}$$

Using a constant specific of 1.09 kJ/(kg K) at $T = 500°C$, the net work is 265.3 kJ/kg.

3.3 Ideal-Cycle Analysis for Turbofan Engines

Most modern aircraft employ turbofans, or some variations thereof, as their primary power plant. As shown in Figure 3.3, turbofans have the "core" section, where the air flows through the main components of the engine, and the "fan" section, for additional thrust. It is possible to add an afterburner, to boost the thrust, but afterburners are exclusively used for high-performance military aircraft, and we will deal with them in the next section. We will consider idealized processes through both core and fan sections of a turbofan engine, and obtain useful relationships that allow us to analyze these engines. Once we are capable of doing these analyses, then

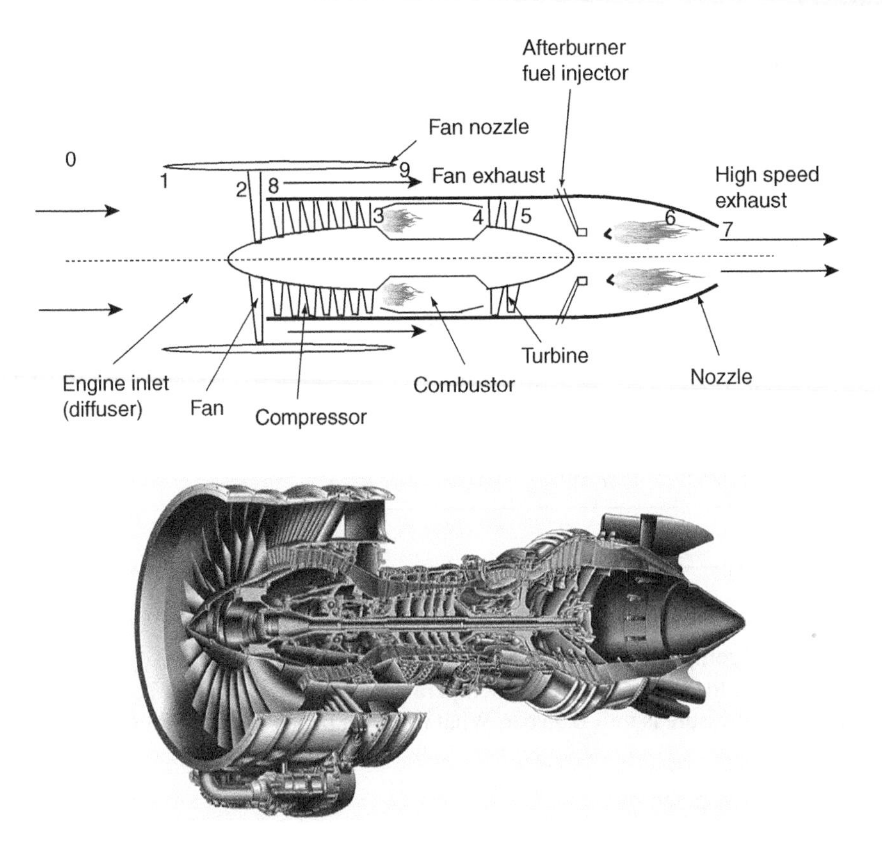

Figure 3.3 Schematic internal view of turbofan Pratt & Whitney PW6000. Copyright United Technologies Corporation.

other variation of turbofan engines, such as turbojets or turbofans with afterburners, can simply be analyzed by removing or adding some of the components (e.g. remove the fan for turbojets or add the afterburner) from the idealized model built in this section. An extreme simplification would be the ramjet, where most of the components are taken out.

We start from an "idealized" cycle of turbofan engines without afterburning. By "idealized", we mean that the flows through all the components of the engine, except for the combustor, are isentropic. That is to say, the flows are assumed to be frictionless and adiabatic, in the diffuser, compressor, fan, turbine and the nozzles. Below we list this and other assumptions for the ideal cycle analysis of turbofans.

Assumptions for Ideal-Cycle Analyses of Turbofans

1. Compressor, fan, and turbine flows are isentropic.
2. Diffuser and nozzle flows are also isentropic, and since they involve no work there are no changes in stagnation pressures and stagnation temperatures through these devices.
3. The "working fluid" through the engine is ideal-gas air, with constant specific heat.
4. The nozzle expansion of air brings the exit pressures to the ambient pressure.

5. Turbine power is used to rotate the compressor and the fan.
6. Combustion occurs at a constant pressure in the main combustor and the afterburner.

Again looking at Figure 3.3, the flow undergoes the following processes in a turbofan.

0–1: External flow. There may be some external acceleration or deceleration if the flight speed is not matched exactly with the required speed of air through the engine (mass flow rate). In either external acceleration or deceleration, the static pressure may change from p_0 to p_1, but there are no changes in the stagnation pressures or temperatures under subsonic flow conditions. In supersonic flows, shocks will alter the stagnation and static pressures.

1–2: Diffuser. The incident air is captured by the inlet and "conditioned" through the diffuser. The function of the diffuser is to bring the air flow to a required speed at the compressor and fan inlet, preferably with a uniform velocity profile, so that all the compressor and fan blades see identical flow. Since the flow is assumed to be isentropic for ideal cycle analyses, there are no changes in the stagnation pressure or temperature.

2–3: The compressor is a rotating device that increases the pressure of the air. The power to turn the compressor comes from the turbine, and since the process is assumed to be isentropic, all of the power is converted to increasing the stagnation enthalpy of the air. In addition, isentropic relationships can be used to relate the changes in stagnation temperatures and pressures.

2–8: Fans are similar to compressors in that some of the power drawn from the turbine is used to turn the fan blades, which again isentropically adds enthalpy to the air flow. For the fan section, this increased enthalpy is converted to kinetic energy in the fan nozzle for thrust enhancement.

3–4: Main combustor. Fuel is injected, mixed with air, and burned in the combustor. This raises the enthalpy of the air (more accurately, the "sensible" enthalpy of the air is increased). This high enthalpy is then available to generate mechanical energy in the turbine and kinetic energy in the nozzle. The combustion is assumed to be constant-pressure.

4–5: Turbines generate mechanical power, by converting the thermal energy (enthalpy) of the air into mechanical energy. Aerodynamically, this is achieved by the high pressure of the air that pushes the turbine blades around their rotational axis. Thermodynamically, this requires high enthalpy that was achieved through combustion in the combustor. The process is again isentropic, and it is also assumed that all of the turbine power goes to power the compressor and the fan.

$$\text{turbine power} = \text{compressor power} + \text{fan power}$$

$$-\dot{W}_t = \dot{W}_c + \dot{W}_{fan} \tag{3.12}$$

In Eq. (3.12), we put a negative sign in front of the turbine power term, since power is defined positive for work output (turbine) and negative for work input (compressor and fan).

5–7: Main nozzle. The flow is accelerated to the exhaust speed. For isentropic acceleration, the stagnation pressure and temperature are constant, but the static pressure will decrease. Assumption 4 above states that this acceleration brings the static pressure to ambient pressure.

$$p_7 = p_o \tag{3.13}$$

8–9: Fan nozzle. This process is the same as the main nozzle, and the exit pressure is equal to the ambient pressure.

$$p_9 = p_o \tag{3.14}$$

As shown in Figure 3.3, the numbers mark different states in the turbofans, for example, 2 for the compressor inlet so that p_2^o and T_2^o represent stagnation pressure and temperature at the compressor inlet, respectively. Parameters without the superscript "o" represent static quantities. We will also often be dealing with the ratios of stagnation pressures and temperatures, so to shorten the writing we will use π_x and τ_x for stagnation pressure and temperature ratios, respectively, across a component "x". So the stagnation pressure ratio and the stagnation temperature ratios across the turbine would be written as follows, with the subscript "t" for turbine.

$$\pi_t = \frac{p_5^o}{p_4^o} \tag{3.15a}$$

$$\tau_t = \frac{T_5^o}{T_4^o} \tag{3.15b}$$

There will also be a few other ratios involving static pressures. The two main ones are the ratio of the stagnation to static parameters in the ambient, and the ratio of the combustion stagnation temperature with respect to the ambient static temperature.

$$\tau_r = \frac{T_0^o}{T_o} \tag{3.16}$$

$$\tau_\lambda = \frac{T_4^o}{T_o} \tag{3.17}$$

Use of these ratios will be clarified in the following.

We begin by looking at the thrust that would be developed by a turbofan, according to Eq. (2.11).

$$F = \dot{m}_C[(1+f)U_{eC} - U_o] + \dot{m}_{fan}(U_{e,fan} - U_o) + (p_{eC} - p_o)A_{eC} + (p_{e,fan} - p_o)A_{e,fan} \tag{2.11}$$

We again assume that the fuel mass flow rate is much smaller than the air mass flow rate ($f \ll 1$), and since the exit pressures in both core and fan nozzles are identical to the ambient pressure, the pressure terms disappear in Eq. (2.11). So the thrust for the ideal-cycle model of a turbofan is

$$F = \dot{m}_C(U_7 - U_o) + \dot{m}_{fan}(U_9 - U_o) \quad (3.18)$$

In Eq. (3.18), we have used U_7 and U_9 as exit velocities in the core and fan nozzle, respectively. Thus, the exit velocities along with the mass flow rates in the core and fan sections generate the thrust. Enthalpy addition in the combustor is needed to generate these exit velocities, and this will require the mass flow rate of the fuel, which can be converted to TSFC. For turbofans, it is customary to use the bypass ratio to relate the mass flow rates of the air through the core and fan sections.

$$\text{bypass ratio} = \alpha = \frac{\dot{m}_{fan}}{\dot{m}_C} \quad (3.19)$$

Then, the total mass flow rate is simply the sum of the mass flow rates through the core and the fan sections.

$$\text{total mass flow} = \dot{m}_o = \dot{m}_C + \dot{m}_{fan} = \dot{m}_C(1+\alpha) = \dot{m}_{fan}\left(\frac{1}{\alpha}+1\right) \quad (3.20)$$

Example 3.3 Bypass ratio

The following nominal data for air flow rates are available.

$$\text{GE90-B4 (used in Boeing 777s)}: a = 8.4, \dot{m}_o = 3037\ \frac{\text{lbm}}{\text{s}}$$

$$\text{F-100-PW-229 (used in F-16s)}: a = 0.4, \dot{m}_o = 112.7\ \frac{\text{kg}}{\text{s}}$$

Using Eq. (3.20), we have

$$\dot{m}_C = \frac{1}{1+\alpha}\dot{m}_o;\ \dot{m}_{fan} = \frac{\alpha}{1+\alpha}\dot{m}_o \quad (\text{E3.3.1})$$

This gives us

$$\text{GE90-B4}: \dot{m}_C = 323\ \frac{\text{lbm}}{\text{s}} \text{ and } \dot{m}_{fan} = 2714\ \frac{\text{lbm}}{\text{s}}$$

$$\text{F-100-PW-229}: \dot{m}_C = 80.5\ \frac{\text{kg}}{\text{s}} \text{ and } \dot{m}_{fan} = 32.2\ \frac{\text{kg}}{\text{s}}$$

Using the bypass ratio, Eq. (3.18) is used to calculate the specific thrust, F_s, of a turbofan engine

$$F_s = \frac{F}{\dot{m}_o} = \frac{1}{1+\alpha}(U_7 - U_o) + \frac{\alpha}{1+\alpha}(U_9 - U_o) \tag{3.21}$$

We can see from Eq. (3.21) that we need the exit velocities, U_7 and U_9. These velocities will be produced by the excess enthalpies in the core and fan channels of the turbofan. So we will start from the front end of a turbofan engine. As noted earlier, we use the stagnation properties, since they include both the static enthalpy (or static pressure) and the kinetic energy (or dynamic pressure) components.

$$\frac{T_o^o}{T_o} \equiv \tau_r = 1 + \frac{\gamma - 1}{2} M_o^2 \tag{3.22a}$$

$$\frac{p_o^o}{p_o} \equiv \pi_r = \tau_r^{\frac{\gamma}{\gamma-1}} = \left(\frac{T_o^o}{T_o}\right)^{\frac{\gamma}{\gamma-1}} \tag{3.22b}$$

M_o = flight Mach number $= U_o/a_o$
a_o = $\sqrt{\gamma R T_o}$ speed of sound based on the ambient temperature
R = specific gas constant for air

For the external flow (0–1) under subsonic conditions, there are no changes in the stagnation pressure or temperature. Likewise, the diffuser flow (1–2) under the assumption of ideal cycle involves no stagnation pressure loss or stagnation temperature change (i.e. the flow is frictionless and adiabatic).

$$T_2^o = T_1^o = T_o^o \tag{3.23a}$$

$$p_2^o = p_1^o = p_o^o \tag{3.23b}$$

Nonetheless, we define the stagnation temperature and pressure ratios for the diffuser for later use.

$$\frac{T_2^o}{T_1^o} \equiv \tau_d = 1 \tag{3.24a}$$

$$\frac{p_2^o}{p_1^o} \equiv \pi_d = 1 \tag{3.24b}$$

As we saw in Section 3.2, for the compressor (2–3) we can relate the inlet and exit stagnation temperatures by applying the energy balance equation (the first law of thermodynamics).

$$-\dot{W}_{CV} = \dot{W}_c = \dot{m}_C c_p (T_3^o - T_2^o) \tag{3.25}$$

Again, we make a sign change, since the work is defined positive for work output, and the compressor work requires power input. Thus, $\dot{W}_{CV}$ in Eq. (3.25) is a negative number, while the compressor power ($\dot{W}_c$) is a positive number, for convenience. We can also divide by the mass flow rate, and deal with the work on a unit mass basis.

$$w_c = \frac{\dot{W}_c}{\dot{m}_C} = c_p(T_3^o - T_2^o) = c_p T_2^o(\tau_c - 1) = c_p T_2^o\left(\pi_c^{\frac{\gamma-1}{\gamma}} - 1\right) \tag{3.26}$$

$$\tau_c \equiv \frac{T_3^o}{T_2^o}$$

$$\pi_c \equiv \frac{p_3^o}{p_2^o} = \text{compression ratio}$$

The temperature ratio across the compressor is related to the compression ratio through the isentropic relationship in the last part of equality in Eq. (3.26).

The fan process is quite analogous in that power input adds enthalpy to the flow, and pressure increase occurs in an isentropic manner.

$$w_f = \frac{\dot{W}_f}{\dot{m}_{fan}} = c_p(T_8^o - T_2^o) = c_p T_2^o(\tau_f - 1) = c_p T_2^o\left(\pi_f^{\frac{\gamma-1}{\gamma}} - 1\right) \tag{3.27}$$

$$\tau_f \equiv \frac{T_8^o}{T_2^o}$$

$$\pi_f \equiv \frac{p_8^o}{p_2^o} = \text{fan pressure ratio}$$

Next, we consider the constant-pressure heat addition in the combustor (3–4). A schematic internal view of a combustor is shown in Figure 3.4. There are two inlet streams, one for the air and another for the fuel.

$$\dot{Q}_{in} = (\dot{m}_C + \dot{m}_f)c_p T_4^o - \dot{m}_C T_3^o \approx \dot{m}_C c_p(T_4^o - T_3^o) = \dot{m}_C c_p T_o\left(\frac{T_4^o}{T_o} - \frac{T_3^o}{T_o}\right) \tag{3.28}$$

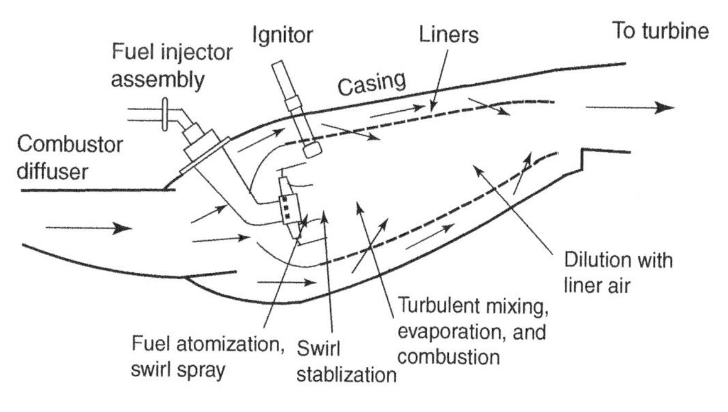

Figure 3.4 Typical gas-turbine combustor.

The mass flow rate of air is much larger than that for the fuel, so we are approximating the total mass flow rate to the mass flow rate of air in Eq. (3.28). The heat input is calculated using the mass flow rate of the fuel and so-called "heating value", h_{PR}. Heating value is defined as the amount of sensible enthalpy that is generated per unit mass of the fuel, and thus has the unit of kJ/kg-fuel or BTU/lbm-fuel. Table 3.2 shows the heating value and other properties of some of the widely used jet fuels. Using this heating value or h_{PR}, Eq. (3.28) becomes

$$\dot{Q}_{in} = \dot{m}_f h_{PR} = \dot{m}_C c_p T_o \left(\frac{T_4^o}{T_o} - \frac{T_3^o}{T_o} \right) = \dot{m}_C c_p T_o (\tau_\lambda - \tau_r \tau_c) \tag{3.29}$$

We can solve for the required mass flow rate of fuel from Eq. (3.29), but we can also write it as a fuel–air ratio, f.

$$f \equiv \frac{\dot{m}_f}{\dot{m}_C} = \frac{c_p T_o}{h_{PR}} (\tau_\lambda - \tau_r \tau_c) = \text{fuel-air ratio} \tag{3.30}$$

Table 3.3 A summary of equations for turbofan without afterburning (ideal-cycle analysis).

Input parameters: M_o, T_o, γ, c_p, h_{PR}, T_4^o, π_c, π_f (fan pressure ratio), α (bypass ratio).
Output parameters: U_7, U_9, F_s, TSFC, η_P, η_T, η_o.
Analysis result:

$$\tau_r = 1 + \frac{\gamma - 1}{2} M_o^2$$

$$\tau_c = (\pi_c)^{(\gamma-1)/\gamma}; \tau_f = (\pi_f)^{(\gamma-1)/\gamma}$$

$$\tau_\lambda = \frac{T_4^o}{T_o}$$

$$\tau_t = 1 - \frac{\tau_r}{\tau_\lambda} \left[\tau_c - 1 + \alpha(\tau_f - 1) \right]$$

$$\left(\frac{U_7}{a_o} \right)^2 = \frac{2}{\gamma - 1} \frac{\tau_\lambda}{\tau_r \tau_c} \left(\tau_r \tau_c \left\{ 1 - \frac{\tau_r}{\tau_\lambda} \left[\tau_c - 1 + \alpha(\tau_f - 1) \right] \right\} - 1 \right)$$

$$\left(\frac{U_9}{a_o} \right)^2 = \frac{2}{\gamma - 1} (\tau_r \tau_f - 1)$$

$$F = \dot{m}_C (U_7 - U_o) + \dot{m}_{fan} (U_9 - U_o)$$

$$F_s = \frac{a_o}{1 + \alpha} \left[\frac{U_7}{a_o} - M_o + \alpha \left(\frac{U_9}{a_o} - M_o \right) \right]$$

$$f = \frac{c_p T_o}{h_{PR}} (\tau_\lambda - \tau_r \tau_c)$$

$$\text{TSFC} = \frac{f}{(1 + \alpha) F_s}$$

$$\eta_T = 1 - \frac{1}{\tau_r \tau_c}; \ \eta_P = 2 \frac{\dfrac{U_7}{U_o} - 1 + \alpha \left(\dfrac{U_9}{U_o} - 1 \right)}{\dfrac{U_7^2}{U_o^2} - 1 + \alpha \left(\dfrac{U_9^2}{U_o^2} - 1 \right)}; \text{ and } \eta_o = \eta_P \eta_T$$

Alternatively, we can use Eq. (3.28) to find the temperature across the combustor.

$$\tau_b \equiv \frac{T_4^o}{T_3^o} = \frac{h_{PR}f}{c_p T_o} \frac{1}{\tau_r \tau_c} + 1 \tag{3.31}$$

The combustion is constant-pressure, so the pressure ratio is one.

$$\pi_b \equiv \frac{p_4^o}{p_3^o} = 1 \tag{3.32}$$

The turbine generates power to run both the compressor and the fan, again in an isentropic expansion process. It is called an expansion process since the power output would lower its stagnation temperature and pressure. The power balance simply states that the turbine power output equals the combined powers of the compressor and the fan.

$$\dot{W}_t = \dot{W}_c + \dot{W}_f \tag{3.33a}$$

$$\dot{m}_C c_p (T_4^o - T_5^o) = \dot{m}_C c_p (T_3^o - T_2^o) + \dot{m}_{fan} c_p (T_8^o - T_2^o) \tag{3.33b}$$

This leads to a simple solution for the turbine exit temperature (Eq. (3.34a)), or we can write the result as the temperature ratio (Eq. (3.34b)), τ_t, for later use.

$$T_5^o = T_4^o - (T_3^o - T_2^o) - \alpha(T_8^o - T_2^o) \tag{3.34a}$$

$$\tau_t = \frac{T_5^o}{T_4^o} = 1 - \frac{\tau_r}{\tau_\lambda}\left[\tau_c - 1 + \alpha(\tau_f - 1)\right] \tag{3.34b}$$

Both Eqs. (3.34a) and (3.34b) state that the stagnation enthalpy is reduced through the turbine by the amount that goes to the compressor and the fan. The pressure ratio is found again through the isentropic relationship.

$$\pi_t = \frac{p_5^o}{p_4^o} = \tau_t^{\frac{\gamma}{\gamma-1}} \tag{3.35}$$

The final process is the nozzle acceleration that converts the available stagnation enthalpy at the turbine exit to kinetic energy. The process is isentropic with no work, so that it is frictionless (stagnation pressure unchanged) and adiabatic (stagnation enthalpy unchanged).

$$\tau_n = \frac{T_7^o}{T_5^o} = 1 \tag{3.36}$$

$$\pi_n = \frac{p_7^o}{p_5^o} = 1 \tag{3.37}$$

We use the energy balance represented in Eq. (3.36) to solve for the exhaust velocity, U_7.

$$h_5^o = h_7^o = h_7 + \frac{U_7^2}{2} \rightarrow U_7 = \sqrt{2c_p(T_7^o - T_7)} \tag{3.38}$$

We can factor out T_7, which converts Eq. (3.38) into an expression for the Mach number at the exit, M_7.

$$U_7 = \sqrt{2c_pT_7\left(\frac{T_7^o}{T_7} - 1\right)} = \sqrt{\frac{2\gamma RT_7}{\gamma - 1}\left[\left(\frac{p_7^o}{p_7}\right)^{\frac{\gamma-1}{\gamma}} - 1\right]} \rightarrow M_7 = \sqrt{\frac{2}{\gamma - 1}\left[\left(\frac{p_7^o}{p_7}\right)^{\frac{\gamma-1}{\gamma}} - 1\right]} \tag{3.39}$$

where $M_7 = \frac{U_7}{a_7} = \frac{U_7}{\sqrt{\gamma RT_7}}$.

We can apply the assumption that the exit pressure is equal to the ambient pressure ($p_7 = p_o$) and write out the pressure ratio in Eq. (3.39).

$$\frac{p_7^o}{p_7} = \frac{p_7^o}{p_o} = \frac{p_7^o p_5^o p_4^o p_3^o p_2^o p_1^o}{p_5^o p_4^o p_3^o p_2^o p_1^o p_o} = \pi_n\pi_t\pi_b\pi_c\pi_d\pi_r = \pi_t\pi_c\pi_r \tag{3.40}$$

$$\pi_n = \pi_b = \pi_d = 1$$

So using the isentropic relationship to the pressure ratio in Eq. (3.40), we can convert it back into temperature ratio and rewrite Eq. (3.39).

$$M_7 = \frac{U_7}{a_7} = \sqrt{\frac{2}{\gamma - 1}(\tau_r\tau_c\tau_t - 1)} \tag{3.41}$$

As we can see in Eq. (3.41), U_7 is related to M_7 through the speed of sound at the exit temperature, T_7.

To find the exit temperature, we use the temperature and pressure ratios previously calculated.

$$T_7 = \frac{T_7}{T_o}T_o = \frac{\frac{T_7^o}{T_o}}{\frac{T_7^o}{T_7}}T_o = \frac{\frac{T_7^o}{T_o}}{\left(\frac{p_7^o}{p_o}\right)^{\frac{\gamma-1}{\gamma}}}T_o \tag{3.42}$$

We have already found the pressure ratio in Eq. (3.42) (see Eq. (3.40)), and we can write out the temperature ratio as well.

$$\frac{T_7^o}{T_o} = \frac{T_7^o T_5^o T_4^o T_3^o T_2^o T_1^o}{T_5^o T_4^o T_3^o T_2^o T_1^o T_o} = \tau_n\tau_t\tau_b\tau_c\tau_d\tau_r = \tau_t\tau_b\tau_c\tau_r \tag{3.43}$$

$$\tau_n = \tau_d = 1$$

Substituting Eqs. (3.40) and (3.43) into Eq. (3.42) results in

$$\frac{T_7}{T_o} = \frac{\tau_t\tau_b\tau_c\tau_r}{\tau_t\tau_c\tau_r} = \tau_b = \frac{T_4^o}{T_3^o} = \frac{\frac{T_4^o}{T_o}}{\frac{T_3^o}{T_2^o}\frac{T_2^o}{T_o}} = \frac{\tau_\lambda}{\tau_c\tau_r} \tag{3.44a}$$

$$\frac{a_7}{a_o} = \frac{\sqrt{\gamma RT_7}}{\sqrt{\gamma RT_o}} = \sqrt{\tau_b} \tag{3.44b}$$

Equation (3.44) can be used to calculate T_7 and therefore a_7, but it also tells us that the static temperature increase is due to the combustion heat input in the combustor.

Finally, combining the results of Eqs. (3.41) and (3.44b) gives us the exit velocity.

$$\frac{U_7}{a_o} = \frac{a_7 M_7}{a_o} = \sqrt{\frac{2\tau_b}{\gamma - 1}(\tau_r \tau_c \tau_t - 1)} = \sqrt{\frac{2}{\gamma - 1}\frac{\tau_\lambda}{\tau_r \tau_c}(\tau_r \tau_c \tau_t - 1)} \tag{3.45}$$

We can find the fan exit velocity, following an analogous route. We can see in Eq. (3.42) that the Mach number (the expression inside the square root sign) consists of all the temperature ratios across the core section of the engine. For the fan section, there are only two temperature ratios to be included, τ_r and τ_f. Also, the ratio of the speed of sound at the exit and ambient was found, for the core section, to be simply, τ_b. Following the same process to find Eq. (3.44), we can obtain $T_9 = T_o$ or $a_9 = a_o$. Combining these observations, we get

$$\frac{U_9}{a_o} = \sqrt{\frac{2}{\gamma - 1}(\tau_r \tau_f - 1)} \tag{3.46}$$

With the exit velocities in the core and fan sections of the engine, we can write the engine thrust.

$$F = \dot{m}_C (U_7 - U_o) + \dot{m}_{fan}(U_9 - U_o) \tag{3.47}$$

The specific thrust is as follows.

$$F_s = \frac{F}{\dot{m}_o} = \frac{a_o}{1+\alpha}\left[\frac{U_7}{a_o} - M_o + \alpha\left(\frac{U_9}{a_o} - M_o\right)\right] \tag{3.48}$$

We can use the fuel–air ratio in the combustor, f in Eq. (3.30), to find TSFC.

$$\text{TSFC} = \frac{\dot{m}_f}{F} = \frac{\frac{\dot{m}_f}{\dot{m}_C}}{\frac{F}{\dot{m}_C}} = \frac{f}{\frac{\dot{m}_o}{\dot{m}_C}\frac{F}{\dot{m}_o}} = \frac{f}{(1+\alpha)F_s} \tag{3.49}$$

As shown in Section 3.2, the thermal efficiency (Eq. (3.11)) of a gas-turbine engine is a function only of the overall compression ratio. This applies to turbofans equally, and in the current notation the thermal efficiency is

$$\eta_T = 1 - \frac{1}{(\pi_r \pi_c)^{\frac{\gamma-1}{\gamma}}} = 1 - \frac{1}{\tau_r \tau_c} \tag{3.50}$$

It can also be shown that the propulsion efficiency is

$$\eta_P = 2\frac{\frac{U_7}{U_o} - 1 + \alpha\left(\frac{U_9}{U_o} - 1\right)}{\frac{U_7^2}{U_o^2} - 1 + \alpha\left(\frac{U_9^2}{U_o^2} - 1\right)} \tag{3.51}$$

Example 3.4 Analysis of large bypass ratio turbofan engine

We can use the information in Table 3.2, for the following turbofan engine.

Input parameters: $M_o = 0.8$ @ $H = 30\,\text{kft}$, $\dot{m}_o = 1125\ \frac{\text{lbm}}{\text{s}}$, $h_{PR} = 18\,400\ \text{BTU/lbm}$, $T_4^o = 1850°\text{R}$. $\pi_c = 15$, $\pi_f = 1.75$, and $\alpha = 6.5$.

First, at H = 30 000 ft, the atmospheric air data are $T_o = 411.8°\text{R}$, $p_o = 629.5\ \text{lbf/ft}^2$, and $a_o = 995\ \text{ft/s}$. $\gamma = 1.4$, $c_P = 0.24\ \frac{\text{BTU}}{\text{lbm}°\text{R}}$. $U_o = M_o a_o = 796\ \text{ft/s}$.

$$\tau_r = 1 + \frac{\gamma - 1}{2} M_o^2 = 1.128, T_2^o = \tau_r T_o = 464.5°\text{R}; p_2^o = \tau_r^{\gamma/(\gamma-1)} p_o = 959.6\ \frac{\text{lbf}}{\text{ft}^2} \tag{E3.4.1}$$

$$\tau_c = (\pi_c)^{(\gamma-1)/\gamma} = 2.17; \tau_f = (\pi_f)^{(\gamma-1)/\gamma} = 1.173 \tag{E3.4.2}$$

$$\tau_\lambda = \frac{T_4^o}{T_o} = 4.49 \tag{E3.4.3}$$

Using the above temperature ratios,

$$\left(\frac{U_7}{a_o}\right)^2 = \frac{2}{\gamma - 1} \frac{\tau_\lambda}{\tau_r \tau_c} \left(\tau_r \tau_c \left\{ 1 - \frac{\tau_r}{\tau_\lambda} \left[\tau_c - 1 + \alpha(\tau_f - 1) \right] \right\} - 1 \right) = 0.337 \tag{E3.4.4}$$

$$\text{Or, } U_7/a_o = 0.577 \rightarrow U_7 = 574.8\ \text{ft/s} \tag{E3.4.5}$$

$$\left(\frac{U_9}{a_o}\right)^2 = \frac{2}{\gamma - 1} (\tau_r \tau_f - 1) = 1.616 \tag{E3.4.6}$$

$$\text{Or, } U_9/a_o = 1.27 \rightarrow U_9 = 1264.8\ \text{ft/s} \tag{E3.4.7}$$

Using Eq. (E3.3.1),

$$\dot{m}_C = \frac{1}{1+\alpha} \dot{m}_o = 150\ \frac{\text{lbm}}{\text{s}}; \dot{m}_{fan} = \frac{\alpha}{1+\alpha} \dot{m}_o = 975\ \frac{\text{lbm}}{\text{s}} \tag{E3.4.8}$$

$$F = \dot{m}_C (U_7 - U_o) + \dot{m}_{fan} (U_9 - U_o) = 424\,680\ \text{lbm}\frac{\text{ft}}{\text{s}^2} \times \frac{1\ \text{lbf}}{32.174\ \text{lbm}\frac{\text{ft}}{\text{s}^2}}$$

$$= 13\,199\ \text{lbf} \tag{E3.4.9}$$

$$F_s = \frac{a_o}{1+\alpha} \left[\frac{U_7}{a_o} - M_o + \alpha \left(\frac{U_9}{a_o} - M_o \right) \right] = \frac{F}{\dot{m}_o} = 11.7\ \frac{\text{lbf}}{\text{lbm/s}} \tag{E3.4.10}$$

$$f = \frac{c_p T_o}{h_{PR}} (\tau_\lambda - \tau_r \tau_c) = \frac{0.24\ \frac{\text{BTU}}{\text{lbm} \cdot °\text{R}} \times 411.8°\text{R}}{18\,400\ \frac{\text{BTU}}{\text{lbm}}} (4.49 - 1.128 \times 2.17) = 0.011 \tag{E3.4.11}$$

$$TSFC = \frac{f}{(1+\alpha)F_s} = 0.45\,\frac{\text{lbm/hr}}{\text{lbf}} \tag{E3.4.12}$$

$$\eta_T = 1 - \frac{1}{\tau_r \tau_c} = 0.59 \tag{E3.4.13}$$

$$\eta_P = 2\frac{\dfrac{U_7}{U_o} - 1 + \alpha\left(\dfrac{U_9}{U_o} - 1\right)}{\dfrac{U_7^2}{U_o^2} - 1 + \alpha\left(\dfrac{U_9^2}{U_o^2} - 1\right)} = 0.37 \tag{E3.4.14}$$

$$\text{and } \eta_o = \eta_P \eta_T = 0.22 \tag{E3.4.15}$$

We can also do this kind of calculation using MATLAB® (MCE34.m; at the end of this chapter), which allows us to see the effects of parameter variations, as shown in the following figures.

3.4 Turbojets, Afterburners and Ramjets

We can take the turbofan analysis in the previous section and add or remove components, to find basic performance parameters for other variants of gas-turbine engines.

3.4.1 Turbojet

Analysis of turbojets can be accomplished by removing the fan component from the turbofan engine. For the current ideal-cycle analysis, simply setting $\alpha = 0$ will remove the fan. Thus, the equations in Table 3.2 can be used, with $\alpha = 0$. From Eq. (3.34), for example, the temperature ratio across the turbine is simplified to

$$T_5^o = T_4^o - (T_3^o - T_2^o) \tag{3.52a}$$

$$\tau_t = \frac{T_5^o}{T_4^o} = 1 - \frac{\tau_r}{\tau_\lambda}(\tau_c - 1) \tag{3.52b}$$

With that change, the exit velocity is computed in exactly the same manner as in Eq. (3.45).

$$\frac{U_7}{a_o} = \frac{a_7 M_7}{a_o} = \sqrt{\frac{2\tau_b}{\gamma - 1}(\tau_r \tau_c \tau_t - 1)} = \sqrt{\frac{2}{\gamma - 1}\frac{\tau_\lambda}{\tau_r \tau_c}(\tau_r \tau_c \tau_t - 1)} \tag{3.45}$$

The thrust and specific thrust become

$$F = \dot{m}_o(U_7 - U_o) \tag{3.53}$$

$$F_s = \frac{F}{\dot{m}_o} = (U_7 - U_o) \tag{3.54}$$

The fuel–air ratio (from Eq. (3.30)), TSFC and efficiency terms also follow.

$$f = \frac{\dot{m}_f}{\dot{m}_o} = \frac{c_p T_o}{h_{PR}}(\tau_\lambda - \tau_r\tau_c) = \text{fuel} - \text{air ratio} \tag{3.30}$$

$$\text{TSFC} = \frac{\dot{m}_f}{F} = \frac{\dfrac{\dot{m}_f}{\dot{m}_o}}{\dfrac{F}{\dot{m}_o}} = \frac{f}{\dfrac{F}{\dot{m}_o}} = \frac{f}{F_s} \tag{3.55}$$

$$\eta_T = 1 - \frac{1}{(\pi_r\pi_c)^{\frac{\gamma-1}{\gamma}}} = 1 - \frac{1}{\tau_r\tau_c} \tag{3.56}$$

$$\eta_P = 2\frac{\dfrac{U_7}{U_o} - 1}{\dfrac{U_7^2}{U_o^2} - 1} \tag{3.57}$$

The above results are summarized in Table 3.4.

Table 3.4 A summary of equations for turbojets without afterburning (ideal-cycle analysis).

Input parameters: M_o, T_o, γ, c_p, h_{PR}, T_4^o, π_c.
Output parameters: U_7, F_s, TSFC, η_P, η_T, η_o.
Analysis result:

$$\tau_r = 1 + \frac{\gamma-1}{2}M_o^2$$

$$\tau_c = (\pi_c)^{(\gamma-1)/\gamma}$$

$$\tau_\lambda = \frac{T_4^o}{T_o}$$

$$\tau_t = 1 - \frac{\tau_r}{\tau_\lambda}(\tau_c - 1)$$

$$\left(\frac{U_7}{a_o}\right)^2 = \frac{2}{\gamma-1}\frac{\tau_\lambda}{\tau_r\tau_c}\left(\tau_r\tau_c\left\{1 - \frac{\tau_r}{\tau_\lambda}(\tau_c - 1)\right\} - 1\right)$$

$$F = \dot{m}_C(U_7 - U_o)$$

$$F_s = a_o\left(\frac{U_7}{a_o} - M_o\right)$$

$$f = \frac{c_p T_o}{h_{PR}}(\tau_\lambda - \tau_r\tau_c)$$

$$TSFC = \frac{f}{F_s}$$

$$\eta_T = 1 - \frac{1}{\tau_r\tau_c};\ \eta_P = 2 \times \frac{\dfrac{U_7}{U_o} - 1}{\dfrac{U_7{}^2}{U_o{}^2} - 1};\ \eta_o = \eta_P\eta_T$$

Example 3.5 Analysis of turbojet without afterburning

For the J79-GE-17 used in the venerable F-4E/G, the following data are given.

$$M_o = 0.8@H = 30\,\text{kft}, \dot{m}_o = 170\ \frac{\text{lbm}}{\text{s}}, h_{PR} = 18\ 400\ \text{BTU/lbm}, T_4^o = 1210°\text{R}.$$

$\pi_c = 13.5$.

At $H = 30\ 000$ ft, the atmospheric air data are $T_o = 411.8°\text{R}$, $p_o = 629/5\ \text{lbf/ft}^2$, and $a_o = 995$ ft/s. $\gamma = 1.4$, $c_P = 0.24\ \frac{\text{BTU}}{\text{lbm°R}}$.

$$\tau_r = 1 + \frac{\gamma - 1}{2} M_o^2 = 1.128, T_2^o = \tau_r T_o = 464.5°\text{R}; p_2^o = \tau_r{}^{\gamma/(\gamma-1)} p_o = 959.6\ \frac{\text{lbf}}{\text{ft}^2} \quad \text{(E3.5.1)}$$

$$\tau_c = (\pi_c)^{(\gamma-1)/\gamma} = 2.1 \rightarrow T_3^o = \tau_c T_2^o = 975.5°\text{R}; p_3^o = \pi_c p_2^o = 12\ 954\ \frac{\text{lbf}}{\text{ft}^2} \quad \text{(E3.5.2)}$$

$$\tau_\lambda = \frac{T_4^o}{T_o} = 4.06 \quad \text{(E3.5.3)}$$

$$\tau_t = 1 - \frac{\tau_r}{\tau_\lambda}(\tau_c - 1) = 0.69 \rightarrow T_5^o = \tau_t T_4^o = 1159°\text{R} \quad \text{(E3.5.4)}$$

Using the above temperature ratios,

$$M_7 = \sqrt{\frac{2}{\gamma - 1}(\tau_r \tau_c \tau_t - 1)} = 1.79 = 0.257 \quad \text{(E3.5.5)}$$

$$U_7 = a_o \sqrt{\frac{2}{\gamma - 1} \frac{\tau_\lambda}{\tau_r \tau_c}(\tau_r \tau_c \tau_t - 1)} = 1745\ \text{ft/s} \quad \text{(E3.5.6)}$$

$$F_s = U_7 - a_o M_o = 949.6\ \frac{\text{ft}}{\text{s}} \rightarrow 29.5\ \frac{\text{lbf}}{\text{lbm/s}} \quad \text{(E3.5.7)}$$

$$f = \frac{c_p T_o}{h_{PR}}(\tau_\lambda - \tau_r \tau_c) = \frac{0.24\ \dfrac{\text{BTU}}{\text{lbm}\cdot°\text{R}} \times 411.8°\text{R}}{18\ 400\ \dfrac{\text{BTU}}{\text{lbm}}}(4.06 - 1.128 \times 2.1) = 0.0091 \quad \text{(E3.5.8)}$$

$$\text{TSFC} = \frac{f}{F_s} = 1.11\ \frac{\text{lbm/hr}}{\text{lbf}} \quad \text{(E3.5.9)}$$

$$\eta_T = 1 - \frac{1}{\tau_r \tau_c} = 0.58 \quad \text{(E3.5.10)}$$

$$\eta_P = 2\frac{\dfrac{U_7}{U_o} - 1}{\dfrac{U_7^2}{U_o^2} - 1} = \frac{2M_o}{\dfrac{U_7}{a_o} + M_o} = 0.63 \quad \text{(E3.5.11)}$$

and $\eta_o = \eta_P \eta_T = 0.36$

3.4.2 Turbojets with Afterburners

Let us now take the turbojet engine and add an afterburner. You may recall that the exit velocity was found from the energy balance from turbine exit (5) to the nozzle exit (7) for the non-afterburning gas-turbine engine (see Eq. (3.38)).

$$\text{non-afterburning: } h_5^o = h_7^o = h_7 + \frac{U_7^2}{2} \rightarrow \frac{(U_7^2)_{non-AB}}{2c_p} = T_7^o\left[1 - \left(\frac{p_7}{p_7^o}\right)^{\frac{\gamma-1}{\gamma}}\right]$$
$$= T_5^o\left[1 - \left(\frac{p_7}{p_7^o}\right)^{\frac{\gamma-1}{\gamma}}\right] \tag{3.58}$$

For the afterburning engine, the stagnation temperature is increased from T_5^o to T_6^o through combustion in the afterburner. A schematic of the afterburner is shown in Figure 3.5. A fuel

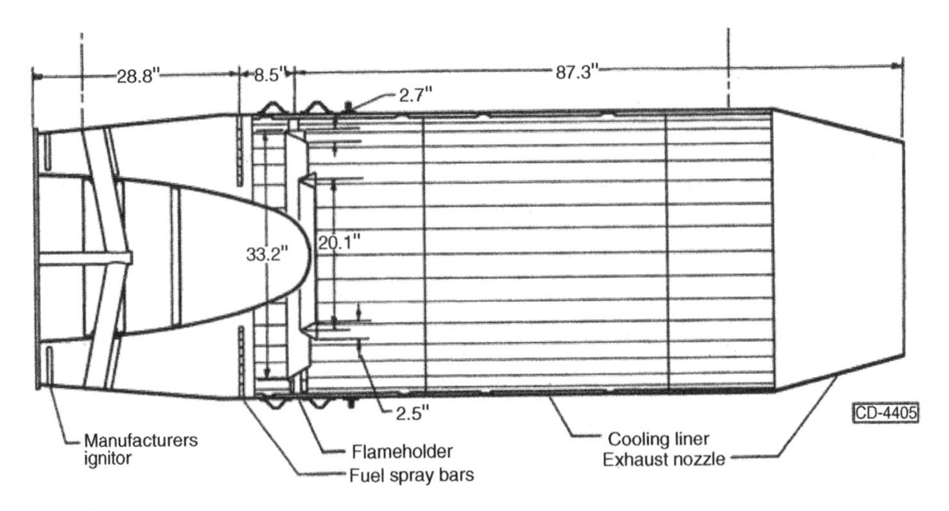

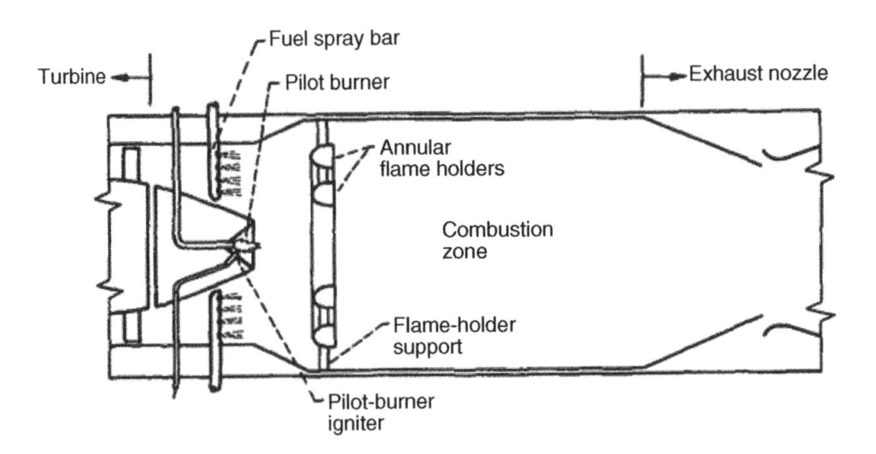

Figure 3.5 Schematic of the afterburner. Courtesy of NASA.

ring adds fuel to the gas, and the flame is anchored in the stabilizer, sustaining this heat addition process. For the afterburning case, we have a slightly changed energy balance.

Afterburning:

$$h_5^o < h_6^o = h_7^o = h_7 + \frac{U_7^2}{2} \rightarrow \frac{(U_7^2)_{AB}}{2c_p} = T_7^o\left[1 - \left(\frac{p_7}{p_7^o}\right)^{\frac{\gamma-1}{\gamma}}\right] = T_6^o\left[1 - \left(\frac{p_7}{p_7^o}\right)^{\frac{\gamma-1}{\gamma}}\right] \tag{3.59}$$

We note that the pressure ratio inside the square brackets in Eqs. (3.58) and (3.59) are the same, since the afterburner combustion is also a constant-pressure process. Comparing Eqs. (3.58) and (3.59) leads to the conclusion that we only need to modify our previous result (Eq. (3.45)) by the ratio of the stagnation temperatures with and without afterburning.

$$\frac{(U_7^2)_{AB}}{(U_7^2)_{non-AB}} = \frac{T_6^o}{T_5^o} = \frac{\frac{T_6^o}{T_o}}{\frac{T_5^o}{T_4^o}\frac{T_4^o}{T_o}} = \frac{\tau_{\lambda,AB}}{\tau_t\tau_\lambda} \tag{3.60}$$

$$\tau_{\lambda,AB} = \frac{T_6^o}{T_o}$$

In Eq. (3.60), $\tau_{\lambda,AB}$ is the stagnation temperature achieved by the afterburner combustion, normalized by the ambient temperature, similar to τ_λ. With this change, Eq. (3.45) is rewritten for the afterburning case.

$$\frac{(U_7)_{AB}}{a_o} = \sqrt{\frac{2}{\gamma-1}\frac{\tau_{\lambda,AB}}{\tau_t\tau_\lambda}\frac{\tau_\lambda}{\tau_r\tau_c}(\tau_r\tau_c\tau_t - 1)} = \sqrt{\frac{2}{\gamma-1}\tau_{\lambda,AB}\left(1 - \frac{1}{\tau_r\tau_c\tau_t}\right)} \tag{3.61}$$

Again, we can use the results for the turbojet thrust (Eq. (3.53)) and specific thrust (Eq. (3.54)), by using the exhaust velocity in Eq. (3.61) for the afterburning case.

For the fuel consumption, the fuel is now fed to both the main combustor and the afterburner. We can consider the combined total mass flow rate of the engine as the source of heat input to the entire engine, and apply the energy balance from the inlet to the exit.

$$\dot{m}_o h_o^o + \dot{m}_{f,total}h_{PR} = \dot{m}_o h_7^o \rightarrow \dot{m}_{f,total}h_{PR} = \dot{m}_o c_p\left(T_7^o - T_o^o\right) \tag{3.62}$$

From this energy balance, we can solve for the total fuel–air ratio in the turbojets with afterburning.

$$f_{total} = \frac{\dot{m}_{f,total}}{\dot{m}_o} = c_pT_o\left(\frac{T_7^o}{T_o} - \frac{T_o^o}{T_o}\right) = c_pT_o\left(\frac{T_6^o}{T_o} - \frac{T_o^o}{T_o}\right) = c_pT_o\left(\tau_{\lambda,AB} - \tau_r\right) \tag{3.63}$$

In Eq. (3.63), only the ratio, $\tau_{\lambda,AB}$ appears, since that ratio represents the combined effect of heat addition in the main combustor and the afterburner. With the total fuel–air ratio and the specific thrust above, the calculation of the TSFC (Eq. (3.55)) is identical to that for turbojet without afterburning.

Table 3.5 A summary of equations for turbojets with afterburning (ideal-cycle analysis).

Input parameters: M_o, T_o, γ, c_p, h_{PR}, T_4^o, T_6^o, π_c.
Output parameters: U_7, F_s, TSFC, η_P, η_T, η_o.
Analysis result:

$$\tau_r = 1 + \frac{\gamma - 1}{2} M_o^2$$

$$\tau_c = (\pi_c)^{(\gamma-1)/\gamma}$$

$$\tau_\lambda = \frac{T_4^o}{T_o}$$

$$\tau_t = 1 - \frac{\tau_r}{\tau_\lambda}(\tau_c - 1)$$

$$\frac{(U_7)_{AB}}{a_o} = \sqrt{\frac{2}{\gamma - 1}\tau_{\lambda,AB}\left(1 - \frac{1}{\tau_r \tau_c \tau_t}\right)}$$

$$f_{total} = c_p T_o(\tau_{\lambda,AB} - \tau_r)$$

$$F_s = a_o\left(\frac{U_7}{a_o} - M_o\right)$$

$$\text{TSFC} = \frac{f_{total}}{F_s}$$

$$\eta_T = \frac{\frac{1}{2}\dot{m}_o\left(U_7^2 - U_o^2\right)}{\dot{m}_{f,total}h_{PR}} = \frac{\left(U_7^2 - U_o^2\right)}{2f_{total}h_{PR}}; \eta_P = 2 \times \frac{\frac{U_7}{U_o} - 1}{\frac{U_7^2}{U_o^2} - 1}; \eta_o = \eta_P \eta_T$$

However, the same thermal efficiency (Eq. (3.56)) cannot be applied since now there is an additional source of heat input. So we return to the definition of the thermal efficiency, where it is defined as the ratio of the change in kinetic energy of the fluid (net outcome of operating the turbojet engine) to the total heat input.

$$\eta_T = \frac{\frac{1}{2}\dot{m}_o\left(U_7^2 - U_o^2\right)}{\dot{m}_{f,total}h_{PR}} = \frac{\left(U_7^2 - U_o^2\right)}{2f_{total}h_{PR}} \tag{3.64}$$

The above results are summarized in Table 3.5.

Example 3.6 Afterburning turbojet

Compare the performance of a turbojet with and without afterburning, for the following engine data.

$$M_o = 2.0@T_o = 411.8°\text{R}.\ h_{PR} = 18\ 400\ \text{BTU/lbm}, T_4^o = 1583°\text{R}, T_6^o = 2000°\text{R}, \text{and}\ \pi_c = 10.$$

$$\gamma = 1.4, c_P = 0.24\ \frac{\text{BTU}}{\text{lbm°R}}.$$

a. *Without afterburning*

$$\tau_r = 1 + \frac{\gamma - 1}{2} M_o^2 = 1.128, T_2^o = \tau_r T_o = 464.5°\text{R}; p_2^o = \tau_r^{\gamma/(\gamma-1)} p_o = 959.6 \frac{\text{lbf}}{\text{ft}^2} \quad \text{(E3.6.1)}$$

$$\tau_c = (\pi_c)^{(\gamma-1)/\gamma} = 2.1 \rightarrow T_3^o = \tau_c T_2^o = 975.5°\text{R}; p_3^o = \pi_c p_2^o = 12,954 \frac{\text{lbf}}{\text{ft}^2} \quad \text{(E3.6.2)}$$

$$\tau_\lambda = \frac{T_4^o}{T_o} = 4.06 \quad \text{(E3.6.3)}$$

$$\tau_t = 1 - \frac{\tau_r}{\tau_\lambda}(\tau_c - 1) = 0.69 \rightarrow T_5^o = \tau_t T_4^o = 1159°\text{R} \quad \text{(E3.6.4)}$$

Using the above temperature ratios,

$$M_7 = \sqrt{\frac{2}{\gamma - 1}(\tau_r \tau_c \tau_t - 1)} = 1.79 = 0.257 \quad \text{(E3.6.5)}$$

$$U_7 = a_o \sqrt{\frac{2}{\gamma - 1} \frac{\tau_\lambda}{\tau_r \tau_c}(\tau_r \tau_c \tau_t - 1)} = 1745 \text{ ft/s} \quad \text{(E3.6.6)}$$

$$F_s = U_7 - a_o M_o = 949.6 \frac{\text{ft}}{\text{s}} \rightarrow 29.5 \frac{\text{lbf}}{\text{lbm/s}} \quad \text{(E3.6.7)}$$

$$f = \frac{c_p T_o}{h_{PR}}(\tau_\lambda - \tau_r \tau_c) = \frac{0.24 \frac{\text{BTU}}{\text{lbm} \cdot °\text{R}} \times 411.8°\text{R}}{18\,400 \frac{\text{BTU}}{\text{lbm}}}(4.06 - 1.128 \times 2.1) = 0.0091 \quad \text{(E3.6.8)}$$

This gives

$$\text{TSFC} = \frac{f}{F_s} = 1.11 \frac{\text{lbm}}{\text{hr} \cdot \text{lbf}} \quad \text{(E3.6.9)}$$

b. *With afterburning*

$$\tau_{\lambda AB} = \frac{T_6^o}{T_o} = 4.86 \quad \text{(E3.6.10)}$$

$$(U_7)_{AB} = a_o \sqrt{\frac{2}{\gamma - 1} \tau_{\lambda AB} (1 - \frac{1}{\tau_r \tau_c \tau_t})} = 3068.5 \text{ ft/s} \quad \text{(E3.6.11)}$$

This generates a specific thrust of

$$F_s = (U_7)_{AB} - a_o M_o = 70.6 \frac{\text{lbf}}{\text{lbm/s}} \quad \text{(E3.6.12)}$$

And,

$$f = \frac{c_p T_o}{h_{PR}}(\tau_{\lambda AB} - \tau_r) = \frac{0.24\,\frac{\text{BTU}}{\text{lbm}\cdot{}^\circ\text{R}} \times 411.8^\circ\text{R}}{18\,400\,\frac{\text{BTU}}{\text{lbm}}}(4.86 - 1.128) = 0.02 \quad \text{(E3.6.13)}$$

$$\text{TSFC} = \frac{f}{F_s} = 1.02\,\frac{\text{lbm}}{\text{hr}\cdot\text{lbf}} \quad \text{(E3.6.14)}$$

3.4.3 Turbofan Engines with Afterburning (Mixed Stream)

Afterburning is also used in turbofan engines for thrust augmentation during take-offs and high-speed maneuvers, almost exclusively in low bypass-ratio, high-performance engines. In these low bypass-ratio turbofan engines, the afterburning takes place in the so-called mixed stream, consisting of the fan air stream and the core gas stream. A schematic of this arrangement is shown in Figure 3.6. If the afterburner is "on", then the mixed-stream turbofan essentially acts like a turbojet, since there is a common exhaust stream generated by the afterburner energy. Therefore, all of the equations developed for turbojets apply for turbofan engines with mixed-stream afterburning, except for the fuel mass flow rate in the afterburner. The reason is that the mixing of the fan and core streams generates a mass-weighted average temperature between T_5^o to T_8^o. In most instances, the stagnation pressure of the two streams is assumed to be equal, so that afterburning in the mixed stream only affects the fuel metering. So we set up a new energy balance for this mixed stream. First, let us define the mass-weighted average temperature between T_5^o and T_8^o, as T_M^o.

$$\dot{m}_C c_p T_5^o + \dot{m}_{fan} c_p T_8^o = \dot{m}_o c_p T_M^o \quad (3.65)$$

We can solve for T_M^o and also write it as a ratio form.

$$\tau_M \equiv \frac{T_M^o}{T_5^o} = \frac{1}{1+\alpha}\left(1 + \alpha\frac{T_8^o}{T_5^o}\right) = \frac{1}{1+\alpha}\left(1 + \alpha\frac{\tau_r\tau_f}{\tau_\lambda\tau_t}\right) \quad (3.66)$$

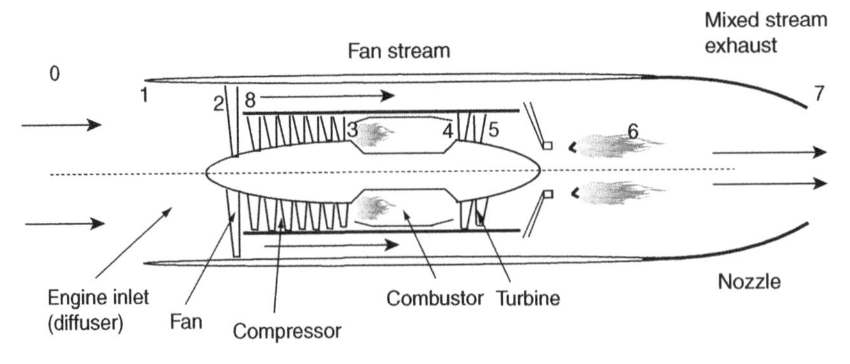

Figure 3.6 Mixed stream afterburner in low bypass-ratio turbofans.

The afterburner adds energy to this mixed stream at stagnation temperature of T_M^o.

$$\dot{m}_{f,AB}h_{PR} + \dot{m}_o c_p T_M^o = \dot{m}_o c_p T_7^o = \dot{m}_o c_p T_6^o \tag{3.67}$$

The last part of the equality in Eq. (3.67) is due to the nozzle being assumed to be adiabatic. The updated fuel–air ratio in the afterburner is

$$f_{AB} \equiv \frac{\dot{m}_{f,AB}}{\dot{m}_o} = \frac{c_p T_o}{h_{PR}}\left(\tau_{\lambda,AB} - \tau_\lambda \tau_t \tau_M\right) \tag{3.68}$$

The combined fuel–air ratio for the main combustor and the afterburner requires a simple addition.

$$f_{total} = \frac{\dot{m}_f + \dot{m}_{f,AB}}{\dot{m}_o} = \frac{f}{1+\alpha} + f_{AB} \tag{3.69}$$

This new total fuel–air ratio is used to calculate the TSFC and the thermal efficiency. As noted above, other aspects of the calculations are identical to those of turbojets.

The above results are summarized in Table 3.6.

Table 3.6 A summary of equations for turbofan with afterburning in the mixed stream (ideal-cycle analysis).

Input parameters: M_o, T_o, γ, c_p, h_{PR}, T_4^o, T_6^o, π_c, π_f, α.
Output parameters: U_7, U_9, F_s, TSFC, η_P, η_T, η_o.
Analysis result:

$$\tau_r = 1 + \frac{\gamma - 1}{2}M_o^2$$

$$\tau_c = (\pi_c)^{(\gamma-1)/\gamma};\ \tau_f = (\pi_f)^{(\gamma-1)/\gamma}$$

$$\tau_\lambda = \frac{T_4^o}{T_o};\ \tau_t = 1 - \frac{\tau_r}{\tau_\lambda}\left[\tau_c - 1 + \alpha(\tau_f - 1)\right]$$

$$\tau_M = \frac{1}{1+\alpha}\left(1 + \alpha\frac{\tau_r \tau_f}{\tau_\lambda \tau_t}\right)$$

$$\left(\frac{U_7}{a_o}\right)^2 = \frac{2}{\gamma - 1}\frac{\tau_\lambda}{\tau_r \tau_c}\left(\tau_r \tau_c\left\{1 - \frac{\tau_r}{\tau_\lambda}\left[\tau_c - 1 + \alpha(\tau_f - 1)\right]\right\} - 1\right)$$

$$\left(\frac{U_9}{a_o}\right)^2 = \frac{2}{\gamma - 1}\left(\tau_r \tau_f - 1\right)$$

$$F = \dot{m}_C(U_7 - U_o) + \dot{m}_{fan}(U_9 - U_o)$$

$$F_s = \frac{a_o}{1+\alpha}\left[\frac{U_7}{a_o} - M_o + \alpha\left(\frac{U_9}{a_o} - M_o\right)\right]$$

$$f = \frac{c_p T_o}{h_{PR}}(\tau_\lambda - \tau_r \tau_c);\ f_{AB} = \frac{c_p T_o}{h_{PR}}\left(\tau_{\lambda,AB} - \tau_\lambda \tau_t \tau_M\right);\ f_{total} = \frac{f}{1+\alpha} + f_{AB}$$

$$\text{TSFC} = \frac{f_{total}}{(1+\alpha)F_s}$$

$$\eta_T = 1 - \frac{1}{\tau_r \tau_c};\ \eta_P = 2 \times \frac{\dfrac{U_7}{U_o} - 1 + \alpha\left(\dfrac{U_9}{U_o} - 1\right)}{\dfrac{U_7^2}{U_o^2} - 1 + \alpha\left(\dfrac{U_9^2}{U_o^2} - 1\right)};\ \eta_o = \eta_P \eta_T$$

3.4.4 Ramjets

At supersonic speeds, there is sufficient kinetic energy in the incident air stream to cause a substantial increase in the pressure when decelerated through the inlet. This so-called ram compression can be used to bypass or discard the compressor altogether, and run a jet engine. This operation is predicated by the presence of ram compression, sufficient only at supersonic speeds. Once such ram compression is achieved, typically with the aid of auxiliary boosters, the weight and complexity of the compressor–turbine assembly can be dispensed with. As shown in Figure 3.7, a ramjet engine looks much like an afterburner with a supersonic inlet. In conventional ramjets, the supersonic stream is decelerated to subsonic speeds through the inlet, prior to the combustion process. This enables relatively effective flame stabilization strategies, similar to afterburners. However, this deceleration incurs losses in stagnation pressure. Nowadays, higher thrust and better efficiency can be achieved by decelerating the flow just enough for flame stabilization, but still maintain combustion at supersonic speeds. These are called scramjet engines, standing for for *s*upersonic *c*ombustion *ramjets*. The choice of fuel and combustor geometry evidently requires a much higher level of intricacy in scramjet engines.

For the time being, we will only consider conventional ramjets under idealized conditions. The thrust and specific thrust are determined in the same manner as for turbojet (Eqs. (3.53) and (3.54)).

$$F = \dot{m}_o(U_7 - U_o) \tag{3.53}$$

$$F_s = \frac{F}{\dot{m}_o} = (U_7 - U_o) \tag{3.54}$$

Determination of the exhaust velocity, U_7, is also similar.

$$\frac{U_7}{a_o} = \frac{a_7}{a_o}M_7 = \sqrt{\frac{T_7}{T_o}}M_7 \tag{3.45}$$

$$M_7 = \sqrt{\frac{2}{\gamma - 1}\left[\left(\frac{p_7^o}{p_7}\right)^{\frac{\gamma-1}{\gamma}} - 1\right]} \tag{3.39}$$

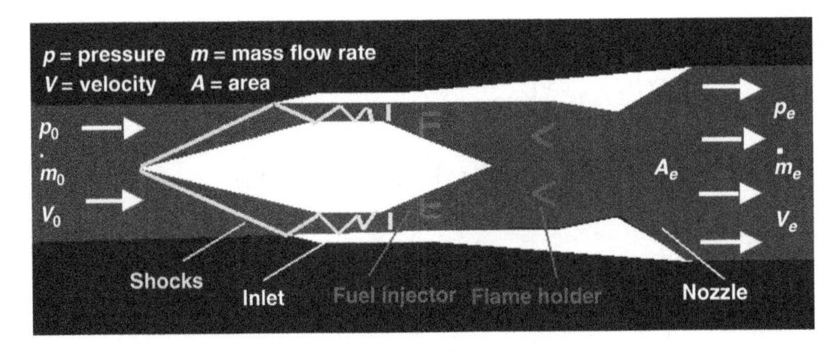

Figure 3.7 Schematic of a ramjet. Courtesy of NASA.

The pressure ratio in Eq. (3.39) is somewhat different from the turbojet due to the stagnation pressure loss across the shock at the inlet.

$$\frac{p_7^o}{p_7} = \frac{p_7^o}{p_o} = \frac{p_7^o}{p_4^o}\frac{p_4^o}{p_3^o}\frac{p_3^o}{p_1^o}\frac{p_1^o}{p_o^o}\frac{p_o^o}{p_o} = \pi_n\pi_b\pi_d\pi_s\pi_r = \pi_s\pi_r \tag{3.70}$$

$$\pi_n = \pi_b = \pi_d = 1$$

A new pressure ratio is used to include the stagnation pressure loss through the shocks at the inlet.

$$\pi_s = \frac{p_1^o}{p_o^o} \tag{3.71}$$

This stagnation pressure loss may be estimated by the following expressions.

$$\begin{aligned} \pi_s &= 1 && \text{for} \quad M_o < 1 \\ \pi_s &= 1 - 0.075(M_o - 1)^{1.35} && \text{for} \quad 1 \le M_o < 5 \\ \pi_s &= \frac{800}{M_o^4 + 935} && \text{for} \quad M_o > 1 \end{aligned} \tag{3.72}$$

The temperature is found in the same manner as Eq. (3.44).

$$\frac{T_7}{T_o} = \frac{\tau_r\tau_s\tau_b}{\tau_r\tau_s} = \tau_b = \frac{\tau_\lambda}{\tau_r} \tag{3.73}$$

Combining the results of Eqs. (3.70) and (3.73) gives us

$$\frac{U_7}{a_o} = \sqrt{\frac{2}{\gamma - 1}\frac{\tau_\lambda}{\tau_r}\left[(\pi_r\pi_s)^{\frac{\gamma-1}{\gamma}} - 1\right]} \tag{3.74}$$

Fuel–air ratio calculation (Eq. (3.30)) is also nearly identical, except that, again, we set the compressor temperature ratio to one in its absence.

$$f \equiv \frac{\dot{m}_f}{\dot{m}_o} = \frac{c_pT_o}{h_{PR}}(\tau_\lambda - \tau_r) \tag{3.75}$$

The above results are summarized in Table 3.7.

Table 3.7 A summary of equations for ramjets.

Input parameters: M_o, T_o, γ, c_p, h_{PR}, T_4^o, π_s.
Output parameters: U_7, F_s, TSFC, η_P, η_T, η_o.
Analysis result:

$$\tau_r = 1 + \frac{\gamma - 1}{2} M_o^2$$

$$\tau_\lambda = \frac{T_4^o}{T_o}$$

$$\frac{U_7}{a_o} = \sqrt{\frac{2}{\gamma - 1} \frac{\tau_\lambda}{\tau_r} \left[(\pi_r \pi_s)^{\frac{\gamma-1}{\gamma}} - 1 \right]}$$

$$f = \frac{c_p T_o}{h_{PR}} (\tau_\lambda - \tau_r)$$

$$F_s = a_o \left(\frac{U_7}{a_o} - M_o \right)$$

$$\text{TSFC} = \frac{f}{F_s}$$

$$\eta_T = \frac{(U_7^2 - U_o^2)}{2 f h_{PR}}; \ \eta_P = 2 \times \frac{\frac{U_7}{U_o} - 1}{\frac{U_7^2}{U_o^2} - 1}; \ \eta_o = \eta_P \eta_T$$

Example 3.7 Ramjet

Find the specific thrust and TSFC for a ramjet operating at $H = 50\,000$ ft, for $M_o = 2$ and 4. $h_{PR} = 45\ 000$ kJ/kg, $T_4^o = 3000$ K and $\gamma = 1.4$.

At 50 000 ft, $T_o = 205$ K, $p_o = 11.6$ kPa, and $a_o = 287$ m/s.

$$M_o = 2$$

First, we can find the stagnation pressure loss at the inlet using Eq. (3.72).

$$\pi_s = 1 - 0.075(M_o - 1)^{1.35} = 1 - 0.075(2 - 1)^{1.35} = 0.925 \quad \text{(E3.7.1)}$$

$$\tau_r = 1 + \frac{\gamma - 1}{2} M_o^2 = 1.8, \pi_r = \frac{p_o^o}{p_o} = (\tau_r)^{\frac{\gamma}{\gamma-1}} = 7.82 \quad \text{(E3.7.2)}$$

$$M_7 = \sqrt{\frac{2}{\gamma - 1} \left[(\pi_r \pi_s)^{\frac{\gamma}{\gamma-1}} - 1 \right]} = 1.93 \quad \text{(E3.7.3)}$$

$$F_s = a_o \left(\frac{U_7}{a_o} - M_o \right) = 1005.4 \frac{\text{N}}{\text{kg/s}}, \text{using } \frac{U_7}{a_o} = \frac{T_7}{T_o} = \sqrt{\tau_b} M_7 = \sqrt{\frac{\tau_\lambda}{\tau_r}} M_7 = 5.5 \quad \text{(E3.7.4)}$$

$$f = \frac{c_p T_o}{h_{PR}}(\tau_\lambda - \tau_r) = \frac{1.004\ \frac{\text{kJ}}{\text{kgK}} \times 205\ \text{K}}{45\ 000\ \frac{\text{kJ}}{\text{kg}}}(14.6 - 1.8) = 0.0586 \quad \text{(E3.7.5)}$$

$$\text{TSFC} = \frac{f}{F_s} = 0.0583\ \frac{\text{kg/s}}{\text{kN}} \quad \text{(E3.7.6)}$$

$$M_o = 4 \quad \text{(E3.7.7)}$$

$$\pi_s = 1 - 0.075(M_o - 1)^{1.35} = 1 - 0.075(4 - 1)^{1.35} = 0.67 \quad \text{(E3.7.8)}$$

$$\tau_r = 1 + \frac{\gamma - 1}{2}M_o^2 = 4.2, \pi_r = \frac{p_o^o}{p_o} = (\tau_r)^{\frac{\gamma}{\gamma-1}} = 151.84 \quad \text{(E3.7.9)}$$

$$M_7 = \sqrt{\frac{2}{\gamma - 1}\left[(\pi_r\pi_s)^{\frac{\gamma}{\gamma-1}} - 1\right]} = 3.47 \quad \text{(E3.7.10)}$$

$$F_s = a_o\left(\frac{U_7}{a_o} - M_o\right) = 1689.9\ \frac{\text{N}}{\text{kg/s}};\ \frac{U_7}{a_o} = \frac{T_7}{T_o} = \sqrt{\tau_b}M_7 = \sqrt{\frac{\tau_\lambda}{\tau_r}}M_7 = 9.89 \quad \text{(E3.7.11)}$$

$$f = \frac{c_p T_o}{h_{PR}}(\tau_\lambda - \tau_r) = \frac{1.004\ \frac{\text{kJ}}{\text{kgK}} \times 205\ \text{K}}{45\ 000\ \frac{\text{kJ}}{\text{kg}}}(14.6 - 4.2) = 0.0476 \quad \text{(E3.7.12)}$$

$$\text{TSFC} = \frac{f}{F_s} = 0.0281\ \frac{\text{kg/s}}{\text{kN}} \quad \text{(E3.7.13)}$$

With the substantial increase in the specific thrust, TSFC decreases.

3.5 Further Uses of Basic Engine Analysis

The above analyses can be used to determine the basic performance parameters of various gas-turbine engines under some idealized assumptions, as we saw in several of the examples in the previous section. In addition, because most of the results are in algebraic equations, we can take these and find optimum conditions for engine operations. Let us, for example, take the turbojet engine result for the exhaust velocity.

$$\frac{U_7}{a_o} = \frac{a_7 M_7}{a_o} = \sqrt{\frac{2\tau_b}{\gamma - 1}(\tau_r\tau_c\tau_t - 1)} = \sqrt{\frac{2}{\gamma - 1}\frac{\tau_\lambda}{\tau_r\tau_c}(\tau_r\tau_c\tau_t - 1)} \quad (3.45)$$

The maximum specific thrust will be obtained at a given flight Mach number, when the exhaust velocity is at a maximum. Then, what should be the compression ratio that generates this maximum specific thrust? Again, we note that, at a given flight Mach number, the maximum specific thrust is found when the exhaust velocity is at maximum. We can take Eq. (3.45), or the square of it, to find where that maximum exhaust velocity occurs as a function of the compression ratio, or τ_c. Under isentropic operation, there is a direct relationship between the compression ratio and τ_c.

$$\frac{\partial}{\partial \tau_c}\left[\left(\frac{U_7}{a_o}\right)^2\right] = \frac{2}{\gamma-1}\frac{\partial}{\partial \tau_c}\left[\frac{\tau_\lambda}{\tau_r\tau_c}(\tau_r\tau_c\tau_t - 1)\right] = 0 \tag{3.76}$$

We can differentiate by parts, while remembering that τ_t is a function of τ_c.

$$-\frac{\tau_\lambda}{\tau_r\tau_c^2}(\tau_r\tau_c\tau_t - 1) + \frac{\tau_\lambda}{\tau_r\tau_c}\left(\tau_r\tau_t + \tau_r\tau_c\frac{\partial \tau_t}{\partial \tau_c}\right) = 0 \tag{3.77}$$

This simplifies to

$$\frac{1}{\tau_r\tau_c^2} + \frac{\partial \tau_t}{\partial \tau_c} = 0 \tag{3.78}$$

The turbine temperature ratio, τ_t, for turbojets is found by setting $\alpha = 1$ in Eq. (3.34b).

$$\tau_t = \frac{T_5^o}{T_4^o} = 1 - \frac{\tau_r}{\tau_\lambda}(\tau_c - 1) \tag{3.79}$$

This is easily differentiated, and the result substituted in Eq. (3.78), to set up the compression condition for maximum specific thrust.

$$(\tau_c)_{\max Fs} = \frac{\sqrt{\tau_\lambda}}{\tau_r} \tag{3.80}$$

This is readily converted to the compression ratio through the isentropic relationship.

$$(\pi_c)_{\max Fs} = \left(\frac{\sqrt{\tau_\lambda}}{\tau_r}\right)^{\frac{\gamma}{\gamma-1}} \tag{3.81}$$

A similar procedure can be applied for turbojets with afterburning. It will be shown in one of the problems at the end of this chapter that the compression ratio that produces the maximum specific thrust in afterburning turbojets is

$$(\pi_c)_{\max\ Fs} = \left[\frac{1}{2}\left(\frac{\tau_\lambda}{\tau_r} + 1\right)\right]^{\frac{\gamma}{\gamma-1}} \quad \text{for afterburning turbojet} \tag{3.82}$$

We can use a similar approach to find, say, the optimum bypass ratio, for minimum TSFC in turbofans. We had an expression for the TSFC in turbofan engines (Eq. (3.49)).

$$\text{TSFC} = \frac{f}{(1+\alpha)F_s} \tag{3.49}$$

Example 3.8 Compression ratio for maximum specific thrust in turbojets

For $T_4^o = 1600$ K and $T_o = 220$, $\tau_\lambda = 7.27$.

The compression ratio for maximum specific thrust is found using Eq. (3.80).

For $M_o = 0.8, \tau_r = 1.128 \rightarrow (\tau_c)_{\max Fs} = \frac{\sqrt{\tau_\lambda}}{\tau_r} = 2.4$, from which $\pi_c = 21.4$

For $M_o = 2.5, \tau_r = 2.25 \rightarrow (\tau_c)_{\max Fs} = \frac{\sqrt{\tau_\lambda}}{\tau_r} = 1.2$, from which $\pi_c = 1.9$

This looks simple enough, but there is much algebra to be done, since the specific thrust contains the bypass ratio in several places. We note that the minimum TSFC will occur when the denominator in Eq. (3.49) is at a maximum. The fuel–air ratio represents the fuel consumption in the core section, and does not involve the bypass ratio in any way.

$$\frac{\partial}{\partial \alpha}\left[(1+\alpha)\frac{F}{\dot{m}_o}\right] = 0 \tag{3.83}$$

The specific thrust has the following form (Eq. (3.48)).

$$F_s = \frac{a_o}{1+\alpha}\left[\frac{U_7}{a_o} - M_o + \alpha\left(\frac{U_9}{a_o} - M_o\right)\right] \tag{3.48}$$

Therefore, Eq. (3.83) is equivalent to

$$\frac{\partial}{\partial \alpha}\left[(1+\alpha)\frac{F}{\dot{m}_o}\right] = 0 \Rightarrow \frac{\partial}{\partial \alpha}\left[\frac{U_7}{U_o} - 1 + \alpha\left(\frac{U_9}{U_o} - 1\right)\right] = 0 \tag{3.84}$$

Note that U_7 is also a function of α, so after much algebra we should arrive at the following result for the bypass ratio, resulting in the minimum TSFC in turbofan engines.

$$(\alpha)_{minTSFC} = \frac{1}{\tau_r(\tau_f - 1)}\left[\tau_\lambda - \tau_r(\tau_c - 1) - \frac{\tau_\lambda}{\tau_r\tau_c} - \frac{1}{4}\left(\sqrt{\tau_r\tau_f - 1} + \sqrt{\tau_r - 1}\right)^2\right] \tag{3.85}$$

A similar approach can be used to find the fan pressure ratio or temperature ratio that minimizes the TSFC in turbofans.

$$(\tau_f)_{\text{minTSFC}} = \frac{\tau_\lambda - \tau_r(\tau_c - 1) - \frac{\tau_\lambda}{\tau_r\tau_c} - \alpha\tau_r + 1}{\tau_r(1+\alpha)} \tag{3.86}$$

Example 3.9 Optimum bypass ratio in turbofan engines

$$\text{For } T_4^o = 1600\text{ K and } T_o = 220, \tau_\lambda = 7.27. \text{ Also, } \pi_c = 12, \text{and } \pi_f = 2.5.$$

The bypass ratio for the minimum TSFC is found using Eq. (3.85).

$$\text{For } M_o = 0.8, \tau_r = 1.128 \rightarrow (\alpha^*)_{\text{minTSFC}} = 7.87.$$

For $M_o = 2.5$, $\tau_r = 2.25 \rightarrow (\alpha^*)_{\text{minTSFC}} = -0.67$, so there are no possible values of bypass ratio at this flight Mach number.

3.6 MATLAB® Program

```
% MCE 34: Calculates the turbofan performance based on the following inputs:
% Mo, To, gamma, cp, hPR, T4o, pc, pf (fan pressure ratio), alpha (bypass ratio).
% OUTPUT PARAMETERS: U7, U9, Fs, TSFC, hP, hT, ho.
% Other variations such as turbojets can be calculated, e.g. simply by setting
% alpha = 0.

% English units are used in this example.
G=32.174; % conversion factor

% INPUT PARAMETERS
Mo = 0.8; To = 411.8; po= 629.5; ao= 995;
gamma = 1.4; cp =0.24; hPR = 18400;
T4o = 1850;, pc =15; pf = 1.75; alpha =6.5;
mo=1125;

%Start of calculations

% Options to vary X(i) = Mo, To, etc.

N = 25;
Xmax=1; Xmin=0; dX=(Xmax-Xmin)/N;
for i = 1:N;
 X(i)= Xmin+dX*(i-1);
 Mo=X(i);

Uo=Mo*ao

tr=1+(gamma-1)/2*Mo^2; T2o=tr*To

tc=pc^((gamma-1)/gamma)
tf=pf^((gamma-1)/gamma)

tL=T4o/To

tt=1-tr/tL*(tc-1+alpha*(tf-1))
```

```
% Exit velocities: U2=(U7/ao)^2; V2=(U9/ao)^2

U2=2/(gamma-1)*tL/(tr*tc)*(tr*tc*tt-1)

U7=sqrt(U2)*ao

V2=2/(gamma-1)*(tr*tf-1);

U9=sqrt(V2)*ao

% Mass flow rates

mc=mo/(1+alpha); mF=mo*alpha/(1+alpha);

% Thrust

F1=mc*(U7-Uo)+mF*(U9-Uo);

F=F1/G

Fs(i) = F/mo

% TSFC

f = cp*To/hPR*(tL-tr*tc);

TSFC(i)=f/((1+alpha)*Fs(i))*3600

% Efficiencies

hT(i) = 1-1/(tr*tc)

N1=U7/Uo-1+alpha*(U9/Uo-1);

D1=(U7/Uo)^2-1+alpha*((U9/Uo)^2-1);

hP(i)=N1/D1

ho(i)=hT(i)*hP(i)

end

plot(X,Fs)
```

3.7 Problems

3.1. Plot the thermal efficiency of the Brayton cycle as a function of the compression ratio, using Eq. (3.11). $\gamma = 1.4$.

$$\eta_T = 1 - \frac{T_o}{T_3^o} = 1 - \frac{1}{\frac{T_3^o}{T_o}} = 1 - \frac{1}{\left(\frac{p_3^o}{p_o}\right)^{\frac{\gamma-1}{\gamma}}} \tag{3.11}$$

3.2. Air enters the compressor of an idealized gas-turbine engine (isentropic compression and expansion) at $p_2 = 0.55$ bar and $T_2 = 275$ K. For the turbine, the inlet pressure and temperature are $p_4 = 12$ bar and $T_4 = 1200$ K, while the exit pressure is $p_5 = 0.95$ bar. What is the net power output per unit mass of air in kJ/kg?

3.3. Compare two turbojet engines. One has a turbine inlet temperature of 1175 K, while the second is identical, except that the turbine is cooled with bleed-off air from the compressor, allowing the turbine inlet temperature to increase to 1550 K. The flight Mach number is 2 at ambient temperature of 225 K. The compressor pressure ratio is 12. In the second engine 10% of the airflow is bled from the exit of the compressor. After cooling the turbine, the bleed air is not used for generation of thrust.

a. What is the specific thrust of the two engines?
b. What is the ratio of the TSFC of the two engines?

3.4. Find the core and fan mass flow rates, for a turbofan engine with $\alpha = 7.5$ and $\dot{m}_o = 1025 \frac{\text{lbm}}{\text{s}}$.

3.5. For a turbojet engine, following specs are known in English units:

$$H = 30\,\text{kft};\ A_o = 2.64\,\text{ft}^2;\ M_o = 0.8;\ \gamma = 1.4;\ c_p = 0.24\,\text{BTU/lbm}\cdot{}^\circ\text{R};$$
$$h_{pr} = 18\,400\,\text{BTU/lbm};$$
$$T_4^o = 3200^\circ\text{R};\ \pi_c = 16.$$

Calculate the TSFC and the specific thrust.

3.6. For a turbojet engine, the following specs are known in imperial units:

$$H = 30\,\text{kft}; A_o = 2.64\,\text{ft}^2\ M_o = 0.8; \gamma = 1.4; c_p = 0.24\,\text{BTU/lbm}^\circ\text{R};$$
$$h_{pr} = 18\,400\ \text{BTU/lbm};$$
$$T_4^o = 3200^\circ\text{R}; \pi_c = 16; T_6^o = 3750^\circ\text{R}.$$

Calculate the TSFC and the specific thrust.

3.7. For a turbofan engine operating at $M_o = 0.75$ $H = 40$ kft, $\dot{m}_o = 2405 \frac{\text{lbm}}{\text{s}}$, $h_{PR} = 18\,400$ BTU/lbm, $T_4^o = 2650^\circ$R, $\pi_c = 12$, $\pi_f = 1.75$, and $\alpha = 6.5$, find the specific thrust, TSFC, thermal and propulsion efficiencies.

3.8. For the J79-GE-17 turbojet, find the specific thrust, TSFC, thermal and propulsion efficiencies.

$$M_o = 0.85@H = 10\,\text{kft}, \dot{m}_o = 170\,\frac{\text{lbm}}{\text{s}}, h_{PR} = 18\,400\ \text{BTU/lbm},$$
$$T_4^o = 1210^\circ\text{R and}\ \pi_c = 13.5.$$

3.9. Calculate the after-burning specific thrust and TSFC for the J79-GE-17 turbojet. $M_o = 0.85$ @ $H = 10$ kft, $\dot{m}_o = 170 \frac{\text{lbm}}{\text{s}}$, $h_{PR} = 18\,400$ BTU/lbm, $T_4^o = 1210^\circ$R, $T_6^o = 1630^\circ$R and $\pi_c = 13.5$.

3.10. For a mixed-stream turbofan engine operating at $M_o = 0.75$ @ $H = 40$ kft, $\dot{m}_o = 125 \frac{\text{lbm}}{\text{s}}$, $h_{PR} = 18\,400$ BTU/lbm, $T_4^o = 2450^\circ$R, $T_6^o = 2610^\circ$R. $\pi_c = 18$, $\pi_f = 1.75$, and $\alpha = 2.25$, find the specific thrust and TSFC.

3.11. Find the specific thrust and TSFC for a ramjet operating at $H = 40$ kft, for $M_o = 2.5$ and 5. $h_{PR} = 45\,000$ kJ/kg, $T_4^o = 2800$ K and $\gamma = 1.4$.

3.12. For a turbofan engine operating at $M_o = 0.75$ @ $H = 40$ kft, $\dot{m}_o = 2405\ \frac{\text{lbm}}{\text{s}}$, $h_{PR} =$ 18 400 BTU/lbm, $T_4^o = 2650°\text{R}$, $\pi_c = 12$, $\pi_f = 1.75$, plot the specific thrust, TSFC, thermal and propulsion efficiencies as a function of the bypass ratio for $\alpha = 0$ to 12.

3.13. For a turbofan engine operating at $H = 40$ kft, $\dot{m}_o = 2405\ \frac{\text{lbm}}{\text{s}}$, $h_{PR} = 18\ 400$ BTU/lbm, $T_4^o = 2650°\text{R}$, $\pi_c = 12$, $\pi_f = 1.75$, and $\alpha = 6.5$, plot the specific thrust, TSFC, thermal and propulsion efficiencies as a function of the flight Mach number from 0.25 to 0.90.

3.14. For a turbofan engine operating at $M_o = 0.75$ at $H = 40$ kft, $\dot{m}_o = 2405\ \frac{\text{lbm}}{\text{s}}$, $h_{PR} =$ 18 400 BTU/lbm, $T_4^o = 2650°\text{R}$, $\pi_f = 1.75$, and $\alpha = 6.5$, plot the specific thrust, TSFC, thermal and propulsion efficiencies as a function of the compression ratio for $\pi_c =$ 10 to 25.

3.15. For a turbofan engine operating at $M_o = 0.75$ at $H = 40$ kft, $\dot{m}_o = 2405\ \frac{\text{lbm}}{\text{s}}$, $h_{PR} =$ 18 400 BTU/lbm, $\pi_c = 10$, $\pi_f = 1.75$, and $\alpha = 6.5$, plot the specific thrust, TSFC, thermal and propulsion efficiencies as a function of the turbine inlet temperature from $T_4^o = 2250$ to $4000°\text{R}$.

3.16. For the J79-GE-17 turbojet, plot the specific thrust, TSFC, thermal and propulsion efficiencies as a function of M_o. $H = 30$ kft, $\dot{m}_o = 170\ \frac{\text{lbm}}{\text{s}}$, $h_{PR} = 18\ 400$ BTU/lbm, $T_4^o = 1510°\text{R}$ and $\pi_c = 13.5$.

3.17. Plot the after-burning specific thrust and TSFC for the J79-GE-17 turbojet as a function of M_o. $H = 30$ kft, $\dot{m}_o = 170\ \frac{\text{lbm}}{\text{s}}$, $h_{PR} = 18\ 400$ BTU/lbm, $T_4^o = 1510°\text{R}$, $T_6^o = 1630°\text{R}$ and $\pi_c = 13.5$.

3.18. For a mixed-stream turbofan engine operating at $M_o = 0.75$ @ $H = 40$ kft, $\dot{m}_o = 125\ \frac{\text{lbm}}{\text{s}}$, $h_{PR} = 18\ 400$ BTU/lbm, $T_4^o = 2450°\text{R}$, $T_6^o = 2610°\text{R}$. $\pi_c = 18$, $\pi_f =$ 1.75, plot the specific thrust and TSFC as a function of α.

3.19. Plot the specific thrust and TSFC for a ramjet as a function of the combustion temperature, $T_4^o = 1800 - 3500$ K, at $H = 40$ kft, for $M_o = 2.5$ and 5. $h_{PR} =$ 45 000 kJ/kg and $\gamma = 1.4$.

3.20. For an *ideal-cycle* turbojet, (a) determine the compressor pressure ratio that gives the maximum specific thrust at altitude of 6 km, for $M_o = 0.80$; $T_4^o = 1650$ K, $\gamma = 1.4$; (b) what is the maximum specific thrust [N/(kg/s)] corresponding to this π_c?

3.21. For an *ideal-cycle turbofan* engine, calculate U_7, U_9, and the specific thrust for the following operating parameters:

$M_o = 0.9$
H = altitude = 40 kft
$\gamma = 1.4$
$c_p = 0.24$ BTU/(lbm · °R)
$h_{pr} = 18\ 400$ BTU/lbm
$T_4^o = 3000°\text{R}$
$\pi_c = 20$; $\pi_f = 4$; $\alpha = 5.158$;
$p_o/p_7 = 1.0$, $p_o/p_9 = 1.0$.

Bibliography

Jack, D. (2005) *Elements of Gas-Turbine Propulsion*, McGraw Hill, Mattingly.

4

Gas-Turbine Components: Inlets and Nozzles

4.1 Gas-Turbine Inlets

Some typical gas-turbine inlets (also called diffusers) are shown in Figure 4.1. For commercial aircraft, the inlet design is relatively simple, in that in most instances the angle of attack of the aircraft and therefore the inlet will not be very large and therefore the flow is able to enter the engine in a "well-behaved" manner. It is at high angles of attack or yaw during high wind or high-performance maneuvers that the flow can depart from the design conditions. Aside from these off-design conditions, the primary functions of inlets, or diffusers, are to (1) deliver the required mass flow rate into the engine at appropriate Mach numbers; (2) to provide this air flow at a nearly uniform velocity profile into the compressor and (3) minimize the stagnation pressure loss through the diffuser. These requirements arise due to the sensitivity of the compressor operation to both the magnitude and uniformity of the incoming flow. There are other design considerations such as cost, weight, nacelle drag and noise reduction, but the two primary functions are those noted above. The control of the flow speed or the mass flow rate is achieved primarily through inlet sizing. Since the aircraft spends most of its time in cruise, the diffuser is sized to provide the optimum mass flow rate at cruise. At Mach numbers away from this cruise condition, external acceleration/deceleration is induced along with other devices in the inlet such as air bleeds or side vents to augment the mass flow rate control.

At supersonic speeds, the flow characteristics can become quite complex, with shocks inside and outside of the inlets. With all these variables, the Mach number, angles of attack and yaw, the design process involves much testing with both computational fluid dynamic (CFD), wind tunnel tests and flight tests. Figure 4.2 shows an example of CFD analysis showing what can happen inside the diffuser at high angles of attack. The flow is not able to negotiate the low pressure on the inner lower surface, and separates. The flow separation can cause recirculation, which is certainly counter to the design target of sending the air flow through. Moreover, the recirculation results in loss of stagnation pressure, and also significant departure from uniform flow pattern going into the compressor. This effect can be amplified

Aerospace Propulsion, First Edition. T.-W. Lee.
© 2014 John Wiley & Sons, Ltd. Published 2014 by John Wiley & Sons, Ltd.
Companion Website: www.wiley.com/go/aerospaceprop

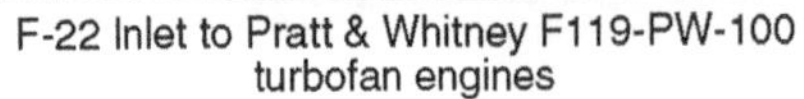

F-22 Inlet to Pratt & Whitney F119-PW-100 turbofan engines

F-22 Inlet to Pratt & Whitney F100-PW-220 turbofan engines

F-16 Inlet to Pratt & Whitney TF-33-P-3/103 Engine on B-52 Stratofortress

Figure 4.1 Gas-turbine engine inlets. Courtesy of US Department of Defence.

through loss in engine thrust, slower speed and even greater flow non-uniformity in the inlet, eventually leading to complete engine stall.

In this chapter, we only consider relatively simple cases of inlet operation, and leave more detailed analyses such as the one shown in Figure 4.2 to CFD studies or projects. With wide availability of CFD packages such as ANSYS Fluent, CFD analysis of inlets can be done at relatively low time and cost expenditures.

4.2 Subsonic Diffuser Operation

Let us first consider what may happen to the flow outside the diffuser in subsonic conditions. You may recall that at subsonic speeds the presence of the inlet is "felt" by the flow ahead, and the incident flow adjusts its streamlines to accommodate the inlet. At high flight speeds or low engine mass flow rates, the inlet presents itself as an obstacle to the flow, and the flow attempts to bypass the inlet by diverging away, as shown in Figure 4.3. There is a certain amount of external flow deceleration as the stream-tube area increases as the flow approaches the inlet. At low speeds or high engine mass flow rates – for example at take-off – the opposite occurs where the engine acts as a sink to the flow and the flow rushes in from a wide stream-tube, again as shown in Figure 4.3. There is then external acceleration in such instances.

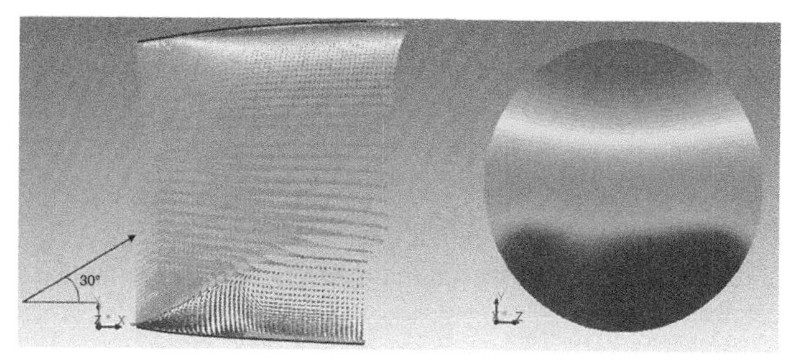

Flow velocity vectors and static pressure, in a circular inlet at angle of attack of 30 deg.

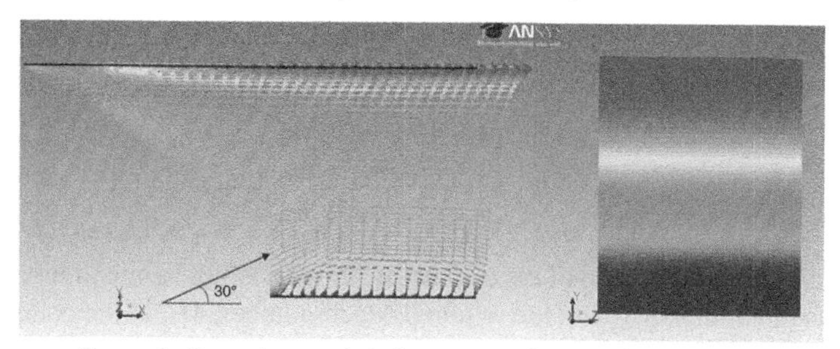

Flow velocity vectors and static pressure, in a rectangular inlet at angle of attack of 30 deg.

Figure 4.2 CFD analysis of inlet flow at a high angle of attack.

As shown in Figure 4.3, the upstream "capture area" can be estimated under ideal, isentropic flow conditions. The ideal conditions mean that the flow is frictionless and that there are no flow disturbances in or out of the diffuser. We start from the mass flow rate balance.

$$\rho_o U_o A_o = \rho_1 U_1 A_1 \rightarrow \frac{A_o}{A_1} = \frac{\rho_1 U_1}{\rho_o U_o} = \frac{\rho_1 a_1 M_1}{\rho_o a_o M_o} = \frac{M_1}{M_o}\frac{\rho_1}{\rho_o}\sqrt{\frac{T_1}{T_o}} \tag{4.1}$$

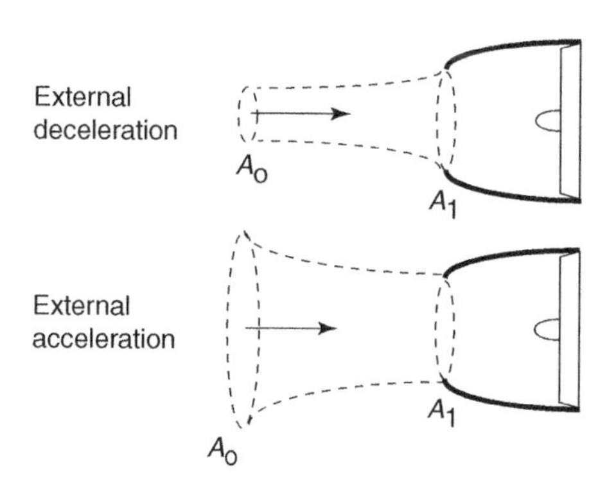

Figure 4.3 External deceleration and acceleration.

We can use the adiabatic condition (the stagnation temperature remains the same) for the temperature ratio, and the isentropic relationship for the density ratio, to find

$$\frac{A_o}{A_1} = \frac{M_1}{M_o}\left[\frac{1+\frac{\gamma-1}{2}M_o^2}{1+\frac{\gamma-1}{2}M_1^2}\right]^{\frac{\gamma+1}{2(\gamma-1)}} \tag{4.2}$$

Equation (4.2) can be used to relate the area ratio between any two points that are connected by isentropic, adiabatic flows.

$$\frac{A_1}{A_2} = \frac{M_2}{M_1}\left[\frac{1+\frac{\gamma-1}{2}M_1^2}{1+\frac{\gamma-1}{2}M_2^2}\right]^{\frac{\gamma+1}{2(\gamma-1)}} \tag{4.3a}$$

$$\frac{A_o}{A_2} = \frac{M_2}{M_o}\left[\frac{1+\frac{\gamma-1}{2}M_o^2}{1+\frac{\gamma-1}{2}M_2^2}\right]^{\frac{\gamma+1}{2(\gamma-1)}} \tag{4.3b}$$

Equation (4.3b) relates the ratio of the capture area to the compressor entrance area, for the given flight Mach number, M_o, and the required Mach number, M_2, at the compressor inlet. The conditions at the compressor entrance are dictated by the target mass flow rate and corresponding velocity or the Mach number. Figure 4.4 shows the plot of Eq. (4.3b) for various M_2. It can be seen that a large capture area is required during low-speed operations (acceleration), as noted above, while the area ratio becomes less than one (deceleration) at normal cruise Mach numbers.

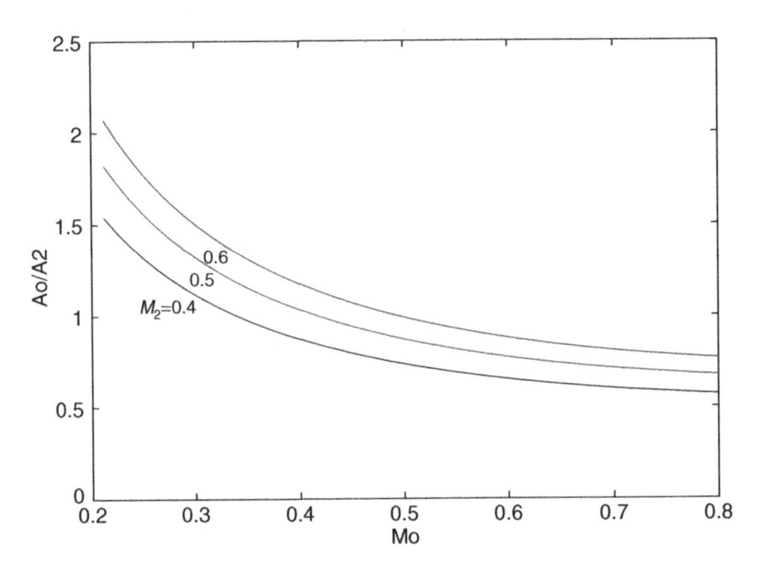

Figure 4.4 Area ratio as a function of the flight Mach numbers, for various M_2, under ideal flow assumption (Eq. (4.3b)).

How would the external acceleration or deceleration affect the flow *inside* the diffuser, under actual conditions (removing the ideal flow assumptions in Eqs. (Eqs. (4.1)–(4.3))? If the flow speed increases (external acceleration), then the static pressure would decrease at "1". For a fixed diffuser with a constant p_2, lower p_1 means that the flow is going through a steeper uphill (up-pressure) climb. A convenient measure of the flow performance of a diffuser is the pressure-recovery coefficient.

$$C_p = \frac{p_2 - p_1}{\frac{1}{2}\rho_1 U_1^2} \tag{4.4}$$

Since the pressure gradient is adverse, $p_2 > p_1$, the pressure is quite sensitive to any increase in the pressure coefficient. If the adverse pressure gradient is too steep, or C_p too large, then the flow can separate and create highly undesirable recirculation within the diffuser (diffuser stall). Thus, external deceleration is highly preferred, by sizing the inlet to minimize external acceleration at low speeds, while allowing for external deceleration at high Mach numbers. This involves oversizing the inlet area, which of course comes at the cost of bulk and increased aerodynamic drag.

Let us now look at how the pressure-recovery coefficient (Eq. (4.4)) may be estimated for simple diffuser geometries. Figure 4.5 shows some characteristics dimensions for three simple

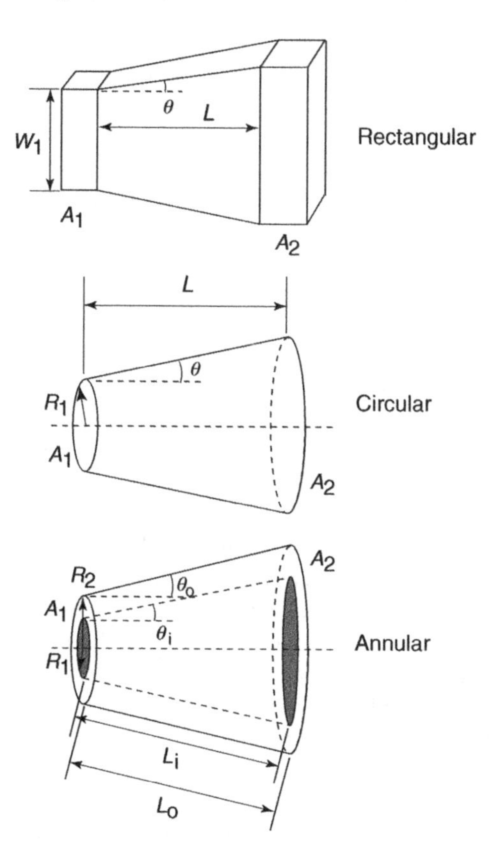

Figure 4.5 Simple diffuser geometries.

diffuser geometries: rectangular, conical and annular. For these geometries, we can use the area ratio, $AR = A_2/A_1$, as our basic diffuser geometrical parameter. We can write the area ratios, as a function of the diffuser dimensions and divergence angles.

$$AR = \frac{A_2}{A_1} = 2\left(1 + \frac{\tan\theta}{W_1}\right) \quad \text{rectangular} \tag{4.5a}$$

$$AR = 1 + \left(1 + \frac{L}{R_1}\tan\theta\right)^2 \quad \text{conical} \tag{4.5b}$$

$$AR = \frac{A_2}{A_1} = \frac{(R_o + L_o \sin\theta_o)^2 + (R_i + L_i \sin\theta_i)^2}{R_o^2 - R_i^2} \quad \text{annular} \tag{4.5c}$$

$$\frac{\bar{L}}{\Delta R} = \frac{L_i + L_o}{2(R_o - R_i)} \quad \text{annular} \tag{4.5d}$$

We can write the "mechanical energy equation" for the inlet flow from 1 to 2.

$$p_1 + \frac{1}{2}\rho U_1^2 - p_{loss} = p_2 + \frac{1}{2}\rho U_2^2 \tag{4.6}$$

We assume low Mach number, incompressible (constant density) flow, where p_{loss} represents the pressure loss through wall friction or flow separation. For steady-state, constant-density flow, the volumetric flow rates at 1 and 2 are equal, so that the velocity is inversely proportional to the area: $U_2/U_1 = A_1/A_2$. Substituting this velocity ratio into Eq. (4.6), and then to Eq. (4.4), gives an expression for the pressure-recovery coefficient.

$$C_p = 1 - \frac{1}{AR^2} - \frac{p_{loss}}{\frac{1}{2}\rho_1 U_1^2} \tag{4.7}$$

We can see that if the area ratio is 1 and the pressure loss is zero, then $C_p = 1$; that is, there is no change in static pressure, as expected. Also, ideal flows with no frictional or other flow losses ($p_{loss} = 0$) will lead to an expression for $C_{p,ideal}$, as a function only of the area ratio.

$$C_{p,ideal} = 1 - \frac{1}{AR^2} \quad \text{(ideal flow)} \tag{4.8}$$

In actual instances, frictional flow recirculation or stall losses will reduce C_p from the ideal cases. Such details of the flow inside can be revealed only through experimental testing or CFD analyses. However, there are some generalized data that can assist in the design process, such as the chart shown in Figure 4.6. Various flow regimes can be identified as a function of the geometrical parameters and the area ratio. It can be seen that "stable" flow – that is, flow with no flow separation – is achieved only for very short nozzles with small divergence angles. The pressure-recovery coefficient itself may be estimated based on test data. Figure 4.7 shows a chart for C_p for conical diffusers.

It should be remembered, however, that the function of the diffuser is to provide not only the necessary mass flow rate at minimum stagnation pressure loss, but also a uniform velocity profile. Data shown in Figure 4.7 only shows the mean pressure-recovery parameter, and

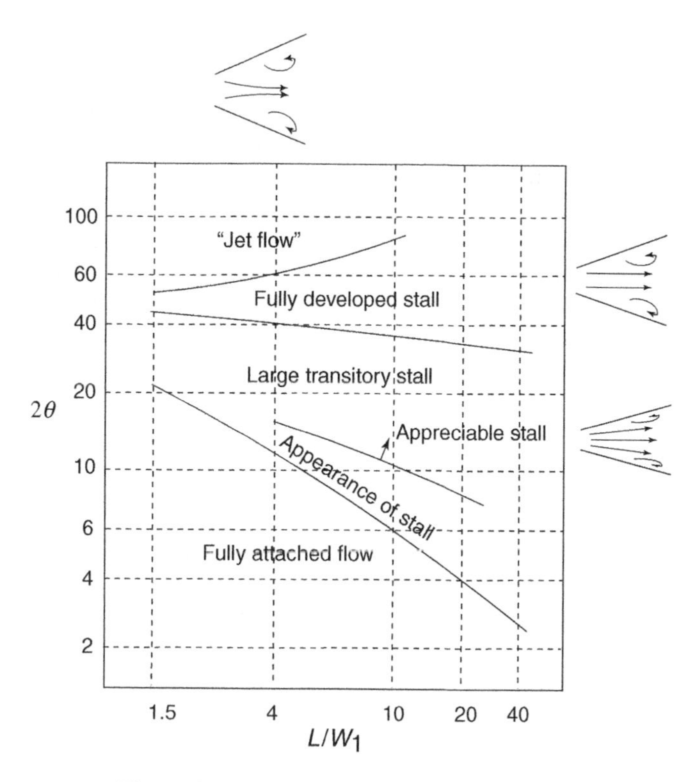

Figure 4.6 Flow regimes in a rectangular duct.

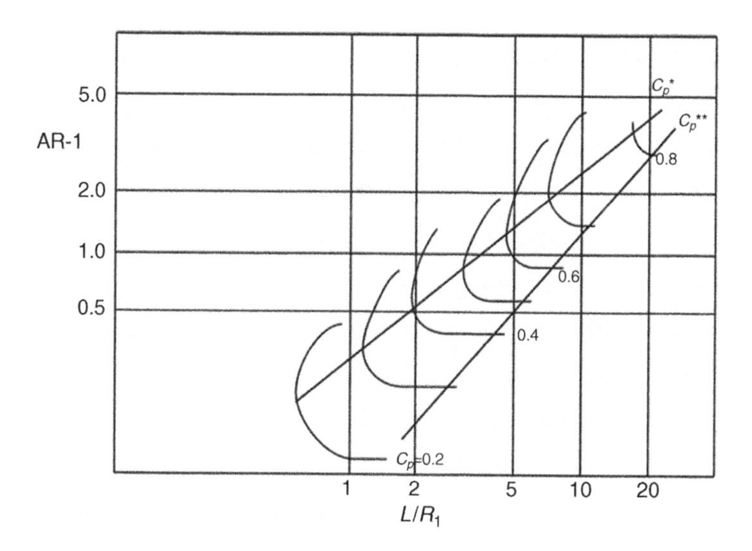

Figure 4.7 Pressure-recovery coefficient for conical inlets.

nothing about the velocity profiles at different angles of attacks. Thus, data such as those shown in Figures 4.6 and Figure 4.7 are more helpful for what to avoid than how to design a good inlet, even for the simplest geometry. For modern inlets, involving three-dimensional curvature and multiple flow turns, the situation can be even more complex. At high angles of attack, the flow can separate internally with recirculation, leading to loss in stagnation pressures (Figure 4.8), in addition to non-uniform flow at the diffuser exit.

The combined effects of the flow inside the diffuser are summarized into the diffuser isentropic efficiency. The diffuser isentropic efficiency is somewhat more difficult to visualize than, say, the compressor isentropic efficiency, because there is no change in the stagnation enthalpy but only a loss in the stagnation pressure. So we need to start by looking at the diffuser process in a *T–s* diagram, as in Figure 4.9. The horizontal axis plots the entropy, while the vertical axis is temperature, or enthalpy when multiplied by the specific heat. We start from the static pressure in

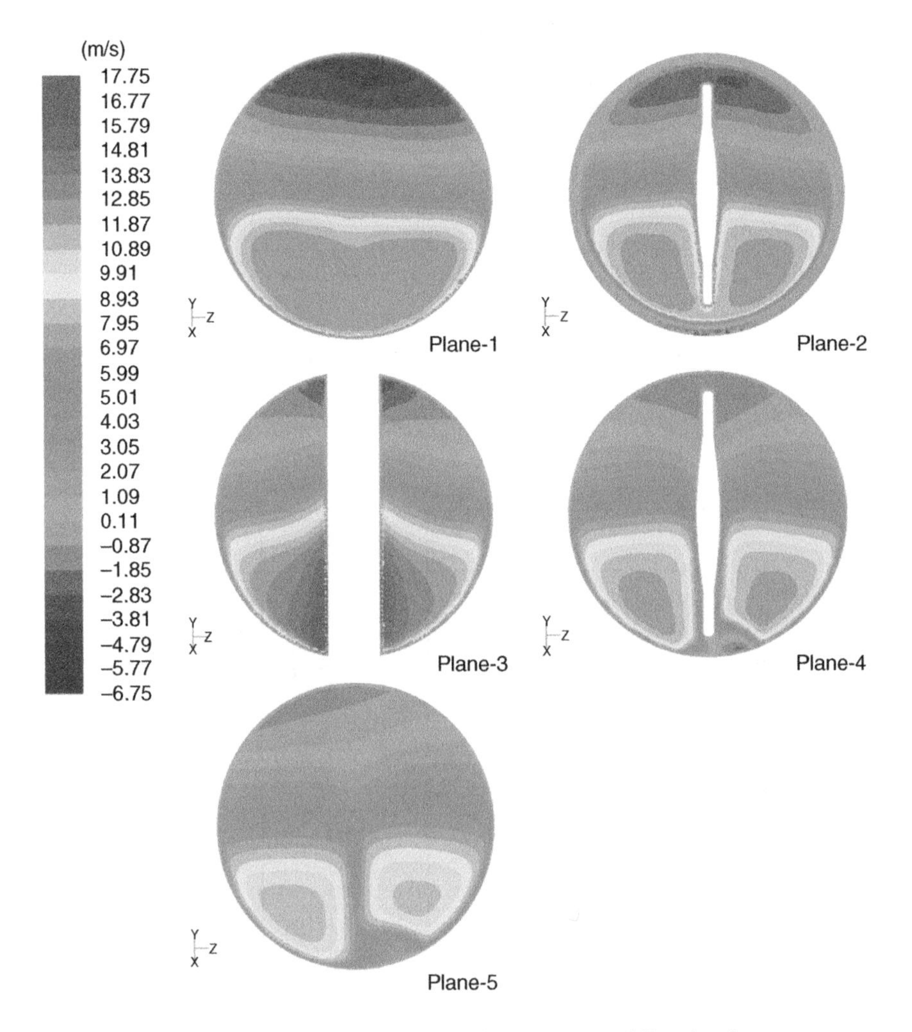

Figure 4.8 Occurrence of swirl downstream of flow bends.

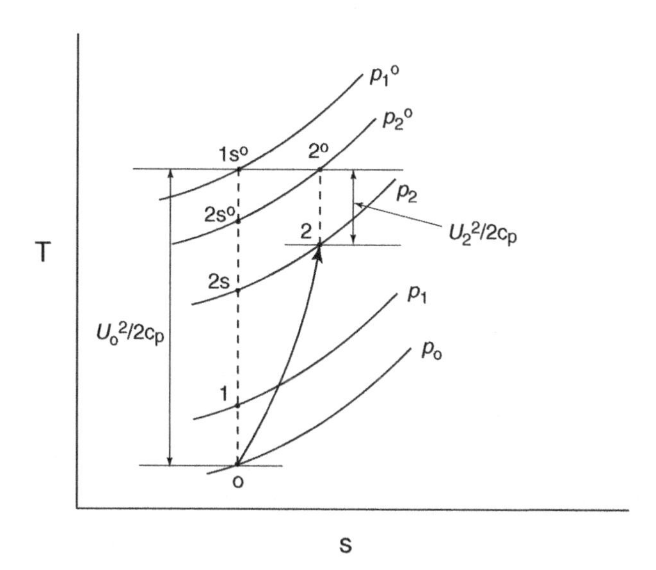

Figure 4.9 The diffuser process shown on *T–s* diagram.

the ambient, p_o. As noted above, the pressure can either decrease (external acceleration) or increase (external deceleration) to p_1. We will consider the more common case of external deceleration and plot a constant-pressure curve for p_1 above p_o in Figure 4.9. Note that constant-pressure lines are exponential curves on a *T–s* diagram. From p_1, if the flow decelerated isentropically in the diffuser, then there is no loss in the stagnation pressure and it will reach the stagnation pressure, $p_1^o = p_o^o$, and this process would follow a vertical line from "1" to "1_s^o", where the subscript "s" represents a state achieved isentropically and the superscript "o" a stagnation state. However, any losses through wall friction, boundary layer separation or shocks, will result in deviation from this isentropic process, in the direction of increasing entropy. Therefore, the actual process will track a curve sketched on Figure 4.9 from "1" to "2". There will be a corresponding stagnation pressure to this actual pressure at 2, which we will call p_2^o. Note that because there is no change in the stagnation enthalpy, the stagnation temperature at p_2^o is exactly the same as that at p_{1s}^o. The definition of diffuser isentropic efficiency is based on the idea that this same stagnation pressure, p_2^o, could have been attained, with less stagnation enthalpy, had the diffuser process gone on isentropically. So in Figure 4.9, we compare the stagnation enthalpy at "2°" with that at "2s°", and use that as our definition of the diffuser isentropic efficiency.

$$\eta_d = \frac{h_{2s}^o - h_o}{h_2^o - h_o} \approx \frac{T_{2s}^o - T_o}{T_2^o - T_o} \tag{4.9a}$$

We can use our standardized temperature ratios to readily calculate this quantity.

$$\eta_d = \frac{T_{2s}^o - T_o}{T_2^o - T_o} = \frac{\dfrac{T_{2s}^o}{T_o^o}\dfrac{T_o^o}{T_o} - 1}{\dfrac{T_o^o}{T_o} - 1} = \frac{\left(\dfrac{p_2^o}{p_o^o}\right)^{\frac{\gamma-1}{\gamma}}\left(\dfrac{p_o^o}{p_o^o}\right)^{\frac{\gamma-1}{\gamma}} - 1}{\tau_r - 1} = \frac{(\pi_d \pi_r)^{\frac{\gamma-1}{\gamma}} - 1}{\tau_r - 1} \tag{4.9b}$$

Equation (4.9b) relates the diffuser isentropic efficiency with the actual pressure ratio in the diffuser, π_d.

We can also relate the (static) pressure-recovery coefficient, C_p, to the diffuser (stagnation) pressure ratio, π_d. Since the pressure losses are represented in Eq. (4.7), we can solve for p_{loss} using Eqs. (4.7) and (4.8).

$$p_{loss} = \frac{1}{2}\rho_1 U_1^2 (C_{p,ideal} - C_p) \tag{4.10}$$

Then, the diffuser pressure ratio becomes

$$\pi_d = \frac{p_2^o}{p_1^o} = 1 - \frac{p_{loss}}{p_1^o} = 1 - \frac{\frac{1}{2}\rho_1 U_1^2}{p_1^o}(C_{p,ideal} - C_p) \tag{4.11}$$

Example 4.1 Diffuser performance

Find the stagnation pressure ratio across a conical diffuser, with the inlet radius, $R_1 = 0.5$ m, divergence half-angle of 10°, and length of 1.0 m at M_o of 0.8 at sea level.

We will make use of Figure 4.7 for this analysis. First, we need Eq. (4.5b).

$$AR = 1 + \left(1 + \frac{L}{R_1}\tan\theta\right)^2 = 2.185 \tag{E4.1.1}$$

$L/R_1 = 2$, so from Figure 4.7 we find $C_p \approx 0.38$ and $C_{p,ideal} = 1 - \frac{1}{AR^2} = 0.79$.

We will assume that there is no external acceleration or deceleration, $M_o = M_1$. At sea level, $a_1 = 340.6$ m/s, $\rho_1 = 1.226$ kg/m^3, so that $U_1 = M_1 a_1 = 272.5$ m/s.

$$p_1^o = p_1\left[1 + \frac{\gamma - 1}{2}M_1^2\right]^{\frac{\gamma}{\gamma-1}} = 101\ 325\left(1 + \frac{1.4 - 1}{1.4}0.8^2\right)^{\frac{1.4}{1.4-1}} = 137\ 626\ \text{Pa} \tag{E4.1.2}$$

From Eq. (4.11),

$$\pi_d = \frac{p_2^o}{p_1^o} = 1 - \frac{\frac{1}{2}\rho_1 U_1^2}{p_1^o}(C_{p,ideal} - C_p) = 1 - \frac{0.5(1.226)(272.5)^2}{137626}(0.79 - 0.38)$$
$$= 0.864 \tag{E4.1.3}$$

The diffuser efficiency is related to the pressure ratio, using Eq. (4.9b).

$$\tau_r = 1 + \frac{\gamma - 1}{2}M_o^2 = 1.128 \rightarrow \pi_r = 1.524$$
$$\eta_d = \frac{(\pi_d \pi_r)^{\frac{\gamma-1}{\gamma}} - 1}{\tau_r - 1} = \frac{(0.864 \times 1.524)^{\frac{1.4-1}{1.4}} - 1}{1.128 - 1} = 0.639 \tag{E4.1.4}$$

So this particular design leads to a rather poor diffuser performance.

4.3 Supersonic Inlet Operation

During the discussion of ramjets in Chapter 3, we briefly considered the stagnation pressure loss in supersonic inlets. One of the correlations that can be used is

$$
\begin{aligned}
\pi_s &= 1 && \text{for} \quad M_o < 1 \\
\pi_s &= 1 - 0.075(M_o - 1)^{1.35} && \text{for} \quad 1 \le M_o < 5 \\
\pi_s &= \frac{800}{M_o^4 + 935} && \text{for} \quad M_o > 5
\end{aligned}
\tag{4.12}
$$

Here again, π_s denotes the external stagnation pressure loss outside of the inlet, due to presence of shocks for $M_o > 1$. Although π_s is simply given as a function of the flight Mach number in Eq. (4.12), with π_s decreasing with the Mach number above M_o of 1, the actual stagnation pressure loss depends on the inlet geometry. A simple comparison between a normal-shock inlet and a wedge inlet shows this fact (see Example 4.2).

Example 4.2 A comparison of the stagnation pressure loss in normal-shock and wedge inlets at M_o of 3

In a normal-shock inlet, the normal-shock relations give the following downstream conditions for $M_o = 3$: $M_1 = 0.475$, $\frac{p_1}{p_o} = 10.33$, and $\frac{p_1^o}{p_o^o} = 0.328$. Thus, a compression ratio of 10.33 has been achieved at the expense of nearly 72% loss in stagnation pressure.

For a two-ramp inlet as shown in Figure E4.2.1, two 15° turns create oblique shocks followed by a weak normal shock at the cowl tip. For $M_o = 3$, we have the following calculations.

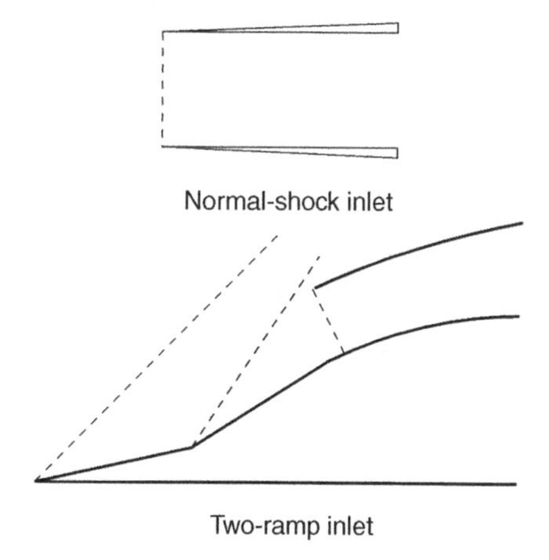

Figure E4.2.1 A normal shock inlet, and a two-ramp inlet.

$M_o = 3, \theta_a = 15°$, the oblique shock angle is $\beta_a = 32°$. $(M_o)_n = M_o \sin\beta_a = 1.59$. From normal shock relations, $(M_a)_n = 0.6715 = M_a \sin(\beta_a - \theta_a) \rightarrow M_a = \dfrac{(M_a)_n}{\sin(\beta_a - \theta_a)} = 2.3$.

And $\dfrac{p_a}{p_o} = 2.783$, and $\dfrac{p_a^o}{p_o^o} = 0.8989$.

We carry the calculation to the next oblique shock. $M_a = 2.3$, $\theta_b = 15°$, the oblique shock angle is $\beta_b = 55°$. $(M_a)_n = M_a \sin\beta_b = 1.48$.

$$(M_b)_n = 0.7083 = M_b \sin(\beta_b - \theta_b) \rightarrow M_b = \frac{(M_b)_n}{\sin(\beta_b - \theta_b)} = 1.67.$$

And $\dfrac{p_b}{p_a} = 2.389$, $\dfrac{p_b^o}{p_a^o} = 0.9360$.

Finally, for $M_b = 1.67$, the normal shock relations give $M_1 = 0.65$, $\dfrac{p_1}{p_b} = 3.087$, and $\dfrac{p_1^o}{p_a^o} = 0.8680$.

The overall compression is found by $\dfrac{p_1}{p_o} = \dfrac{p_a p_b p_1}{p_o p_a p_b} = 20.5$ and $\dfrac{p_1^o}{p_o^o} = \dfrac{p_a^o p_b^o p_1^o}{p_o^o p_a^o p_b^o} = 0.73$. So a much higher static pressure rise is achieved with much lower stagnation pressure loss.

As with the subsonic inlets, the details of the flow for supersonic flows can only be determined through computational analyses or experimental testing. However, we can use the oblique shock and other gas-dynamic relations (see Appendices C and D) to find the "bulk" flow characteristics. First, we are interested in basic properties such as the static and stagnation pressure under supersonic flow conditions. Figure 4.10 shows a schematic of a supersonic inlet, and approximate locations of the sequence of shocks. For each of the shocks, we can apply the oblique and normal shock relations (see Example 4.2), to find the ratio of static and stagnation pressures and other parameters. Depending on the flight condition, the ramps in variable-geometry inlets can be adjusted to optimize the flow. In addition, various bleed, bypass or inlet ports are opened or closed to provide the right mass flow rate into the engine.

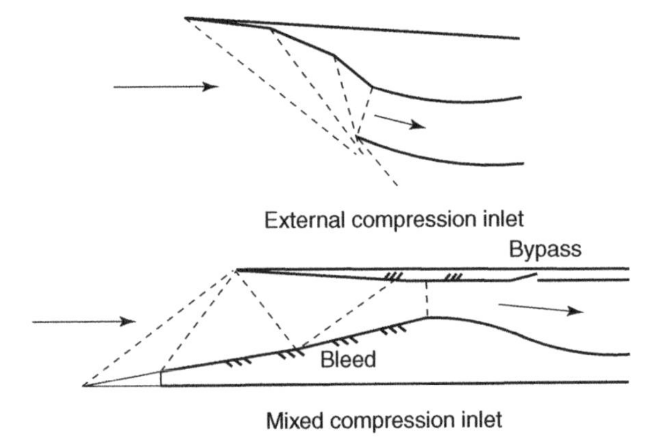

Figure 4.10 Supersonic inlets: external-compression and mixed-compression inlets.

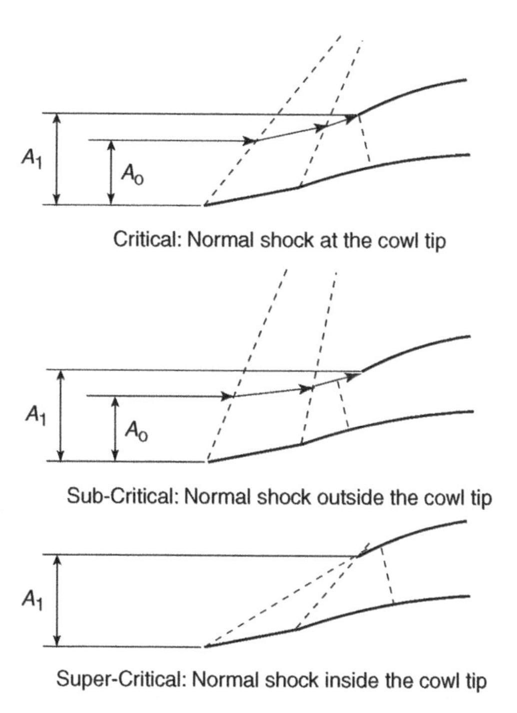

Figure 4.11 Mass flow rate consideration for supersonic inlets: supercritical, critical, and subcritical operations.

The ramps turn the flow direction, through the oblique shocks, as shown in Figure 4.10. This will determine the upstream flow capture area, as shown in Figure 4.11. Supercritical operation refers to the normal shock residing within the inlet. Under these conditions, the oblique shock angle can be more acute than the hypothetical extension line from the wedge tip to the "top" of the inlet ($\beta < \delta$). This results in so-called "full flow", where the upstream capture area is equal to the inlet blockage area ($A_o = A_1$). Although this provides the most mass flow rate per the inlet blockage area, acute angles mean weak oblique shocks, inevitably leading to a stronger normal shock. Therefore, the overall stagnation pressure loss is large. For critical or subcritical operations, the normal shock is at or outside of the cowl tip, so that the oblique shock angle is greater than or equal to the angle for the extension line ($\beta \geq \delta$), and flow spillage occurs due to diversion of the flow. This results in so-called "restricted flow", where the upstream capture area is less than the inlet blockage area ($A_o < A_1$). Let us consider the rectangular inlet with a wedge, as shown in Figure 4.11, to figure out the capture area A_o. The geometry of the inlet gives the following relationships.

$$\begin{aligned} A_o &= (L_N - l_N)\tan\beta, \\ A_1 &= L_N \tan\delta \end{aligned} \quad \rightarrow \frac{A_o}{A_1} = \frac{(L_N - l_N)\tan\beta}{L_N \tan\delta} \tag{4.13}$$

Using the fact that the flow behind the shock is parallel to the wedge surface, we can also write

$$A_1 - A_o = l_N \tan\theta \rightarrow 1 - \frac{A_o}{A_1} = \frac{l_N \tan\theta}{L_N \tan\delta} \tag{4.14}$$

Eliminating L_N and l_N in Eqs. (4.13) and (4.14) gives us a simple relation for A_o/A_1.

$$\frac{A_o}{A_1} = \frac{\cot\theta - \cot\delta}{\cot\theta - \cot\beta} \quad \text{for wedge} \tag{4.15}$$

For the cone geometry, we can apply a similar method to find

$$\frac{A_o}{A_1} = \frac{\cot^2\theta - \cot^2\delta}{\cot^2\theta - \cot^2\beta} \quad \text{for cone} \tag{4.16}$$

For a sequence of ramps as shown in Figure 4.11, we can simply keep multiplying Eq. (4.15).

$$\frac{A_o}{A_1} = \frac{\cot\theta_a - \cot\delta_a}{\cot\theta_a - \cot\beta_a}\frac{\cot\theta_b - \cot\delta_b}{\cot\theta_b - \cot\beta_b} \quad \text{for two-ramp wedge} \tag{4.17}$$

The modes (supercritical, critical, subcritical) shown in Figure 4.11 are determined by the flight Mach number, the ramp (or cone) angles and the "back pressure", which is the static pressure at the compressor inlet (p_2) in this case. The flow is subsonic at the compressor inlet, so the pressure increases with decelerating flow in the diffuser, as shown in Figure 4.12. The location of the normal shock is determined by the pressure curve that must arrive at p_2. If the

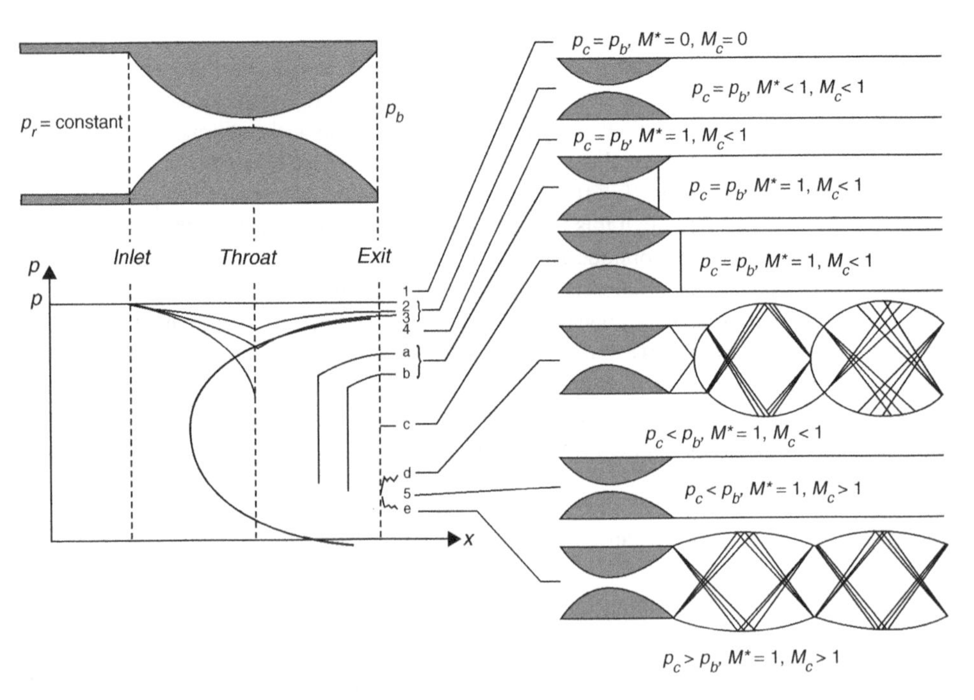

Figure 4.12 Pressure in a convergent-divergent geometry.

back pressure is high, then the normal shock will have to relocate upstream so that the pressure curve again matches p_2. The critical operation is the "target design" where the compressor mass flow rate and static pressure places the normal shock at the cowl tip. If p_2 were higher at a given flight Mach number, then the normal shock would travel upstream onto the external ramp, as shown in Figure 4.11. As shown in the previous analysis of capture area, the subcritical mode results in higher flow spillage, and also the normal shock sitting on the external ramp causes some flow problems. The static pressure undergoes a big jump across the normal shock, and as far as the flow is concerned this is an adverse pressure gradient, so that flow separation can lead to poor flow stability. The flow separation slows the flow speed in the "recirculation bubble", and thus the net available flow area is effectively decreased. Higher flow speed at the net available flow area will move the pressure curves downward, so that the normal shock moves downstream, which will eliminate the flow separation and then the normal shock will be moved upstream again. This process can repeat itself at high frequencies, and this problem is referred to as "buzz instability". At high engine mass flow rates and low back pressures (p_2), the normal shock is moved downstream into the inlet, and this mode is called supercritical. The presence of a strong normal shock (since it is preceded by weak oblique shocks) causes larger losses in stagnation pressure, and also flow non-uniformity due to adverse pressure gradient inside the diffuser. Therefore, the compromise is to design the supersonic inlets for subcritical or critical operations by oversizing the inlet, and then use bleeds to control the mass flow rate and boundary layer controls in the external ramp to suppress flow separation.

4.4 Gas-Turbine Nozzles

The function of nozzles is to attain the maximum kinetic energy at the exit, with the available thermal energy of the incoming flow. Other functions include mixing of the fan and core streams in turbofans, thrust vectoring or augmentation, noise and infrared signature reductions. For high bypass turbofans, the incoming thermal energy is relatively small and a simple convergent nozzle may suffice to accelerate the flow. For high-performance engines, convergent-divergent nozzles are used to accelerate the exhaust to supersonic speeds. The fundamental measure of performance is the nozzle isentropic efficiency, defined as

$$\eta_n = \frac{h_6^o - h_7}{h_6^o - h_{7s}} = \frac{\frac{U_7^2}{2}}{\frac{U_{7s}^2}{2}} \tag{4.18}$$

As shown in Figure 4.13, the nozzle efficiency is the ratio of the actual speed attained as the flow accelerates from p_6^o to p_7, to the speed that would have been attained if the flow had been isentropic. On a T–s diagram, the vertical distance between the stagnation and static temperatures represents the kinetic energy, which explains the definition in Eq. (4.18). It also shows that the actual expansion would involve a reduction in the stagnation pressure even though the stagnation temperature remains the same.

For isentropic flows, the flow in the nozzle follows the one-dimensional gas-dynamic relations. The stagnation to static pressure ratio at an arbitrary location in the nozzle, x, is given by

$$\frac{p_6^o}{p_x} = \left(1 + \frac{\gamma - 1}{2} M_x^2\right)^{\frac{\gamma}{\gamma-1}} \tag{4.19}$$

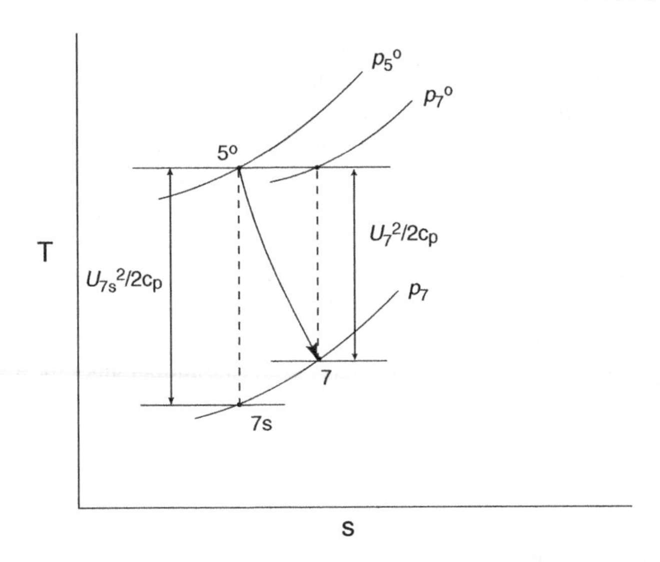

Figure 4.13 *T–s* diagram for the nozzle expansion process.

We can find the pressure ratio for the flow to go supersonic ($M_x > 1$) by setting $M_x = 1$ (at the throat), which gives the pressure ratio of 1.893. Thus, if the stagnation pressure is 1.893 times greater than the exit pressure, then the flow can attain supersonic speeds through a convergent-divergent nozzle. The local Mach number can then be calculated as a function of the area ratio ($A_x/A*$), where A_x is the nozzle cross-sectional area at x and $A*$ the throat area.

$$\left(\frac{A_x}{A*}\right)^2 = \frac{1}{M_x^2}\left[\frac{2}{\gamma+1}\left(1+\frac{\gamma-1}{2}M_x^2\right)\right]^{\frac{\gamma+1}{2(\gamma-1)}} \tag{4.20}$$

There are two possible values of Mach number for a given area ratio in Eq. (4.20), one for the supersonic flow in the divergent section and another for the subsonic section. Knowing the local Mach number from Eq. (4.20) allows us to calculate the local static pressure, using Eq. (4.19).

If the nozzle area and the length are designed so that $p_7 = p_o$, then the flow will undergo "smooth" isentropic expansion. However, let us consider more realistic cases. In simple convergent nozzles for low-speed gas-turbine engines, the nozzle ends at the throat and thus has the minimum area at the exit. The Mach number will be equal to 1 at the exit, as long as the pressure ratio is greater than 1.893, as given by Eq. (4.19). If the nozzle is extended to include a short divergent section, the flow is allowed to accelerate further with accompanying decrease in static pressure, as given by Eqs. (4.19) and (4.20). If the exit static pressure is higher than the ambient pressure, then the nozzle is said to under-expanded, and expansion waves at the exit will lower the pressure to the ambient pressure outside of the nozzle. Further downstream, the expansion waves and jet streamlines interact with the free jet boundary to create a complex flow pattern as shown in Figure 4.14 (also often visible in photos of high-speed aircraft), which goes on until dissipated by viscous effects. In the other extreme case, where the nozzle length is longer than required, then the pressure undershoots the ambient

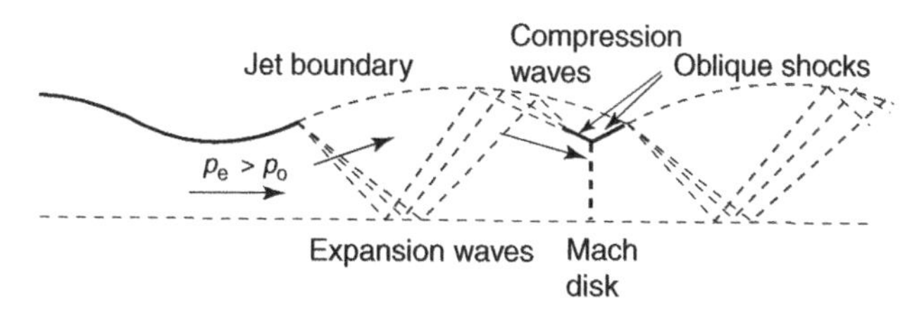

Figure 4.14 Flow pattern in under-expanded nozzle jets.

pressure (over-expanded jet). In order to adjust to the higher ambient pressure, oblique shocks form at the nozzle exit, which create so-called shock-diamonds outside of the nozzle as shown in Figure 4.15. If the overexpansion proceeds further, then a stronger normal shock may form at the exit or move inside the nozzle, with adverse effects such as loss in flow acceleration and boundary layer separation. However, since this would require grossly overdesigned nozzle, overexpansion is not a frequent problem in gas-turbine nozzles.

Let us return to the nozzle efficiency, given by Eq. (4.18), and how we may use this parameter to estimate the nozzle performance.

$$\eta_n = \frac{\frac{U_7^2}{2}}{\frac{U_{7s}^2}{2}} = \frac{h_6^o - h_7}{h_6^o - h_{7s}} = \frac{T_6^o - T_7}{T_6^o - T_{7s}} = \frac{1 - T_7/T_6^o}{1 - T_{7s}/T_6^o} = \frac{1 - T_7/T_7^o}{1 - T_{7s}/T_6^o} \tag{4.21}$$

We have used the adiabatic nozzle condition in the last part of the equality. Using isentropic relationships between the appropriate points (Figure 4.13), we obtain

$$\eta_n = \frac{1 - (p_7/p_7^o)^{\frac{\gamma-1}{\gamma}}}{1 - (p_7/p_6^o)^{\frac{\gamma-1}{\gamma}}} \tag{4.22}$$

We can use Eq. (4.22) to solve for the nozzle pressure ratio, π_n, to the nozzle efficiency.

$$\pi_n = \frac{p_7^o}{p_6^o} = \frac{p_7}{p_6^o} \frac{1}{\left\{1 - \eta_n \left[1 - \left(\frac{p_7}{p_6^o}\right)^{\frac{\gamma-1}{\gamma}}\right]\right\}^{\frac{\gamma}{1-\gamma}}} \tag{4.23}$$

The nozzle efficiency itself may be a complex function of the nozzle geometry, Mach number and operating pressures, but at this point we assume it is given through testing or

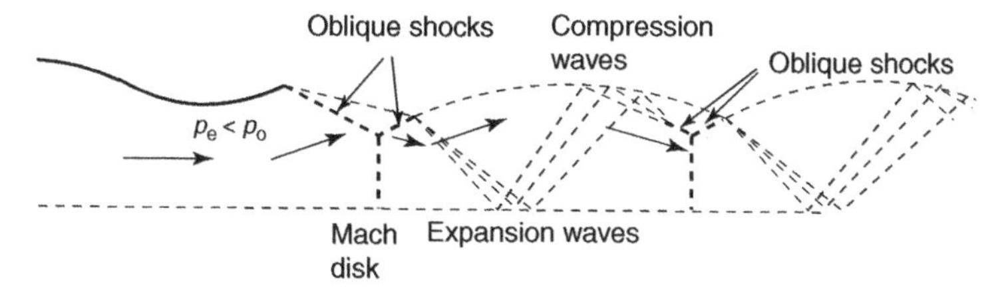

Figure 4.15 Flow pattern in over-expanded nozzle jets.

computational fluid dynamic data. The nozzle efficiency leads to a more accurate estimate of the exit velocity, U_7.

$$U_7 = \sqrt{2(h_6^o - h_7)} = \sqrt{2\eta_n(h_6^o - h_{7s})} = \sqrt{\frac{2\gamma}{\gamma - 1}\eta_n RT_6^o\left[1 - \left(\frac{p_7}{p_6^o}\right)^{\frac{\gamma-1}{\gamma}}\right]} \tag{4.24}$$

4.5 Problems

4.1. Find the stagnation pressure ratio across a conical diffuser, with the inlet radius, $R_1 = 0.75$ m, divergence half-angle of 5°, and length of 2.0 m at M_o of 0.8 at sea level.

4.2. If the stagnation pressure loss is 12% of flight dynamic pressure, $p_o^o - p_2^o = 0.1\frac{1}{2}\rho U_o^2$ in an inlet, what are the pressure ratio, π_d, and the diffuser efficiency, η_d?

4.3. The mass flow rate through an inlet is 115 kg/s at $M_o = 0.8$ and $A_1 = 5\ \text{m}^2$. The diffuser efficiency is 0.92, and the Mach number at the diffuser exit is 0.4. The ambient temperature and pressure are 225 K and 10.5 kPa, respectively. Find (a) the inlet static pressure, p_1 and (b) the static pressure at the diffuser exit, p_2.

4.4. Plot the area ratio, AR, for a conical inlet as a function of the angle, θ, $L = 1.2$ m and R_1 and 0.25 m.

4.5. Add the pressure recovery coefficient in the plot for Problem 4.3 for five values of θ.

4.6. Plot the area ratio, AR, for a conical inlet as a function of the diffuser length, L for $\theta = 15°$, and R_1 and 0.25 m.

4.7. Add the pressure recovery coefficient in the plot for Problem 4.5 for five values of L.

4.8. Determine the stagnation pressure loss across a single wedge of 15° inclination, for $M_o = 1.5$, 2, 4 and 6, and compare with Eq. (4.12). The flow downstream of the wedge has a normal shock.

4.9. A contoured inlet, like the one on F-16s, can be considered as a multiple, weak oblique shock inlet with compression wave emanating from each contoured turn. We need to simplify the analysis by approximating the number of oblique shocks to two in this case, as shown below (assume that there is a normal shock following the two turns and there is no change in the flow in the straight section). For an incident Mach number of 2.25 at a pressure of 0.25 atm and T = 225 K, determine at the end of the inlet ("1"), (a) stagnation pressure; (b) static temperature; (c) Mach number (d) flow velocity and (e) the upstream "capture" area (or height, assuming the inlet to be rectangular).

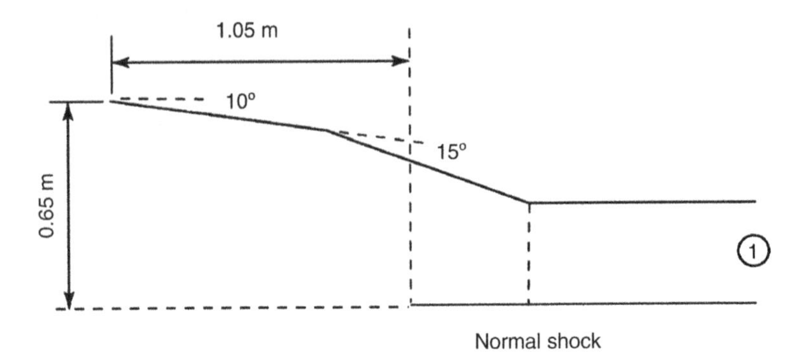

4.10. Determine the stagnation pressure loss across two wedges of 15° inclination each, for $M_o = 1.5$, 2, 4 and 6, and compare with Eq. (4.12). The flow downstream of the wedges has a normal shock.

4.11. Plot the stagnation pressure ratio as a function of the Mach number, using Eq. (4.12). Insert as data points, the stagnation pressure ratio across two wedges of 10° inclination each, for $M_o = 1.5$, 2, 4 and 6.

4.12. Find approximate dimensions for the F-15 aircraft inlet. Calculate the stagnation pressure ratio across the inlet for flight Mach number of 0.8, 1.5 and 2.0.

4.13. For $p_7 = p_o$, plot the nozzle efficiency as a function of the nozzle stagnation pressure ratio, π_n at altitudes of 0, 10 000, 30 000 and 40 000 ft.

4.14. A isentropic convergent-divergent nozzle operates at a pressure of $p_o = 5$ kPa. The nozzle stagnation pressure and temperature are $p^o = 101$ kPa and $T^o = 1800$ K, respectively. Calculate (a) the nozzle area expansion ratio $A_7/A*$, for perfect expansion; (b) nozzle exit Mach number M_7 for perfect expansion and (c) nozzle exit velocity, U_7. $\gamma = 1.4$.

4.15. Air at 1500 K, 300 kPa and $M = 0.5$ enters a nozzle with an inlet area of 0.5 m^2 and exits the nozzle at 75 kPa. The flow is assumed to be isentropic with a constant specific heat of $\gamma = 1.4$. Determine the following:

a. The velocity and mass flow rate of the entering air.
b. The temperature and Mach number of the exiting air.
c. The exit area, exit velocity and magnitude and direction of the net force on the nozzle (assume the ambient is at 75 kPa).

Bibliography

Hill, P. and Petersen, C. (1992) *Mechanics and Thermodynamics of Propulsion*, 2nd edn, Addison-Wesley Publishing Company.

Saeed Farokhi (2009) *Aircraft Propulsion*, Wiley.

5

Compressors and Turbines

5.1 Introduction

Aerodynamically, compressors and turbines operate in a similar manner to airfoils or propellers/wind turbines. The airfoil generates lift by being propelled in the forward direction, and the contour of the airfoil and the angle of attack causes higher speed and lower pressure on the top side than the bottom. The pressure difference between the top and bottom surfaces leads to the lift. For compressors, the blades move with a circular motion; nonetheless the forward motion of the compressor blades generates higher pressure on the back surface, which when stacked in multiple stages provides compression in gas-turbine engines. Turbine blades, on the other hand, rotate based on the pressure difference between the forward and back surfaces. This conversion of air momentum into blade motion also involves conversion of air enthalpy into mechanical work, or vice versa for the compressors.

Figure 5.1 shows some schematics of the axial compressors. The air enters the compressor and flows through the inlet guide vane. A "stage" refers to a rotor and stator combination. The rotor is the element that is attached to the hub, which rotates powered by the turbine. The stator is a stationary element attached to the casing, which guides the flow at an optimal angle and also serves to provide uniform flow toward the next stage. Modern compressors comprise upwards of 12 stages, and for high-efficiency engines can be divided into separate low- and high-pressure compressors. Modern engines can achieve overall pressure ratio of 40 or higher with 15 stages.

We can see that the annular area around the compressor hub decreases, going downstream. The reason is that the density will increase with increasing pressure across the compressor, and in order to keep a nearly constant axial velocity the flow area needs to decrease. Since the mass flow rate is fixed under steady-state operation, in order to keep the axial velocity the same, the compressor flow area needs to scale inversely proportional to the air density, $A \sim 1/\rho$. There are several choices to reduce the cross-sectional area, constant outer diameter (COD), constant mean diameter (CMD), and constant hub diameter (CHD). However, these are more based on fabrication requirements, and for optimal flows an alternative design may be chosen.

Before looking at the aero-thermodynamics of compressor stages, let us look at the thermodynamic aspect of the compressor operation. Figure 5.2 shows the compressor process on a T–s diagram. The function of the compressor is to raise the pressure from "2" to "3" with as

Aerospace Propulsion, First Edition. T.-W. Lee.
© 2014 John Wiley & Sons, Ltd. Published 2014 by John Wiley & Sons, Ltd.
Companion Website: www.wiley.com/go/aerospaceprop

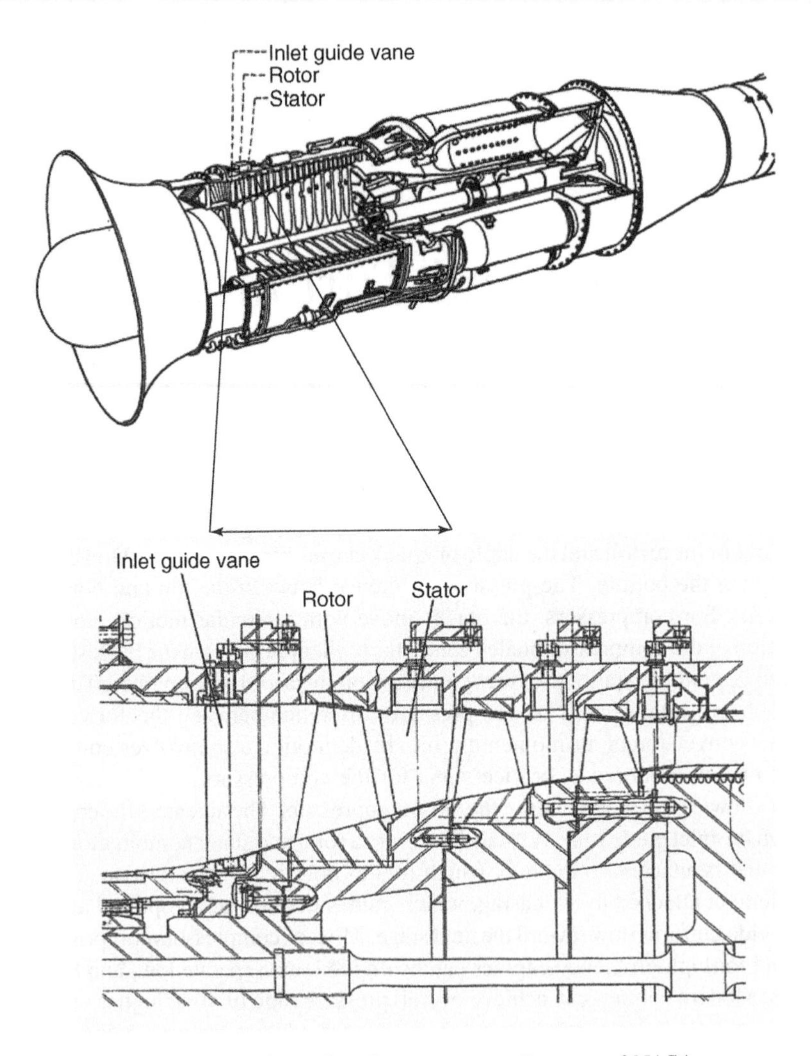

Figure 5.1 Schematics of compressors. Courtesy of NASA.

small losses as possible (minimum entropy increase). An ideal, isentropic process would be a vertical line in Figure 5.2, but the real process involves entropy increase so that the pressure increase is obtained at the cost of some work input, or the enthalpy difference between "2" and "3". The loss terms include the usual frictional losses at the blade surfaces and flow separation near the trailing edges of the compressor blades. The compressor isentropic efficiency is defined as the ratio of the isentropic work to the actual work required to generate the pressure increase.

$$\eta_c = \frac{h_{3s}^o - h_2^o}{h_3^o - h_2^o} \approx \frac{\dfrac{T_{3s}^o}{T_2^o} - 1}{\dfrac{T_3^o}{T_2^o} - 1} = \frac{\left(\dfrac{p_3^o}{p_2^o}\right)^{\frac{\gamma-1}{\gamma}} - 1}{\dfrac{T_3^o}{T_2^o} - 1} \tag{5.1}$$

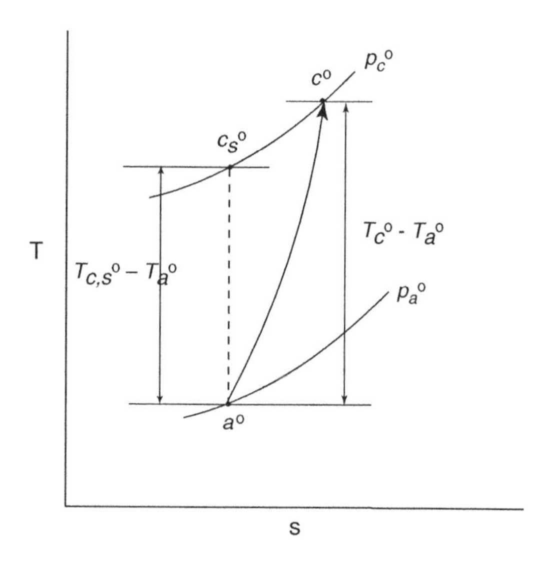

Figure 5.2 Compressor process sketched on a *T–s* diagram.

The same expression can be applied to a single stage in a compressor, by defining the compressor stage efficiency, η_{st}.

$$\eta_{st} = \frac{h^o_{c,s} - h^o_a}{h^o_c - h^o_a} \approx \frac{\frac{T^o_{c,s}}{T^o_a} - 1}{\frac{T^o_c}{T^o_a} - 1} = \frac{\left(\frac{p^o_c}{p^o_a}\right)^{\frac{\gamma-1}{\gamma}} - 1}{\frac{T^o_c}{T^o_a} - 1} \tag{5.2}$$

The subscripts "a" and "c" refer to upstream and downstream of the compressor stage, respectively.

5.2 Basic Compressor Aero-Thermodynamics

Figure 5.3 shows a schematic of the compressor aerodynamics. Consider a cross-sectional element of a compressor blade, rooted a few centimeters into the page and rotating "downward" around a horizontal axis. The air enters this compressor stage from left to right. The inlet guide vane, or the stator in the prior stage, guides the flow at a small downward angle, and the absolute velocity, denoted by vector $\vec{c}$, shows this flow direction. The rotating motion, depicted as downward in Figure 5.3, is represented by vector $\vec{u}$. The resultant relative velocity to the compressor blade is found by taking the difference between the two vectors.

$$\vec{w} = \vec{c} - \vec{u} \tag{5.3}$$

This vector relationship is valid both upstream ("a") and downstream ("b") of the rotor, so that individual velocity components can be obtained at these locations. The stator ("*b*" to "*c*") is a stationary element, so we just need the absolute velocities. The angles are denoted by Greek letters, α and β, with respect to the horizontal direction, for the absolute and relative velocities,

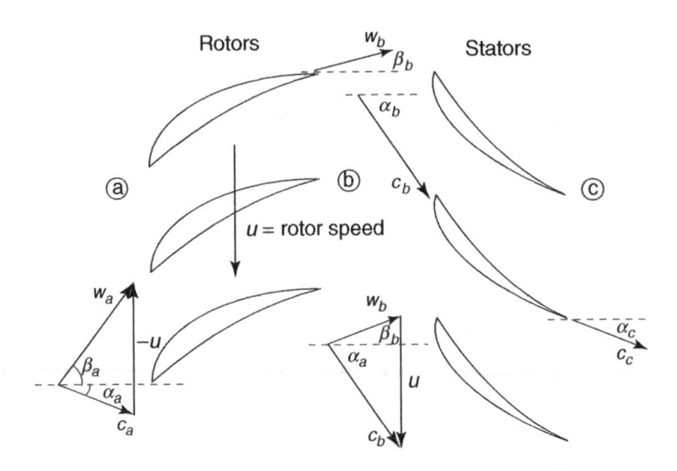

Figure 5.3 Compressor aerodynamics, with velocity triangle.

respectively. The velocity triangle connecting the vectors in Eq. (5.3) is typically superimposed on another one, as shown in Figure 5.3.

We can see in Figure 5.3 that there needs to be a balance between the velocity vectors for the relative velocity toward the blade, w_a, to within acceptable range of angle of attack. The angle, β_a, can change either due to the absolute velocity, c_a, or the rotational velocity, u. Similarly, at the trailing edge of the rotor ("b"), the angle, β_b, should be reasonably well-aligned with the rotor, otherwise the flow can separate prematurely, leading to large losses in efficiency. Since $u\,(=\Omega r)$ is a function of the radial velocity (Ω) and the radial location (r), the velocity triangle will change as a function of the radial location as well. We can also realize that some rotational (angular) momentum is imparted by the rotor to the fluid, as the tangential component at "b" is greater than at "a", that is, $c_{\theta b} > c_{\theta a}$. We will see shortly that this adds both momentum and kinetic energy to the fluid, which is translated into an increase in stagnation pressure across the rotor.

From velocity triangles, we can also relate the velocity components to one another. For example,

$$u = c_{\theta a} + w_{\theta a} = c_{za} \tan \alpha_a + c_{za} \tan \beta_a = c_{\theta b} + w_{\theta b} = c_{zb} \tan \alpha_b + c_{zb} \tan \beta_b \tag{5.4}$$

Sometimes, it is reasonable to assume that the axial velocity remains the same ($c_{za} = c_{zb}$), in which case the relationship between the angles becomes quite simple.

$$\tan \alpha_a + \tan \beta_a = \tan \alpha_b + \tan \beta_b \quad (\text{if } c_{za} = c_{zb}) \tag{5.5}$$

In order to relate the velocities across the compressor stage to the thermodynamic parameters, let us consider the dynamics in a cylindrical coordinate system. The Newton's second law for a fluid of mass, δm, in a rotating system is given as follows:

$$\begin{aligned} F_r &= \delta m\left(\frac{dc_r}{dt} - c_\theta \frac{d\theta}{dt}\right) \\ F_\theta &= \delta m\left(\frac{dc_\theta}{dt} + c_r \frac{d\theta}{dt}\right) \\ F_z &= \delta m\left(\frac{dc_z}{dt}\right) \end{aligned} \tag{5.6}$$

The left-hand sides represent the forces in the radial (r), tangential (θ) and axial directions (z), while the right-hand sides are the mass times the acceleration terms in respective directions. For the time being, we will use the middle equation for the force balance in the tangential direction. First, multiplying both sides by the radius, r, will result in the moment or torque, T, on the left-hand side.

$$rF_\theta = T = \delta m\left(r\frac{dc_\theta}{dt} - c_r r\frac{d\theta}{dt}\right) = \delta m\frac{d(rc_\theta)}{dt} \approx \frac{\delta m}{\Delta t}\Delta(rc_\theta) \tag{5.7}$$

In the last part of the equality in Eq. (5.7), we consider the fluid mass (δm) to traverse the rotor in time of Δt, the ratio of which is the mass flow rate. So we can write the torque that needs to be applied for a given rotor element as

$$rF_\theta = T = \dot{m}\left[(rc_\theta)_b - (rc_\theta)_a\right] \tag{5.8}$$

The torque can be converted to power, by multiplying by the angular speed, Ω.

$$P_s = T\Omega = \dot{m}\Omega\left[(rc_\theta)_b - (rc_\theta)_a\right] = \dot{m}u(c_{\theta b} - c_{\theta a}) \tag{5.9}$$

We have used the fact that $u = \Omega r$. The power can be converted back to work per unit mass of the fluid crossing the rotor element.

$$w_s = P_s \dot{m} = u(c_{\theta b} - c_{\theta a}) = u\Delta c_\theta \tag{5.10}$$

Equation (5.10) shows that the rotor work is proportional to the rotational speed, u, and the change in the tangential velocity of the fluid. Equation (5.10) is sometimes referred to as the Euler (turbine) equation, and is applicable for both compressors and turbines.

To relate the velocities to the compressor stage pressure increase, we need to use Eq. (5.2), which can be rewritten as

$$\frac{p_c^o}{p_a^o} = \left[1 + \eta_{st}\frac{\Delta T^o}{T_a^o}\right]^{\frac{\gamma}{\gamma-1}} \tag{5.11}$$

ΔT^o is the stagnation temperature difference from "a" to "c". This stagnation temperature difference times the specific heat of the fluid (c_p) is equal to the work input to the fluid, which is given by Eq. (5.10). Therefore, the pressure ratio can be determined as

$$\frac{p_c^o}{p_a^o} = \left[1 + \eta_{st}\frac{u\Delta c_\theta}{c_p T_a^o}\right]^{\frac{\gamma}{\gamma-1}} \tag{5.12}$$

Determination of the stage efficiency, η_{st}, is not an easy task, and depends on several operating conditions, such as the incoming air velocity, rotational speed and blade design. Usually, sophisticated computational fluid dynamic analyses and/or experimental testing are required. Equation (5.12) shows that higher u and Δc_θ will result in higher stage pressure ratio. However, beyond a certain rotational speed, the flow may approach transonic speeds, which

Example 5.1 Construction of velocity triangle for compressor state analysis

For a compressor stage at $r = 0.75$ m, the rotational speed is 2500 rpm. $T_a^o = 350\,K$, $c_{za} = c_{zb} = 150$ m/s, $\alpha_a = 35°$ and $\beta_b = 10°$. Complete the velocity triangle and calculate the stage pressure ratio for $\eta_{st} = 0.9$ (Figure E5.1.1).

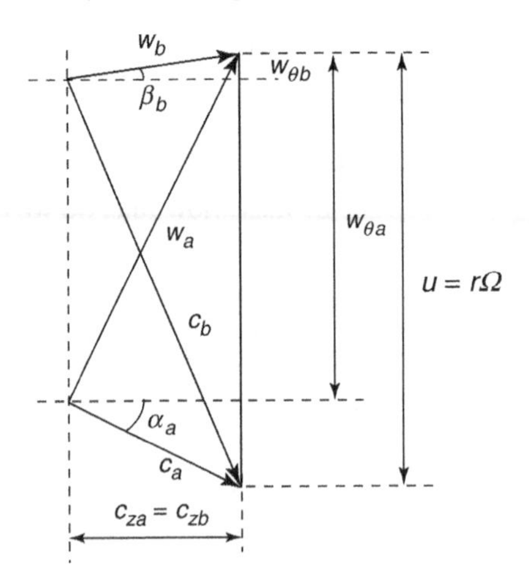

Figure E5.1.1 Velocity triangle.

$$u = 2\pi rn = 2\pi(0.75)\frac{2500\,\frac{\text{rpm}}{\text{s}}}{60\,\frac{\text{s}}{\text{min}}} = 196\,\frac{\text{m}}{\text{s}}$$

$$c_z = c_a \cos\alpha_a \rightarrow c_a = \frac{c_z}{\cos\alpha_a} = 183\,\frac{\text{m}}{\text{s}}$$

$$c_{\theta a} = c_a \sin\alpha_a = 105\,\frac{\text{m}}{\text{s}};\; w_{\theta a} = u - c_{\theta a} = 91\,\frac{\text{m}}{\text{s}};\; w_a = \sqrt{c_{za}^2 + w_{\theta a}^2} = 175\,\frac{\text{m}}{\text{s}}$$

$$\beta_a = \tan^{-1}\frac{w_{\theta a}}{c_{za}} = 31.3°$$

Similarly, $w_b = \dfrac{c_{zb}}{\cos\beta_a} = 152\,\dfrac{\text{m}}{\text{s}}$

$$w_{\theta b} = w_b \sin\beta_a = 26\,\frac{\text{m}}{\text{s}};\; c_{\theta b} = u - w_{\theta b} = 170\,\frac{\text{m}}{\text{s}};\; c_b = \sqrt{c_{zb}^2 + c_{\theta b}^2} = 226\,\frac{\text{m}}{\text{s}}$$

$$\alpha_b = \tan^{-1}\frac{c_{\theta b}}{c_{zb}} = 48.6°$$

The stage work is$w_s = u\Delta c_\theta = u(c_{\theta b} - c_{\theta a}) = 12.7\,\frac{\text{kJ}}{\text{kg}}$.

We can use Eq. (5.12) to find the compression ratio at this location, with $c_p = 1.04$ kJ/(kg K).

$$\frac{p_c^o}{p_a^o} = \left[1 + \eta_{st}\frac{u\Delta c_\theta}{c_p T_a^o}\right]^{\frac{\gamma}{\gamma-1}} = 1.114$$

must be avoided. Also, one must be careful to maintain a balance in the velocity triangles, to avoid rotor stalls (boundary layer separation). Similarly, there is a limit on Δc_θ for stable rotor operation. Due to the increasing pressure in the direction of the flow (adverse pressure gradient) in compressors, the pressure ratio that can be achieved in a single stage is limited to about 1.3 ~ 1.4; that is, the additive term in Eq. (5.12) is relatively small compared to 1. In that case, we can approximate the pressure ratio as

$$\frac{p_c^o}{p_a^o} = \left[1 + \eta_{st}\frac{u\Delta c_\theta}{c_p T_a^o}\right]^{\frac{\gamma}{\gamma-1}} \approx 1 + \frac{\gamma}{\gamma-1}\eta_{st}\frac{u\Delta c_\theta}{c_p T_a^o} = 1 + \eta_{st}\frac{u\Delta c_\theta}{RT_a^o} \tag{5.13}$$

R = specific gas constant for the fluid (air)

5.2.1 Compressor Stage Performance

At this point, we can examine the parameters that will affect the stage pressure ratio and efficiency. The aerodynamics of the flow depends on many parameters.

$$p_c^o, \eta_{st} = f(\dot{m}, p_a^o, T_a^o, \Omega, D, \gamma, R, \nu, \ldots) \tag{5.14}$$

As is customary in aerodynamic analyses, we can group the variables into non-dimensional sets of parameters.

$$\frac{p_c^o}{p_a^o}, \eta_{st} = f\left(\frac{\dot{m}\sqrt{RT_a^o}}{p_a^o D^2}, \frac{\Omega D}{\sqrt{RT_a^o}}, \frac{\Omega D^2}{\nu}\right) \tag{5.15}$$

We are down to three non-dimensional groups of numbers, for the independent variables. The final parameter in Eq. (5.15) is the Reynolds number, which does not cause significant changes in the performance parameters as long as the flow is turbulent and the Reynolds number sufficiently high. The two remaining parameters are non-dimensional mass flow rate number and rotational speed. For fixed D of the stage and gas constant R, we can omit these two factors.

$$\frac{p_c^o}{p_a^o}, \eta_{st} = f\left(\frac{\dot{m}\sqrt{T_a^o}}{p_a^o}, \frac{N}{\sqrt{T_a^o}}\right) \tag{5.16}$$

We can see that the compressor performance depends essentially on the mass flow rate and rotational speed (N = rpm) parameters. Figure 5.4 shows typical variations in the compressor stage pressure ratio and efficiency as a function of these two parameters. We can see in Figure 5.4 that there is an optimum mass flow rate for each rotation rate, N, and the performance diminishes as the mass flow rate is altered from this point. The reason is illustrated in Figure 5.4. The mass flow rate is proportional to the axial velocity (c_z) and rotation rate (N) to the rotational speed (u). For a given u, there is only a single value of c_z that will align the relative velocity vector to the orientation of the blade. If the mass flow rate and therefore c_z deviates from this value in either direction, then the alignment is offset, and the boundary layer behavior across the blade is also off from the optimum. This results in loss of performance. If the departure from the optimum is large, then the compressor blades can stall all together, which is referred to as compressor stalls. Compressor stall can occur for too large

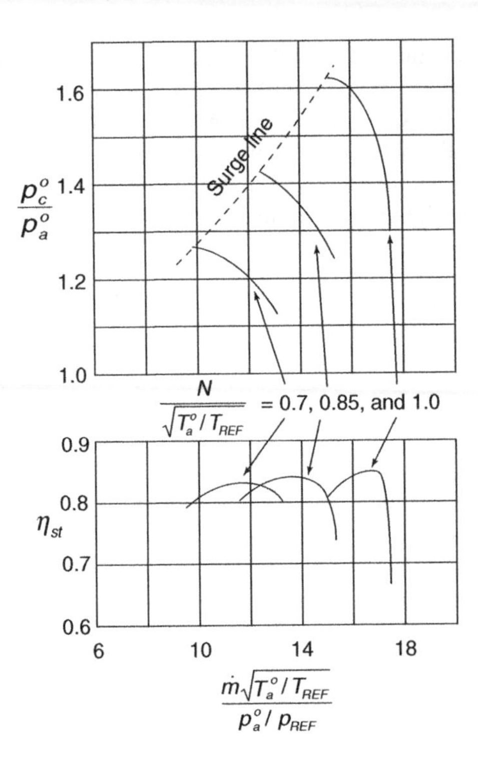

Figure 5.4 Compressor stage pressure ratio and stage efficiency.

(positive) or small (negative) angles of attack, and the stall conditions are plotted as "surge" lines in the performance curves, as shown in Figure 5.4.

Since the compressor stage consists of a large number of rotors/stators, during testing and analyses a cascade of blades is considered, as shown in Figure 5.5. Some definitions of the cascade arrangement are also shown in Figure 5.5. The blade angle typically refers to the blade

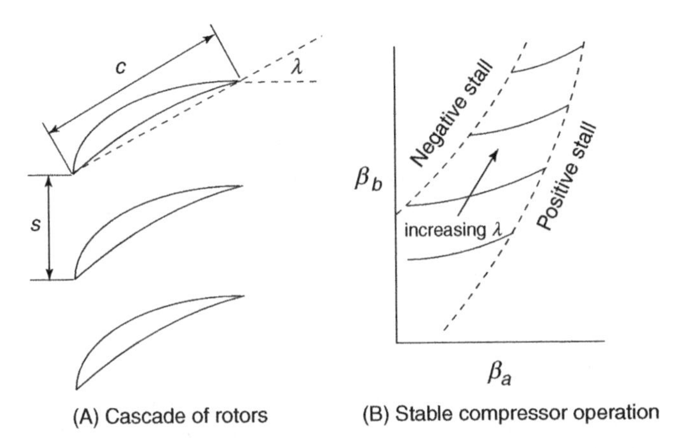

Figure 5.5 Cascade of rotors: (a) cascade geometry; (b) a chart of blade angles for stable compressor operational; and (c) effect of the mass flow rate and rotational velocity on the velocity triangle.

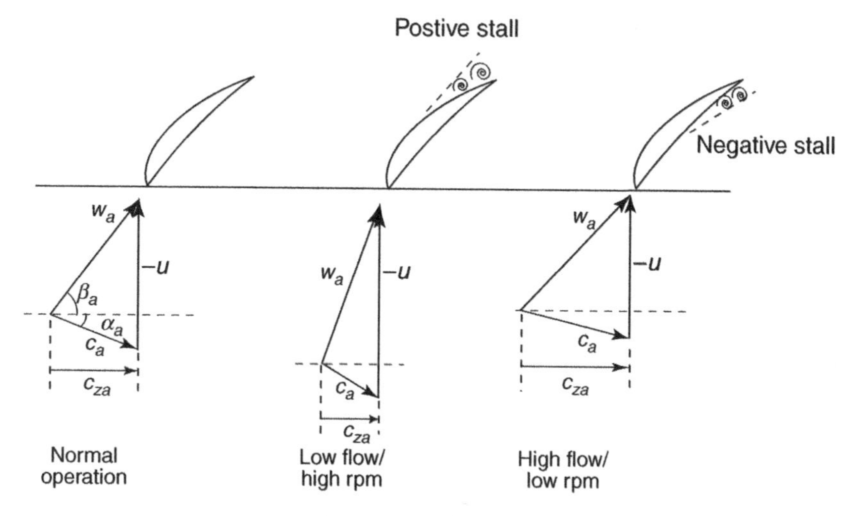

Figure 5.5 *(Continued)*

orientation at the trailing edge, with respect to the horizontal direction. As noted above, stalls can occur if the incoming flow angle, βa, is too large or too small. The effect of the mass flow rate and rotational velocity on the velocity triangle is also shown in Figure 5.5.

5.2.2 Pressure Coefficient and Boundary Layer Separation

Compressor stalls can occur due to imbalance of the axial air speed and the rotational speed, leading the flow misalignment, as shown in Figures 5.4 and Figure 5.5. Even for well-aligned flow, boundary layer separation can occur if the flow curvature over the blade surface is too large. We saw in Eq. (5.12) that U and Δc_θ are the main factors in increasing the stage pressure ratio. We cannot continue to increase Δc_θ in an attempt to maximize the pressure ratio, as this will lead to large flow turn over the blade surface. A measure of this flow curvature and the corresponding likelihood of boundary layer separation is the pressure coefficient, C_p, which is defined in the same manner as in airfoil aerodynamics.

$$C_p = \frac{p_b - p_a}{\frac{1}{2}\rho w_a^2} \tag{5.17}$$

If we assume constant density across the rotor, then the pressure difference can be related to the relative velocities, using Bernoulli's law.

$$p_b - p_a = \frac{1}{2}\rho(w_a^2 - w_b^2) \tag{5.18}$$

This leads to a simple relationship for the pressure coefficient.

$$C_p = 1 - \left(\frac{w_b}{w_a}\right)^2 \tag{5.19}$$

So the information from the velocity triangles – for example, the magnitude of the relative velocities – is useful in a variety of ways. Furthermore, if we consider a simple case of constant axial velocity across the rotor ($w_{za} = w_{zb}$), then Eq. (5.19) can be simplified once more.

$$C_p = 1 - \frac{\cos^2\beta_a}{\cos^2\beta_b} \tag{5.20}$$

Depending on a number of upstream flow parameters, boundary layer separation is expected for C_p above about 0.4. The pressure coefficient at the time of boundary layer separation is referred to as the critical pressure coefficient, C_p^{crit}. For a given β_a and C_p^{crit}, Eq. (5.20) can be used to find the limiting blade angle at the trailing edge (i.e. the allowable curvature of the blade).

$$\beta_b^{\min} = Cos^{-1}\left(\frac{\cos\beta_a}{\sqrt{1 - C_p^{crit}}}\right) \tag{5.21}$$

5.2.3 *de Haller Number and the Diffusion Factor*

An alternative measure of the blade flow quality is the de Haller number, which is a measure of the flow deceleration over the rotor. The logic is that the flow deceleration will lead to pressure increase, and therefore to a more adverse pressure gradient. The de Haller number (dH) is defined as the ratio of the exit to incident relative velocities for rotors. A similar definition exists for stators. The de Haller number should be at least 0.72, for acceptable flow qualities (no boundary layer separation).

$$dH = \frac{w_b}{w_a} \quad \text{(rotor)} \tag{5.22a}$$

$$dH = \frac{c_c}{c_b} \quad \text{(stator)} \tag{5.22b}$$

A more precise measure of the flow deceleration is the "diffusion factor", which is defined as follows:

$$\text{DF} = (\text{diffusion factor}) = \frac{(\text{maximum velocity}) - (\text{exit velocity})}{(\text{average velocity})} \tag{5.23}$$

In practice, it is customary to approximate the denominator (average velocity) as the incident velocity, and the maximum velocity as a function of the change in the tangential speed and the solidity. The solidity, σ, is a cascade parameter, defined as the ratio of the chord (c) and the blade spacing (s), that is, $\sigma = c/s$. Using these ideas, the diffusion factor can be calculated as

$$\text{DF} = \frac{w_{max} - w_b}{w_a} = \frac{\left(w_a + \dfrac{\Delta w_\theta}{2\sigma}\right) - w_b}{w_a} = 1 - \frac{w_b}{w_a} + \frac{|w_{\theta a} - w_{\theta b}|}{2\sigma w_a} \quad \text{(rotors)} \tag{5.24a}$$

A similar expression can be developed for stators.

$$\mathrm{DF} = \frac{c_{max} - c_c}{c_a} = \frac{\left(c_b + \dfrac{\Delta c_\theta}{2\sigma}\right) - c_c}{c_b} = 1 - \frac{c_c}{c_b} + \frac{|c_{\theta b} - c_{\theta a}|}{2\sigma c_b} \quad \text{(stators)} \tag{5.24b}$$

Diffusion factor is limited to approximately 0.5 or below, for stable aerodynamics, although in modern designs slightly higher values can be withstood.

5.2.4 Mach Number Effect

Even though the incoming velocity toward to blade (c_a) and the rotational speed (u) are individually subsonic, the relative velocity arising from the two components can approach the transonic range. As can be seen in velocity triangles, the vector sum can be larger than the individual components. Since the blade design uses subsonic airfoil cross-sections, transonic speeds are to be avoided. However, due to the high rotational speeds in modern compressors, some sections of the blade (toward the tip) may reach transonic speeds, and therefore low efficiency. A measure of the compressibility of the flow is the blade Mach number, based on w_a.

$$M_a = \frac{w_a}{\sqrt{\dfrac{\gamma p_a}{\rho_a}}} \tag{5.25}$$

If the incoming flow angle is altered, then the blade Mach number can be reduced. This can be achieved with inlet guide vanes (IGV) and the stators. Figure 5.6 shows a comparison of relative velocities with and without inlet guide vanes. By directing the flow in the same direction as the rotor rotation, the magnitude of the relative velocity is reduced and therefore also the blade Mach number.

$$\text{Without IGV}: \quad w_a = \sqrt{c_a^2 + u^2} = \sqrt{c_{za}^2 + u^2} \tag{5.26a}$$

$$\text{With IGV}: \quad w_a = \sqrt{c_{za}^2 + (u - c_{za}\tan\alpha_a)^2} \tag{5.26b}$$

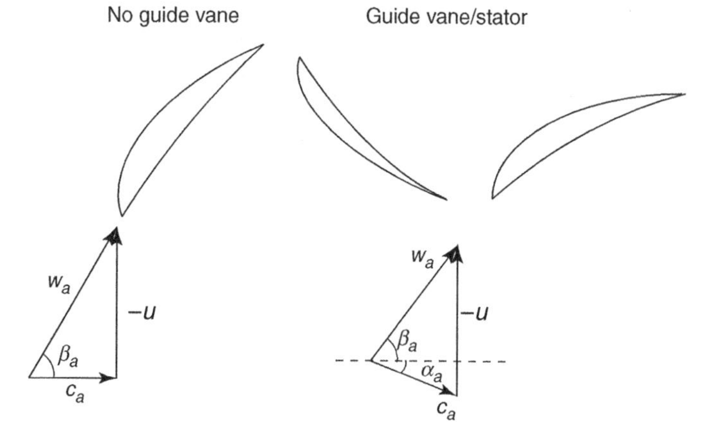

Figure 5.6 Inlet guide vanes/stators and resulting velocity triangles.

5.2.5 *Degree of Reaction*

We saw in above discussions that the rotor stall is an important concern during design and analyses of compressor stages. Poor flow characteristics can occur across both the rotors and the stators if the pressure gradient is too adverse. For this reason, we need to keep the static pressure increase across the rotor and stator elements about equal. A measure of this division of the static pressure increase across rotor and stator is called the "degree of reaction", R, and is defined as the ratio of the static enthalpy change across the rotor ("a" to "b") to the stagnation enthalpy change across the stage ("a" to "c").

$$R = \frac{h_b - h_a}{h_c^o - h_a^o} \approx \frac{p_b - p_a}{p_c - p_a} \tag{5.27}$$

Before estimating the degree of reaction with Eq. (5.27), we should prove how the ratio of enthalpy changes are related to the ratio of static pressure changes. We start by writing the first law of thermodynamics in the following form.

$$Tds = dh - \frac{dp}{\rho} \approx 0 \tag{5.28}$$

We set the entropy change to nearly zero, since we are dealing with nearly isentropic flows without any major flow problems. For nearly constant-density flows, the right-hand side furnishes us with a relationship between enthalpy and pressure, which may be applied for both static and stagnation quantities.

$$h_b - h_a \approx \frac{1}{\rho}(p_b - p_a) \tag{5.29a}$$

$$h_c^o - h_a^o \approx \frac{1}{\rho}(p_c^o - p_a^o) \approx \frac{1}{\rho}(p_c - p_a) \tag{5.29b}$$

The stagnation pressure on the right-hand side of Eq. (5.29b) consists of the static (p) and dynamic pressure ($1/2\rho c^2$). Since the stages are designed to keep the same absolute velocities going from one stage to the next ($c_c \approx c_a$), the dynamic pressure contributions are about the same. Thus, the stagnation pressure difference nearly equals the static pressure difference. This proves that the degree of reaction, as defined with the enthalpy terms, is indeed a good measure of the static pressure increases. An optimum balance of the static pressure increase across the rotor and the stator, an equal share of the increase, would be achieved if the degree of reaction is 0.5. This is the quantity that is sought during stage design and analyses.

Now, we can estimate the degree of reaction, again based on velocities. First, we note that the denominator of Eq. (5.27) is equal to the rotor work, since the stator does not add to the stagnation enthalpy of the flow.

$$h_c^o - h_a^o = u(c_{\theta b} - c_{\theta a}) \tag{5.30}$$

In order to assess the static enthalpy change across the rotor, we consider the energy balance in a coordinate frame attached to the rotor blade.

$$h_a + \frac{w_a^2}{2} = h_b + \frac{w_b^2}{2} \tag{5.31}$$

In this coordinate frame (attached to the rotor), there is no change in the stagnation enthalpy since the rotational velocity relative to the rotor is zero. Thus, static enthalpy difference is simply the difference of the kinetic energy (of the relative velocities) across the rotor.

$$h_b - h_a = \frac{w_a^2}{2} - \frac{w_b^2}{2} \tag{5.32}$$

So Eq. (5.27) becomes

$$R = \frac{h_b - h_a}{h_c^o - h_a^o} = \frac{w_a^2 - w_b^2}{2u(c_{\theta b} - c_{\theta b})} \tag{5.33}$$

This expression may be simplified further if we assume that $c_{za} = c_{zb}$, that is, constant axial velocity across the rotor, as shown in Figure 5.7. We can break down the magnitude of the relative velocities.

$$\begin{aligned} w_a^2 &= w_{\theta a}^2 + w_{za}^2 \\ w_b^2 &= w_{\theta b}^2 + w_{zb}^2 \end{aligned} \tag{5.34}$$

From the examination of the velocity triangle in Figure 5.7, and a little algebra, the following expression for R is valid for $c_{za} = c_{zb}$.

$$R = \frac{w_{\theta a} + w_{\theta b}}{2u} \quad (\text{for } c_{za} = c_{zb}) \tag{5.35}$$

The velocity terms can be converted to angles.

$$\begin{aligned} R &= \frac{c_z}{u}\left(\frac{\tan\beta_a + \tan\beta_b}{2}\right) \\ &= \frac{1}{2} + \frac{c_z}{u}\left(\frac{-\tan\alpha_a + \tan\beta_b}{2}\right) \quad (\text{for } c_{za} = c_{zb}) \\ &= \frac{1}{2} + \frac{c_z}{u}\left(\frac{\tan\beta_a - \tan\alpha_b}{2}\right) \end{aligned} \tag{5.36}$$

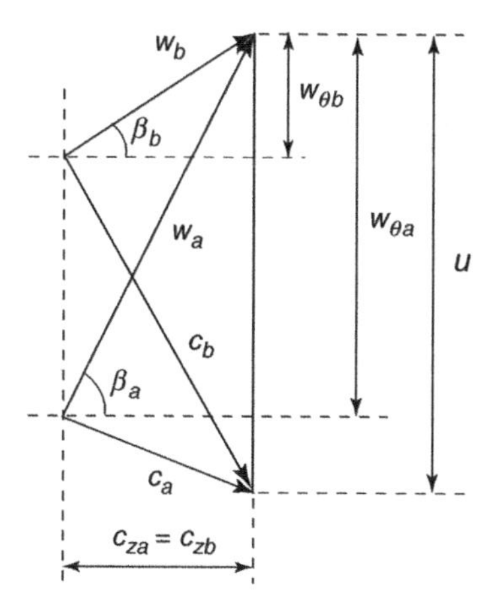

Figure 5.7 Velocity triangle for determining the degree of reaction.

The latter two equalities are obtained after noting the following from the velocity triangle.

$$\frac{u}{c_z} = \tan\alpha_a + \tan\beta_a = \tan\alpha_b + \tan\beta_b \tag{5.37}$$

During compressor design, it is customary to target $R = 0.5$ at the mid-radius, $r_m = (r_t + r_h)/2$, where the subscript "t" and "h" refer to the tip and hub (root) locations, respectively.

Example 5.2 Calculation of the compressor stage parameters

For a compressor stage, the stagnation temperature and pressure at "a" are 380K and 195 kPa, respectively. $c_{za} = 85\,\text{m/s}$, $c_{\theta a} = 45\,\text{m/s}$, $w_{zb} = 65\,\text{m/s}$, $w_{\theta b} = 15\,\text{m/s}$, and $u = 175\,\text{m/s}$. Calculate the specific work, degree of reaction, Mach number and the pressure coefficient.

$$c_{\theta b} = u - w_{\theta b} = 160\frac{\text{m}}{\text{s}}, \text{ so that } \Delta c_\theta = c_{\theta b} - c_{\theta a} = 115\frac{\text{m}}{\text{s}}.\ w_s = u\Delta c_\theta$$
$$= 20.125\,\text{kJ/kg}.$$

Since c_{za} and c_{zb} are not equal, we need to use Eq. (5.33) for the degree of reaction.

$$w_a^2 = w_{za}^2 + w_{\theta a}^2 = c_{za}^2 + (u - c_{\theta a})^2 = 24125\left(\frac{\text{m}}{\text{s}}\right)^2$$

$$w_b^2 = w_{zb}^2 + w_{\theta b}^2 = 4450\left(\frac{\text{m}}{\text{s}}\right)^2$$

$$R = \frac{h_b - h_a}{h_c^o - h_a^o} = \frac{w_a^2 - w_b^2}{2u(c_{\theta b} - c_{\theta a})} = \frac{24125 - 4450}{2(175)(115)} = 0.489$$

As shown in Eq. (5.25), the Mach number is based on the relative velocity.

$$M_a = \frac{w_a}{\sqrt{\dfrac{\gamma p_a}{\rho_a}}}$$

The local speed of sound is $a_a = \sqrt{\dfrac{\gamma p_a}{\rho_a}} = \sqrt{\gamma R T_a} = 388\,\text{m/s}$, based on $T_a = T_a^o - \dfrac{c_a^2}{2c_p} = 375.6\,\text{K}$.

$$M_a = \frac{w_a}{a_a} = 0.4$$

The pressure coefficient follows as $C_p = 1 - \left(\dfrac{w_b}{w_a}\right)^2 = 0.816$.

5.3 Radial Variations in Compressors

Thus far, discussions have focused on compressor parameters for the stage as a whole, or at a fixed radial location. Due to rotation of the compressor blades, properties vary over the radial length of the rotor. Certainly, the rotational velocity varies as $u = \Omega r$, as do the other velocities, such as c_θ and c_z, as a function of r. We can start our examination of this variation from the force balance in the radial direction as shown in Figure 5.8. The dynamical equation in the radial direction (Eq. (5.6)) allows us to relate the acceleration with the net force acting on a fluid volume in the radial direction.

$$F_r = \delta m\left(\frac{dc_r}{dt} - c_\theta\frac{d\theta}{dt}\right) = -\delta m\frac{c_\theta^2}{r} \tag{5.6}$$

The acceleration term on the right-hand side of Eq. (5.6) is called the centripetal acceleration. What are the forces acting in the radial direction? For a fluid element "suspended" in the compressor channel, we only need to consider the pressure forces as shown in Figure 5.8.

$$F_r = prd\theta dz + \left(p + \frac{\partial p}{\partial r}dr\right)(r + dr)d\theta dz + 2(pdrdz)\frac{d\theta}{2} \tag{5.38}$$

In the above equation, we are making the small-angle approximation ($\sin\frac{d\theta}{2} \approx \frac{d\theta}{2}$) for the pressure forces acting on the slanted sides. Using Eq. (5.38) in (5.6) gives us

$$\frac{\partial p}{\partial r} = \rho\frac{c_\theta^2}{r} \tag{5.39}$$

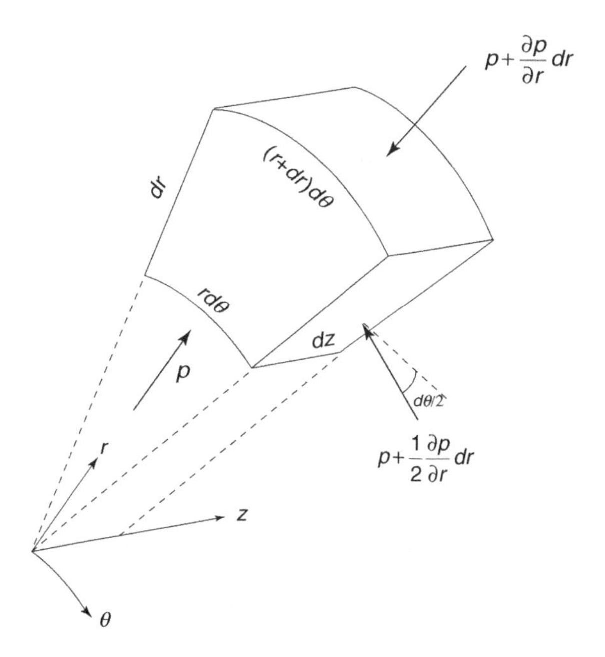

Figure 5.8 Force balance in the radial direction.

At this point, we look at the energy balance in the radial direction, using the first law of thermodynamics.

$$T\frac{\partial s}{\partial r} = \frac{\partial h}{\partial r} - \frac{1}{\rho}\frac{\partial p}{\partial r} \tag{5.40}$$

We can convert the static enthalpy to stagnation enthalpy after noting that the radial component of the velocity is very small. The flow is bounded in the radial direction by the hub and the casing, and this forces the radial velocity to be negligible outside of undesirable leakage flow or recirculation.

$$h^o = h + \frac{c_\theta^2 + c_z^2}{2} \tag{5.41}$$

Using Eqs. (5.41) in (5.40) leads to

$$T\frac{\partial s}{\partial r} = \frac{\partial h^o}{\partial r} - \frac{\partial}{\partial r}\left(\frac{c_\theta^2 + c_z^2}{2}\right) - \frac{1}{\rho}\frac{\partial p}{\partial r} \tag{5.42}$$

Using the result for the pressure gradient (Eq. (5.39)) in the above equation gives us

$$T\frac{\partial s}{\partial r} = \frac{\partial h^o}{\partial r} - \frac{\partial}{\partial r}\left(\frac{c_\theta^2 + c_z^2}{2}\right) - \frac{c_\theta^2}{r} \tag{5.43}$$

One of the major objectives of the compressor design is to attain uniform pressure increase along the radial length of the rotor. This means that both the rotor work (as given by the stagnation enthalpy change) and the loss terms (as given by the entropy change) should be the same at all radii. Thus, the condition for "radial equilibrium" in compressors is achieved by setting the radial gradients of entropy and stagnation enthalpy to zeros in Eq. (5.43). This furnishes the "equation for radial equilibrium".

$$\frac{\partial}{\partial r}\left(c_z^2\right) = -\frac{2c_\theta^2}{r} - \frac{\partial}{\partial r}\left(c_\theta^2\right) = -\frac{1}{r^2}\frac{\partial}{\partial r}(rc_\theta)^2 \tag{5.44}$$

There are several combinations of c_z and c_θ that will satisfy Eq. (5.44). The simplest, and most often used, combination is if c_z and rc_θ are held constant, which would set both sides of Eq. (5.44) to zero.

$$\text{Free-vortex flow}: \quad c_z = \text{const.}, rc_q = \text{const.} \tag{5.45}$$

In the free-vortex design, $c_\theta \sim 1/r$, so the tangential velocity decreases from the hub to the root. This kind of velocity distribution is attained by adjusting the blade orientation angles. Since c_θ changes, the velocity triangles also change as a function of the radius, as shown in Figure 5.9.

Free vortex is not the only possible option for the swirl distribution, that is, the radial distribution of tangential speeds. Although often used due to some desirable properties, free vortex does involve excessive blade twist from the root to the tip, as shown in the previous

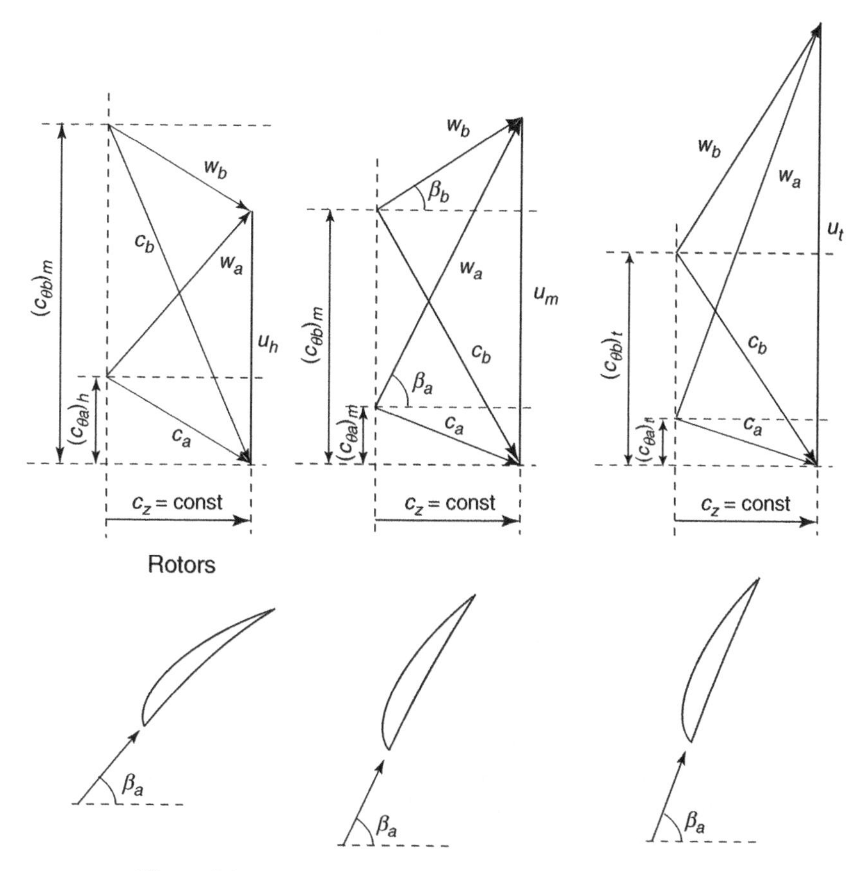

Figure 5.9 Velocity triangles for the free-vortex flow design.

example. The reason is that the tangential velocity is inversely proportional to the radius, while the rotation speed increases linearly with the radius. In general, the swirl distribution can be specified at the inlet and exit of the rotor as follows:

$$c_{\theta a} = A\left(\frac{r}{r_m}\right)^n - B\frac{r_m}{r} \tag{5.46a}$$

$$c_{\theta b} = A\left(\frac{r}{r_m}\right)^n + B\frac{r_m}{r} \tag{5.46b}$$

Specified in this manner, the constants A and B can be directly related to the stagnation enthalpy rise and the degree of reaction. For example, using Eq. (5.46) in Eq. (5.10) gives us

$$w_s = u\Delta\theta = 2\Omega rB\frac{r_m}{r} = 2\Omega Br_m \tag{5.47}$$

The free-vortex distribution is a case where $n = -1$ in Eq. (5.46), which again produces $c_\theta \sim 1/r$.

$$c_{\theta a} = (A - B)\frac{r_m}{r} \tag{5.48a}$$

$$c_{\theta b} = (A + B)\frac{r_m}{r} \tag{5.48b}$$

The constants A and B are related to the $c_{\theta a}$ and $c_{\theta b}$ at the mid-radius.

$$(c_{\theta a})_m = A - B \tag{5.49a}$$

$$(c_{\theta b})_m = A + B \tag{5.49b}$$

Thus,

$$A = \frac{(c_{\theta a})_m + (c_{\theta b})_m}{2} \tag{5.50a}$$

$$B = \frac{(c_{\theta b})_m - (c_{\theta a})_m}{2} \tag{5.50b}$$

How would we achieve free-vortex swirl distribution? Control of the tangential speeds is obtained through the blade orientation of the preceding stator and the rotor. If the trailing edge of the preceding stator or the inlet guide vane generates $c_{\theta a}$ distribution that follows $c_{\theta a} = (rc_{\theta a})_m/r$ and the rotor trailing edge twisted to produce $c_{\theta b} = (rc_{\theta b})_m/r$, then we would have free-vortex swirl distribution.

5.3.1 *Stage Work and Degree of Reaction for Free-Vortex Swirl Distribution*

For free-vortex, $rc_\theta = const.$ or $c_\theta = c_{\theta m}\frac{r_m}{r}$. We can confirm the constant stage work at all radial locations.

$$w_s = u(c_{\theta b} - c_{\theta a}) = \Omega(rc_{\theta b} - rc_{\theta a}) = \Omega[(r_m c_{\theta m})_b - (r_m c_{\theta m})_a] = \text{const.}$$

For $c_{za} = c_{zb}$, the degree of reaction varies steeply with the radius. From Eq. (5.35), we have

$$R = \frac{w_{\theta a} + w_{\theta b}}{2u} = \frac{(u - c_{\theta a}) + (u - c_{\theta b})}{2u} = 1 - \frac{c_{\theta a} + c_{\theta b}}{2u}$$

For free-vortex, this becomes

$$R = 1 - \frac{(r_m c_{\theta m})_a + (r_m c_{\theta m})_b}{2\Omega r^2}$$

Thus, the degree of reaction varies as inverse of the r^2.

Some of the other choices for the radial variations correspond to $n=0$ and $n=1$ in Eq. (5.46).

$$\text{Exponential } (n=0): \quad c_{\theta a} = A - B\frac{r_m}{r} \tag{5.51a}$$

$$c_{\theta b} = A + B\frac{r_m}{r} \tag{5.51b}$$

Substitution of this distribution in Eq. (5.44) and integration, give us the corresponding axial velocity as a function of the radius, after some algebraic work.

$$(c_{za})^2 = (c_{za})_m^2 - 2\left(A^2 \ln\frac{r}{r_m} + \frac{AB}{r/r_m} - AB\right) \tag{5.52a}$$

$$(c_{zb})^2 = (c_{zb})_m^2 - 2\left(A^2 \ln\frac{r}{r_m} - \frac{AB}{r/r_m} + AB\right) \tag{5.52b}$$

$$\text{Forced-vortex } (n=1): \quad c_{\theta a} = A\frac{r}{r_m} - B\frac{r_m}{r} \tag{5.53a}$$

$$c_{\theta b} = A\frac{r}{r_m} + B\frac{r_m}{r} \tag{5.53a}$$

$$(c_{za})^2 = (c_{za})_m^2 - 2\left(A^2\left(\frac{r}{r_m}\right)^2 + AB\ln\frac{r}{r_m} - A^2\right) \tag{5.54a}$$

$$(c_{zb})^2 = (c_{zb})_m^2 - 2\left(A^2\left(\frac{r}{r_m}\right)^2 - AB\ln\frac{r}{r_m} - A^2\right) \tag{5.54b}$$

5.4 Preliminary Compressor Analysis/Design

During design or analyses of compressors, we can follow the mean streamline at the mid-radius, apply radial variations and estimate the stage and compressor performance. This would require the use of the parameters discussed in previous sections. However, loss terms can only be estimated and incorporated into the stage efficiency, or separately using available correlations. Correlations for losses and other effects are useful for initial estimates, but at some point more detailed computational and/or testing are needed to improve the designs. Nonetheless, an analysis serves as a useful starting point, and also places the performance in a broad perspective by accounting for several different effects in a simple manner. There are methods of compressor analysis of varying sophistication, but we consider a basic approach in this section. Figure 5.10 shows the streamline that follows the locus of the mid-radius. We can consider the velocity triangles and various performance parameters for a given stage, assess the radial variations and sum these effects, and then move on the next stage. Overall addition of the pressure rise and efficiency terms would then provide an estimate of the compressor performance.

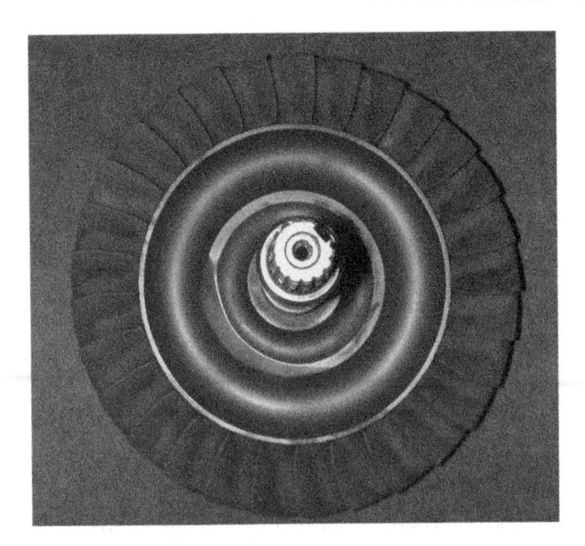

Figure 5.10 Mean streamline along the line connecting the mid-radius.

Example 5.3 Stage analysis starting from the mid-radius

Let us consider a stage analysis for free-vortex swirl distribution. The inlet conditions are as follows: $r_m = 0.75$ m, hub-to-tip ratio $= \zeta = r_h/r_t = 0.8$, rotational speed $= 3000$ rpm, $\eta_{st} = 0.9$; $T_a^o = 350$ K, $p_a^o = 195$ kPa, $c_{za} = c_{zb} = 150$ m/s, $\alpha_a = 35°$, and $\beta_b = 10°$. Following the swirl distribution, the velocity triangles are constructed at the tip and the hub. The average stage parameters are calculated by area-weighting with 1 : 0.5 : 0.5 ratios used for the mid-radius, tip and the hub, respectively, using MATLAB® program, MCE53 at the end of this chapter).

The main results are:

Average pressure ratio: 1.275

Average degree of reaction: 0.36

Average pressure coefficient: 0.454

Tip Mach number: 0.644.

5.5 Centrifugal Compressors

Centrifugal compressors can be made compact, and therefore are useful in small gas-turbine engines (e.g. auxiliary power units, helicopter power plants). They are also useful as turbo-pumps in liquid-propellant rocket engines. However, they tend to have lower efficiency, since the flow is forced radially outward, and has to be eventually turned in the axial direction toward the combustor. This turning leads to some loss in stagnation pressure, as does less

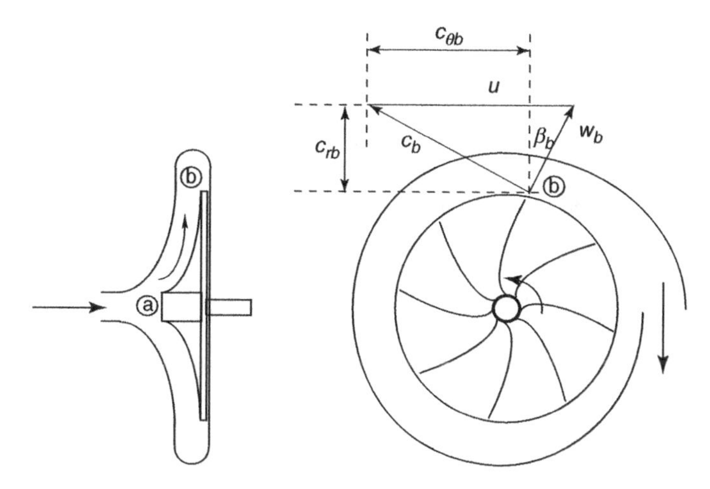

Figure 5.11 A schematic and velocity triangle for a centrifugal compressor.

control over the impeller blades. A schematic of the centrifugal compressor is shown in Figure 5.11. The flow enters normal to the plane of rotation and is forced radially outward by the blades. We can still make use of the Euler equation (Eq. (5.10)) and velocity triangles, except of course that the velocity components are different from axial compressors. For example, the change in the radial component of the velocity contributes to the overall momentum increase, and also the incident flow is at small radius while the exiting flow is at large radius. This requires us to consider the change in the rotational velocity from inlet to exit. Thus, we start from the Euler equation in the following form.

$$w_s = (uc_\theta)_b - (uc_\theta)_a \tag{5.55}$$

The work can also be related to the change in the stagnation enthalpy.

$$w_s = h_b^o - h_a^o = h_b - h_a + \frac{c_b^2}{2} - \frac{c_a^2}{2} \tag{5.56}$$

We can solve for the static enthalpy difference, while substituting Eq. (5.56) for the work term.

$$h_b - h_a = (uc_\theta)_b - (uc_\theta)_a + \frac{c_b^2}{2} - \frac{c_a^2}{2} \tag{5.57}$$

We can use the velocity triangle to relate the absolute with the relative velocity components.

$$\begin{aligned} w_\theta &= c_\theta - u \\ w_r &= c_r \\ w_z &= c_z \end{aligned} \tag{5.58}$$

The magnitude of the velocities has contributions from all three components.

$$\begin{aligned} c^2 &= c_r^2 + c_\theta^2 + c_z^2 \\ w^2 &= w_r^2 + w_\theta^2 + w_z^2 \end{aligned} \tag{5.59}$$

Using Eq. (5.59) in Eq. (5.57), while making use of Eq. (5.58), we obtain

$$h_b - h_a = \frac{u_b^2}{2} - \frac{u_a^2}{2} - \left(\frac{w_b^2}{2} - \frac{w_a^2}{2}\right) = \Omega^2\left(\frac{u_b^2}{2} - \frac{u_a^2}{2}\right) - \left(\frac{w_b^2}{2} - \frac{w_a^2}{2}\right) \tag{5.60}$$

Since the angular velocity, Ω, is a constant for a given operating condition, we can rewrite Eq. (5.60) in the following form.

$$\Delta h = \Delta\left(\frac{\Omega^2 r^2}{2}\right) - \Delta\left(\frac{w^2}{2}\right) \tag{5.61}$$

We may relate this enthalpy difference to the pressure difference, using the first law.

$$T\Delta s = \Delta h - \frac{\Delta p}{\rho} \rightarrow \frac{\Delta p}{\rho} = \Delta\left(\frac{\Omega^2 r^2}{2}\right) - \Delta\left(\frac{w^2}{2}\right) - T\Delta s \tag{5.62}$$

For nearly isentropic flows, this becomes

$$\frac{\Delta p}{\rho} \approx \Delta\left(\frac{\Omega^2 r^2}{2}\right) - \Delta\left(\frac{w^2}{2}\right) \tag{5.63}$$

Equation (5.63) shows the mechanism of the pressure increase in centrifugal compressors. It is the rotational motion imparted to the fluid that generates the pressure rise. The higher the rotational speed or the exit radius, the more dramatic is the pressure increase ($\sim\Omega^2$, r^2). The relative velocity at the exit is excess momentum escaping the compressor, and thus has a negative effect on the compression.

We can obtain the pressure increase, again by relating it to the work done on the fluid. Using Eqs. (5.55) and (5.56), we can write

$$w_s = h_b^o - h_a^o \approx c_p(T_b^o - T_a^o) \rightarrow \frac{T_b^o}{T_a^o} = 1 + \frac{(uc_\theta)_b - (uc_\theta)_a}{c_p T_a^o} = 1 + (\gamma - 1)\frac{(uc_\theta)_b - (uc_\theta)_a}{(a_a^o)^2} \tag{5.64}$$

For most centrifugal compressors, $c_{\theta a} \approx 0$, so we have

$$\frac{T_b^o}{T_a^o} = 1 + (\gamma - 1)\left(\frac{u_b}{a_a^o}\right)^2 \frac{c_{\theta b}}{u} \tag{5.65}$$

Example 5.4 Centrifugal compressor

For a centrifugal compressor rotating at 4200 rpm, find the compression ratio for the following specifications. $T_a^o = 320K$; $\eta_{st} = 0.8$; $r_b = 0.8$; $r_a = 0.025$; $\beta_b = 60°$, (volumetric flow rate) = 6.05 m³/s and t (the thickness of the flow section at the exit) = 0.16 m.

Using the volumetric flow rate, $c_{rb} = \dfrac{\dot{m}/\rho}{2\pi r_b t} = w_{rb} = 8.0\,\text{m/s}$.

$$\tan\beta_b = \frac{w_{b\theta}}{w_{br}} \rightarrow w_{\theta b} = w_{rb}\tan\beta_b = 13.9\,\text{m/s}.$$

$$c_{\theta b} = u_b - w_{\theta b} = 316\,\text{m/s}.$$

Equation (5.66) can now be used.

$$\frac{p_b^o}{p_a^o} = \left[1 + \eta_{st}\frac{u_b c_{\theta b}}{c_p T_a^o}\right]^{\frac{\gamma}{\gamma-1}} = 1.094$$

The stagnation temperature can again be converted to the compression ratio, using the stage efficiency.

$$\frac{p_b^o}{p_a^o} = \left[1 + \eta_{st}\frac{u_b c_{\theta b}}{c_p T_a^o}\right]^{\frac{\gamma}{\gamma-1}} \tag{5.66}$$

From the velocity triangle in Figure 5.11, we can evaluate the compression ratio.

$$\frac{p_b^o}{p_a^o} = \left[1 + \eta_{st}\frac{u_b(u_b - w_{rb}\tan\beta_b)}{c_p T_a^o}\right]^{\frac{\gamma}{\gamma-1}} \tag{5.67}$$

5.6 Turbine

We can view the turbine process as the opposite of the compressor, where the enthalpy of the fluid is converted to the mechanical power output of the turbine. This process is sketched in Figure 5.12. The isentropic efficiency of the turbine or turbine stage is defined as

$$\eta_t = \frac{h_4^o - h_5^o}{h_4^o - h_{5s}^o} \approx \frac{1 - \dfrac{T_5^o}{T_4^o}}{1 - \left(\dfrac{p_5^o}{p_4^o}\right)^{\frac{\gamma-1}{\gamma}}} \tag{5.68}$$

Again, η_t is less than one, due to similar loss terms as in compressors, such as frictional, flow separation and end losses, although the flow separation issues are much less severe in turbines

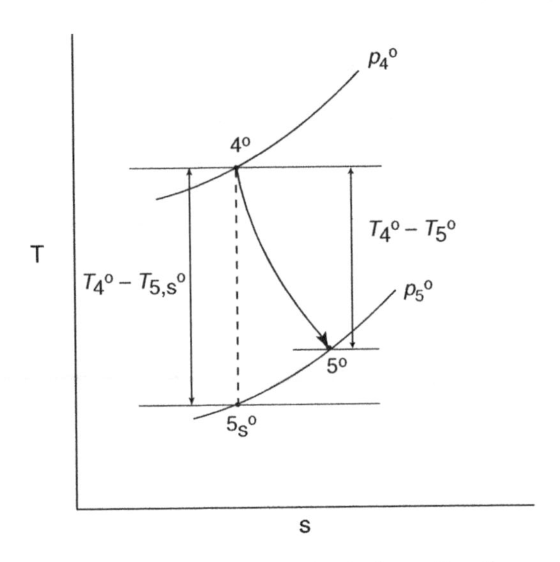

Figure 5.12 Turbine process sketched on a *T–s* diagram.

due to the favorable pressure gradients. Equation (5.68) is useful in defining the performance of turbines, simply by measuring the input and exit temperatures and pressures. Once measured at a range of operating conditions (mass flow rate and rpm), it is also obviously useful for estimating the turbine power output based on the inlet and exit pressure conditions.

Figure 5.13 shows some typical turbines and velocity triangles. The tangential velocity decreases going across the rotor blade, and this momentum difference is used to "lift" the turbine blades, to generate turbine rotation and power. This is the opposite of compressor process, where the action of the rotor generates higher tangential velocity. Thus, the configuration of the rotor and the corresponding velocity triangles easily identify whether the process is power input (compressor) or extraction (turbine). The stators in the turbine is sometimes referred to as "nozzles" as they direct the flow and also lower the static pressure. Other than this reversal in Δc_θ, we can again use the Euler turbine equation to calculate the turbine stage work based on velocity triangle data.

$$w_s = u\Delta c_\theta = u(c_{\theta a} - c_{\theta b}) \tag{5.69}$$

Again, larger work per stage would result for higher u and Δc_θ. Similar to compressors, the rotational velocity (u) is limited by Mach number effects, and the maximum allowable stress, which is much lower in turbines due to the higher operating temperatures. The turbine is downstream of the combustor and the starting operating temperature is that of the combustor exit. On the other hand, much higher Δc_θ can be achieved in turbines, without incurring flow problems, due to favorable pressure gradient that exist in the turbine. The velocity triangle in Figure 5.13 is typical, where the flow direction is reversed in the tangential direction across the rotor.

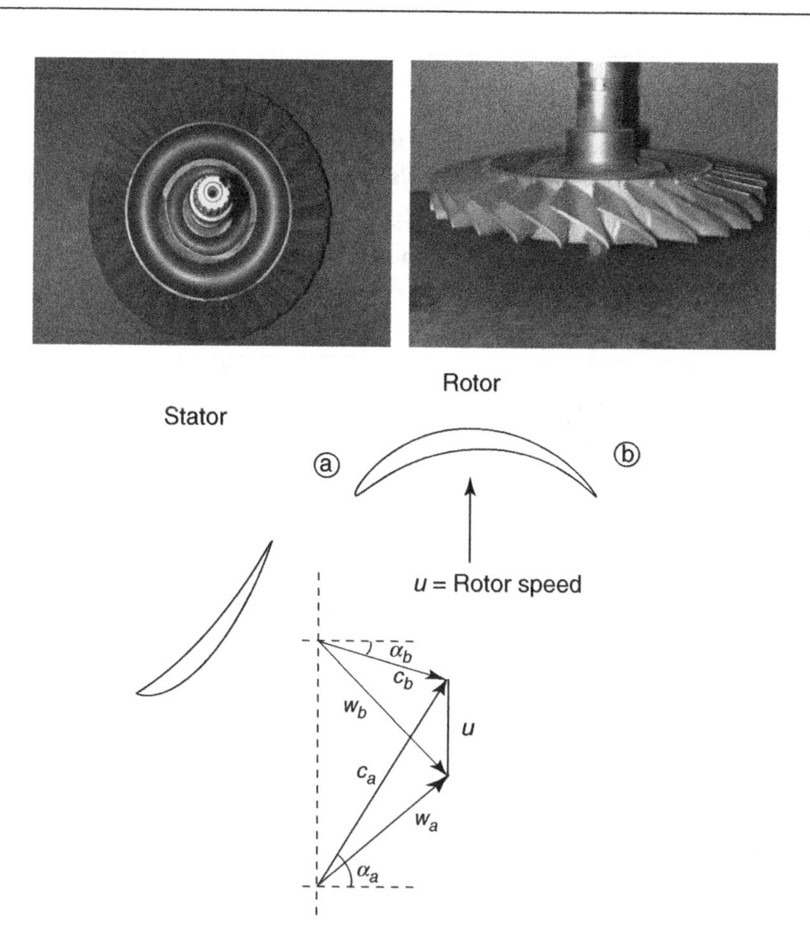

Figure 5.13 Typical turbines and velocity triangles.

A measure of performance of turbine stage is the turbine work ratio.

$$\frac{w_s}{u^2} = \frac{\Delta c_\theta}{u} = \frac{c_{\theta a} - c_{\theta b}}{u} = \frac{c_{za} \tan \alpha_a - c_{zb} \tan \alpha_b}{u} \tag{5.70}$$

For a given mass flow rate (c_{za}), the stage work is maximized by increasing α_a and decreasing α_b. However, similar to compressors, both are limited by some flow constraints. For example, increase in α_a results in high incident Mach number. The incident absolute velocity is related to α_a.

$$c_a = \frac{c_{za}}{\cos \alpha_a} \tag{5.71}$$

For a given c_{za}, as α_a approaches $\pi/2$, c_a approaches infinity. In practical turbines, α_a is limited to about 70°. Also, as shown in Figure 5.13, decrease in α_b requires a large β_a, which involves a severe flow turning, which may cause flow separation in spite of the favorable pressure gradients in the turbine.

Example 5.5 Turbine stage analysis

At mid-radius of a turbine stage, $c_{za} = c_{zb}; R_m = 0.5; \frac{\Delta c_\theta}{u} = 1$; and $\frac{c_z}{u} = 0.4$. Construct the velocity triangle at the mid-radius and also at the hub for $rc_\theta = ar + b$ and $R_h = 0$.

At the mid-radius, $R = 0.5$ means a symmetric velocity triangle with $\alpha_a = \beta_b; \alpha_b = \beta_a$. Also, since $\Delta c_\theta = u_m$, $c_{\theta a} = u_m$; $c_{\theta b} = 0$, this leads to the velocity triangle shown in Figure E5.5.1.

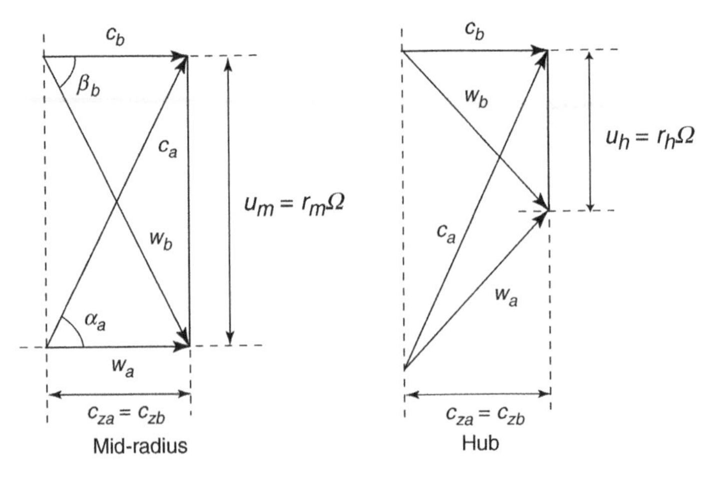

Figure E5.5.1 Turbine stage velocity triangles.

For the swirl distribution, $rc_\theta = ar + b$, $(rc_{\theta b})_m = 0 \rightarrow c_{\theta b} = 0$ at all radii. Also, $R = 0$ means $w_{\theta a} = w_{\theta b}$. This leads to the velocity triangle shown in Figure E5.5.1, at the hub.

5.6.1 *Estimation of the Blade Stagnation Temperature*

The turbine blade is subject to a harsh thermal environment, due to its downstream location from the combustor. For typical alloys used for turbines, the maximum allowable stress decreases steeply with increasing temperature. For modern turbines, active cooling is used to cover the blade surfaces with a thin film of cooling air. First, we need to estimate the temperature encountered by the blades. The relative motion of the blade results in higher temperature, particularly at the leading edge of the blades. We start from the static temperature entering the turbine stage.

$$T_4 = T_4^o - \frac{c_{\theta a}^2 + c_{za}^2}{2c_p} \tag{5.72}$$

We can normalize by the stagnation temperature at the turbine inlet, and multiply and divide by $\Omega = u_t/r_t = u/r$.

$$\frac{T_4}{T_4^o} = 1 - \frac{\gamma - 1}{2}\left(\frac{u_t}{a_4^o}\right)^2\left(\frac{r}{r_t}\right)^2\left[\left(\frac{c_{\theta a}}{u}\right)^2 + \left(\frac{c_{za}}{u}\right)^2\right] \tag{5.73}$$

Now, the blade temperature is estimated by the stagnation temperature caused by the relative velocity w_a, with the recovery factor, r_f, quantifying the stagnation effect. For turbulent boundary layer flow over the turbine blades, the recovery factor can be estimated as $\sqrt[3]{Pr}$, where Pr stands for the Prandtl number $Pr = \mu c_p/k$. This estimate typically gives $0.9 < r_f < 0.95$ for air under turbine operating conditions.

$$T_{blade} = T_4 + r_f \frac{w_a^2}{2c_p} \tag{5.74}$$

Using Eq. (5.73) in (5.74) we get

$$\frac{T_{blade}}{T_4^o} = 1 - \frac{\gamma - 1}{2}\left(\frac{u_t}{a_4^o}\right)^2\left(\frac{r}{r_t}\right)^2\left[\left(\frac{c_{\theta a}}{u}\right)^2 + \left(\frac{c_{za}}{u}\right)^2 - r_f\left\{\left(1 - \frac{c_{\theta a}}{u}\right)^2 + \left(\frac{c_{za}}{u}\right)^2\right\}\right] \tag{5.75}$$

For high blade temperature, active cooling is deployed, with passages within the blade element to deliver cooler air (directly from the compressor). A thin film of cooling air provides a boundary layer to impede heat transfer onto the blade surfaces. Film cooling is the subject of active research, and its effectiveness is a function of injection angle, locations, mass flow rates and cooling air temperature.

Example 5.6 Estimate of the turbine blade surface temperature

Estimate the blade surface temperature for a turbine stage at $r_m = 0.5$ m, $T_a^o = 1400$ K; $c_{za} = 125$ m/s; $\Omega = 500$ rad/s; $r_f = 0.9$; $\alpha_a = 60°$, and $\beta_a = -\beta_b$.

$c_{\theta a} = c_{za} \tan\alpha_a = 216.5$ m/s, and $u = 2pr_m\Omega = 175$ m/s.

$$w_a = \sqrt{c_{za}^2 + (c_{\theta a} - u)^2} = 131.7 \text{ m/s}$$

First, the static temperature is $T_a = T_a^o - \frac{c_{\theta a}^2 + c_{za}^2}{2c_p} = 1569$ K.

Then, the blade surface temperature is

$$T_s = T_a + r_f \frac{w_a^2}{2c_p} = 1577 \text{ K}.$$

We can see that the thermal energy dominates, and the velocity terms make a relatively small impact on the blade surface temperature.

5.6.2 Turbine Blade and Disk Stresses

Figure 5.14 shows a schematic of the turbine rotor attached to the turbine disk. This disk itself would be attached to the rotating hub. The rotors and disk are subjected to high centrifugal stress in a high-temperature environment. For alloys used for turbine materials, the maximum allowable stress may be reduced by half, with a temperature increase of 200 ~ 300 °C. Of course, the mechanical stress on the turbine blade should be below the maximum allowable stress of the material, with a margin of safety. We can estimate the rotor and disk stress, using the basic equations of mechanics. Considering an element on the rotor (Figure 5.14) at radius, r, the force balance in the radial direction (Eq. (5.6)) is

$$F_r = -\delta m \frac{c_\theta^2}{r} = -(\rho_b A_b dr)\frac{(\Omega r)^2}{r} \tag{5.76}$$

The rotational velocity is that of the rotor, and the mass of the element is given by the blade material density (ρ_b) and its volume. The net force on the element can be calculated using the stress in the radial direction, σ_r. Then, Eq. (5.76) becomes

$$\frac{d\sigma_r}{dr} = -\rho_b \Omega^2 r \tag{5.77}$$

We can integrate, by noting that $\sigma_r = 0$ at $r = r_t$.

$$\sigma_r = \rho_b \frac{u_t^2}{2}\left[1 - \left(\frac{r}{r_t}\right)^2\right] \tag{5.78}$$

Thus, the blade stress is the most critical at the hub, and is a function of the blade material density and the rotational speed.

The disk stress can be evaluated in a similar manner, again considering an element shown in Figure 5.14. The force balance gives the following equation for the disk stress, where $b = b(r)$ is the disk width.

$$\frac{d}{dr}(\sigma b r) - \sigma b + \rho_b b r^2 \Omega^2 = 0 \tag{5.79}$$

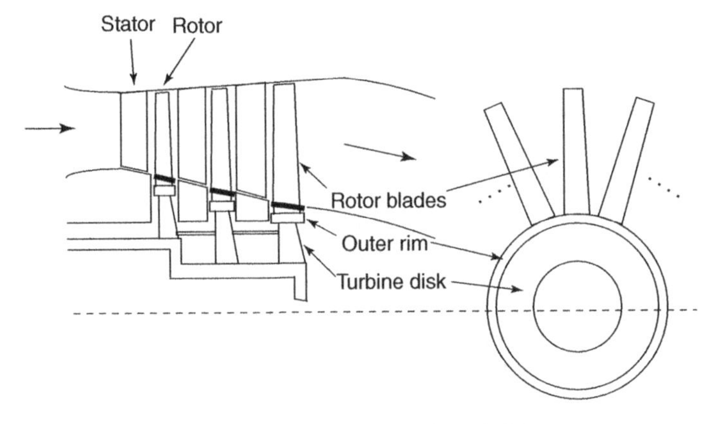

Figure 5.14 Turbine blade and disk.

For a given width distribution, $b = b(r)$, the stress can be found by numerically integrating Eq. (5.79). The simplest case would be $b = \text{const}$, in which case we have

$$\frac{d}{dr}(\sigma r) - \sigma + \rho_b r^2 \Omega^2 = 0 \quad (b = \text{const.}) \tag{5.80}$$

Alternatively, we may wish to keep the stress, σ, constant in the disk, and find the corresponding width distribution. Then Eq. (5.79) becomes

$$\frac{db}{b} = -\frac{\rho_b \Omega^2}{\sigma} r dr \quad (\sigma = \text{const.}) \tag{5.81}$$

This can easily be integrated to give an exponential width distribution, with b_i the width at the inner rim radius, r_i.

$$\ln \frac{b}{b_i} = -\frac{\rho_b \Omega^2}{2\sigma}(r^2 - r_i^2) \tag{5.82}$$

The corresponding stress throughout the disk is

$$\frac{\sigma}{\rho_b u_t^2} = \frac{\left(\frac{r_h}{r_t}\right)^2 \left[1 - \left(\frac{r_i}{r_h}\right)^2\right]^2}{2 \ln\left(\frac{b_i}{b_h}\right)} \tag{5.83}$$

5.7 MATLAB® Programs

```
% MCE 53: Stage analysis using area-weighted average for three blade
% segments.

%Input parameters
Tao=350;
pao=195;
g=1.4; R =287;
cp=1040;
hst=0.9;
rm = 0.75;
z=0.8;
RPM = 3000;
cza = 150;
aa = 35*pi/180;
bb=10*pi/180;

u=RPM*2*pi*rm/60

Ta=Tao-cza^2/(2*cp);
ssa=sqrt(g*R*Ta);
%Calculation of the mid-radius
```

```
cqa=cza*sin(aa);
wqa=u-cqa;
wqb=cza*sin(bb);
cqb=u-wqb;
delc=cqb-cqa;

wsm=u*delc;
prm=(1+hst*u*delc/(cp*Tao))^(g/(g-1))

wa2=wqa^2+cza^2;
wb2=wqb^2+cza^2;

Rm=(wa2-wb2)/(2*u*delc)
Cpm=1-wb2/wa2
Mm=sqrt(wa2)/ssa
% Hub parameters
rmrh=0.5*(1/z+1);
rh=rm/rmrh
uh=u/rmrh;
cqah=cqa*rmrh;
cqbh=cqb*rmrh;
delch=cqbh-cqah;

wsh=uh*delch;
prh=(1+hst*uh*delch/(cp*Tao))^(g/(g-1))

wqah=uh-cqah;
wqbh=uh-cqbh;
wah2=wqah^2+cza^2;
wbh2=wqbh^2+cza^2;

Rh=(wah2-wbh2)/(2*uh*delch)
Cph=1-wbh2/wah2
Mh=sqrt(wah2)/ssa
% Tip parameters
rmrt=0.5*(z+1);
rt=rm/rmrt
ut=u/rmrt;
cqat=cqa*rmrt;
cqbt=cqb*rmrt;
delct=cqbt-cqat;

wst=ut*delct;
prt=(1+hst*ut*delct/(cp*Tao))^(g/(g-1))

wqat=ut-cqat;
wqbt=ut-cqbt;
```

```
wat2=wqat^2+cza^2;
wbt2=wqbt^2+cza^2;
Rt=(wat2-wbt2)/(2*ut*delct)
Cpt=1-wbt2/wat2
Mt=sqrt(wat2)/ssa

% Average parameters with 2:1:1=m:t:h
prA=0.5*prm+0.25*prh+0.25*prt
RA=0.5*Rm+0.25*Rh+0.25*Rt
CpA=0.5*Cpm+0.25*Cph+0.25*Cpt
```

5.8 Problems

5.1. For $\Omega = 5000$ rpm and radial location of $r = 0.65$ m, find α_a and α_b, if $\beta_a = 60°$ and $\beta_b = 30°$. $c_{za} = c_{zb} = 105$ m/s. Draw the velocity triangle.

5.2. For $u = 175$ m/s, find $w_{\theta a}$ and $w_{\theta b}$ if $c_{\theta a} = 35$ m/s and $c_{\theta b} = 125$ m/s, $c_{za} = 120$ m/s and $c_{zb} = 105$ m/s. What are the magnitudes of the velocities, w_a, w_b, c_a, and c_b? Draw the velocity triangle.

5.3. For an axial compressor stage, the following velocity diagram is provided (note the asymmetric axial velocity from a to b). The stagnation temperature and pressure at "a" are 380 K and 195 kPa, respectively. Also, $c_p = 1.04$ kJ/(kg K), $\gamma = 1.4$

Calculate (a) specific work (work per unit mass of air flowing) of the compressor stage; (b) degree of reaction; (c) stagnation pressure ratio if the stage efficiency is 0.90; and (d) the pressure coefficient.

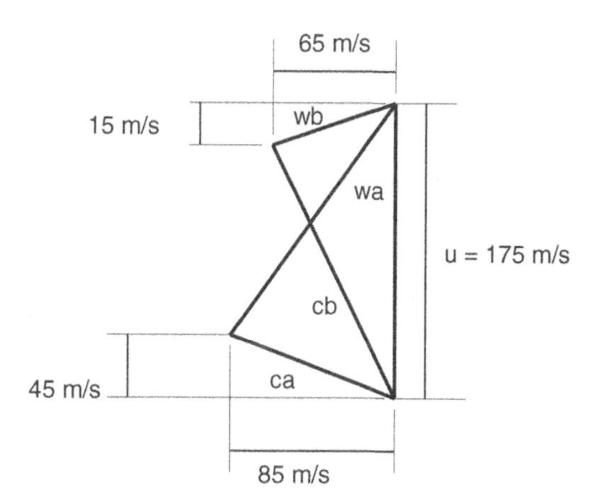

5.4. For a compressor, the tangential velocity distribution is given as follows:

$$(rc_\theta)_a = K_a;\ (rc_\theta)_b = K_b,$$

where the subscript "a" and "b" denote upstream and downstream of the rotor, respectively.

$$r_{hub} = 0.5 \text{ m}, r_{tip} = 0.6 \text{ m}.$$

a. Show that for the above velocity distribution, the work done by the compressor is independent of radius.

b. For the given velocity diagram in Problem 5.3 at the mid-radius, construct velocity diagrams at the hub and the tip. Include all relevant velocities and angles (u, c_a, c_b, w_a, w_b, α_a, β_a, α_b, β_b).

5.5. Analyze the axial compressor stage flow at $r_m = 0.6$ m. The rotational speed of the rotor at this point is $u_m = 250$ m/s, while $c_{za} = 150$ m/s and $c_{\theta a} = 30$ m/s. If c_z = constant, $\alpha_c = \alpha_a$, and the degree of reaction is 0.5. Calculate the following parameters.

a. Rotor angular speed, Ω, in rpm.
b. Rotor exit swirl speed, $c_{\theta b}$.
c. Specific work, w_s.
d. Relative velocity at the rotor exit, w_b.
e. Rotor and stator torques per unit mass flow rate.

5.6. Plot compressor adiabatic efficiency η_c as a function of the compression ratio, $\pi_c = 1$ to 50, for the polytrophic efficiencies of $e_c = 0.8$, 0.85 and 0.9.

5.7. For air at stagnation temperature of 350 K and $c_{za} = 95$ m/s, plot (a) the stagnation pressure ratio as a function of the η_{st}, for $u = 125$ m/s and $\Delta c_\theta = 90$ m/s; (b) the stagnation pressure ratio as a function of the u for $\eta_{st} = 0.89$ and $\Delta c_\theta = 90$ m/s; and (c) the stagnation pressure ratio as a function of the Δc_θ, for $\eta_{st} = 0.89$ and $u = 125$ m/s. Discuss the limits of the parameters, η_{st}, u and Δc_θ.

5.8. For a compressor stage at $r = 0.55$ m, the rotational speed is 7500 rpm. $T_a^o = 350\,K$, $c_{za} = c_{zb} = 150$ m/s, $\alpha_a = 30°$, and $\beta_b = 20°$. Complete the velocity triangle, and calculate the stage pressure ratio for $\eta_{st} = 0.89$.

5.9. Discuss the causes for decrease in stagnation pressure ratio and stage efficiency in Figure 5.4.

5.10. For the conditions in Problems 5.1, 5.2 and 5.4, find C_p (pressure coefficient), dH (de Haller number), M_a (incident Mach number) and R (degree of reaction).

5.11. For a compressor stage, the stagnation temperature and pressure at "a" are 350 K and 105 kPa, respectively. $c_{za} = 115$ m/s, $c_{\theta a} = 45$ m/s, $w_{zb} = 75$ m/s, $w_{\theta b} = 15$ m/s, and $u = 175$ m/s. Complete the velocity triangle and calculate the specific work, degree of reaction, Mach number and pressure coefficient.

5.12. If the data in Problem 5.7 are for $r_m = 0.45$ m, construct the velocity triangles at the hub and the tip using the free-vortex distribution, for $\zeta = r_h/r_t = 0.8$, and find all the relevant parameters including the specific work, degree of reaction, Mach number and pressure coefficient.

5.13. Using the result of Problem 5.1, construct the velocity triangles at the hub and the tip, for an exponential distribution with $A = 180$ and $B = 15$, for $\zeta = r_h/r_t = 0.75$, and find all the relevant stage parameters.

5.14. Consider a compressor stage where the following distributions are to be applied for upstream ("a") and downstream ("b") of the rotor:

$$(rc_\theta)_a = 240\,r^2 + 75; \; (rc_\theta)_b = A_b\,r^2 + B_b$$

a. Determine A_b for the stage work to be constant with respect to radius.
b. Determine B_b for the degree of reaction to be constant with respect to the radius.

c. If the rotational speed of the rotor is 5000 rpm, determine the condition for the degree of reaction to be 0.5 ($R = 0.5$) at all radial locations.

d. Using radial equilibrium equation, determine the radial variation of c_{zb} in terms of r, r_m, $(c_{zb})_m$, A_b, and B_b.

e. Sketch the velocity triangles at $r_h = 0.5$ m, $r_m = 0.625$ m, and $r_t = 0.75$ m. $(c_{zb})_m =$ 100 m/s.

5.15. For an axial compressor stage, the incoming air stagnation temperature is 300 K. At the mid-radius (r_m), $(c_z/u)_m = 0.5$, the degree of reaction is 0.5, and $(c_{\theta b})_m = 0$ ($\alpha_b = 0$). The rotational speed at the tip, u_t, is 400 m/s, and the ratio of hub to tip radius, $r_h/r_t = 0.8$.

a. What is r_m/r_t and the rotational speed at the mid-radius?

b. What is the stagnation pressure ratio at the mid-radius, if the polytropic efficiency is 0.9 and γ is 1.33?

5.16. For a centrifugal compressor rotating at 2500 rpm, find the compression ratio for the following specifications. $T_a^o = 300\,\text{K}; \eta_{st} = 0.85; r_b = 0.55; r_a = 0.015; \beta_b = 60°$, (volumetric flow rate) = 1.25 m^3/s and t (the thickness of the flow section at the exit) = 0.15 m.

5.17. At mid-radius of a turbine stage, $c_{za} = c_{zb}; R_m = 0.5; \frac{\Delta c_\theta}{u} = 1$; and $\frac{c_z}{u} = 0.45$ Construct the velocity triangle at the mid-radius and also at the hub for $rc_\theta =$ const. and $R_h = 0$.

5.18. A turbine stage has a constant axial velocity of 225 m/s, and zero velocity tangential component at the rotor exit. For $u_m = 500$ m/s, determine the following: (a) α_b, for $R = 0.0$; and (b) the rotor specific work for $R = 0.0$ and 0.5.

5.19. Estimate the blade surface temperature for a turbine stage at $r_m = 0.75$ m, $T_a^o = 1200\,\text{K}; c_{za} = 135\,\text{m/s}; \Omega = 400\,\text{rad/s}; r_f = 0.9; \alpha_a = 60°$, and $\beta_a = -\beta_b$.

5.20. Plot the turbine blade stress as a function of r, for an alloy with material density of 4500 kg/m^3 for $\Omega = 2000$, 4000 and 6000 rpm. The blade tip radius is 0.35 m.

5.21. Plot the thickness of the turbine disk, b, as a function of r, for $\Omega = 2000$, 4000 and 6000 rpm. The disk material density of 4500 kg/m^3, and the disk thickness is 25 mm at the inner rim radius of 0.25 m. The allowable stress is 35 000 psi.

Bibliography

Hill, P. and Petersen, C. (1992) *Mechanics and Thermodynamics of Propulsion*, 2nd edn, Addison-Wesley Publishing Company.

6

Combustors and Afterburners

6.1 Combustion Chambers

Some typical combustors and afterburners are shown in Figure 6.1. The primary function of the combustors is to add enthalpy to the flow, by burning fuel. There are many secondary issues, arising from the fact that fuel is injected in the liquid form and combustion needs to proceed completely and stably in a relatively compact volume (for the main combustor). The liquid fuel needs to be atomized and mixed with the air within a short distance from the injector, and stabilized for a wide range of flow conditions. In addition, the outgoing mixture of combustion products and air must be at the required level and uniformity in temperature, so that the turbine blades can withstand the thermal content of the flow. Combustors need to do all of the above with minimal loss in stagnation pressure, and withstand the thermal and pressure loading at the same time.

Although the primary function is identical for the main combustor and the afterburner, structurally and functionally they are quite different. Due to the compact volume afforded for the main combustor, the fuel is injected and burned immediately in so-called non-premixed combustion mode. This is similar to diesel engine combustion where the fuel is injected directly into the combustion chamber, as opposed to the fuel injection into the intake manifold in the spark-ignition engines. In order to accomplish mixing and burning within a short distance, swirls are added to both fuel and air streams. Swirl adds tangential momentum to spread out the fuel, and also forces much more rapid mixing of fuel and air. Only a small portion ($\sim$20%) of the core mass flow rate of air is injected directly into the combustor. The rest is sent into the annular region between the liner and the casing, and sent into the combustor at downstream locations (Figure 6.1) as cooling air. Non-premixed combustion with jet fuel and air produces flame temperatures in excess of 2200 K, and few materials are available to withstand this kind of thermal loading for an extended period of time. Therefore, relatively cooler air from the compressor is sent through a series of liner holes back into the combustor to reduce the overall temperature. Although the temperature decreases during this "dilution" process, the total stagnation enthalpy remains the same. Also, there must not be any "hot" spots in the final temperature profile.

Aerospace Propulsion, First Edition. T.-W. Lee.
© 2014 John Wiley & Sons, Ltd. Published 2014 by John Wiley & Sons, Ltd.
Companion Website: www.wiley.com/go/aerospaceprop

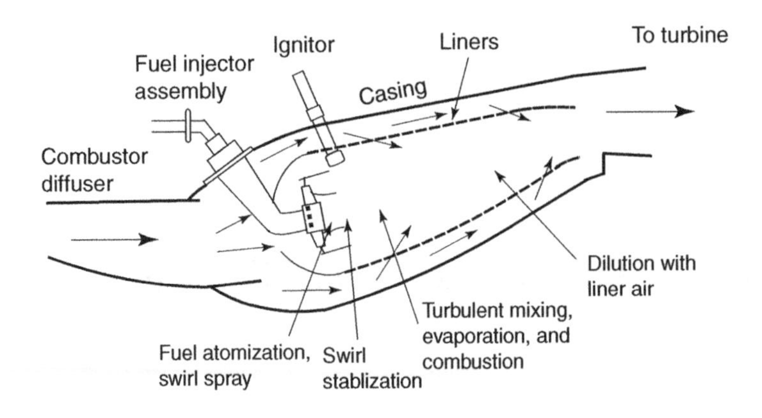

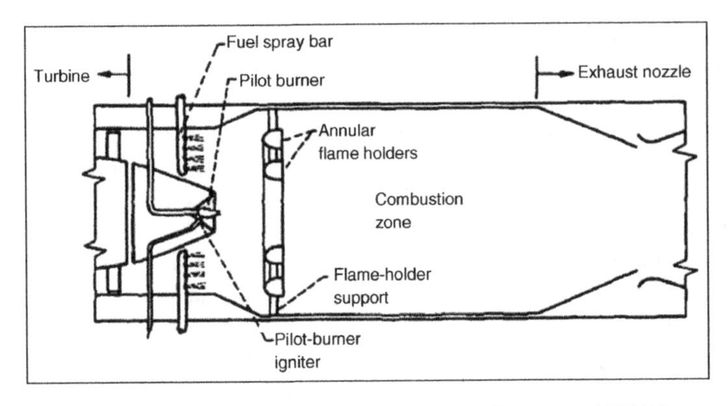

Figure 6.1 Combustors and afterburners. Courtesy of NASA.

Three types of liner-casing geometries are used: annular, can-annular and can combustors. In the annular combustors, the liner and the casing form a common annular region for the air flow, into which fuel is injected at several locations. It has the simplest geometry, but lacks the control of air flow and fuel–air mixing. For example, there is no mechanism to block the large-scale circumferential motion of the air, and therefore the flow uniformity is not ensured across the entire annular region. A compromise is a can-annular combustor, where the casing encloses can-shaped liners. A more precise control of the flow can be achieved in the can combustors where the flow is separated into tubular casing-liner combinations.

In contrast to the main combustor, the afterburner flames are anchored at the flame stabilizers downstream of the fuel injectors. The injected fuel now has some distance and "residence" time to mix with the surrounding air flow, so that the combustion proceeds in the so-called premixed mode. This is similar to the spark-ignition engine combustion, where the fuel is injected into the intake manifold and allowed to mix with air, prior to ignition. In the afterburner, the flame is anchored at the flame stabilizer, which acts as a low-speed zone so that the flame speed (which is in the direction of upstream fuel–air mixture) is balanced with the flow speed moving downstream. We will look at the details of these processes in the following sections.

Table 6.1 Characteristics of some of the jet fuels.

	Jet-A	JP-4	JP-5
Vapor pressure [atm] at 100°F	0.007	0.18	0.003
Initial boiling point [°C]	169	60	182
Flashpoint [°C]	52	−25	65
Heating value [kJ/kg]	43000	42800	42800
Specific gravity	0.710	0.758	0.818
Approx. annual US consumption [billion gallons]	12	5	1
Main use	Commercial airliners	USAF	US Navy

6.2 Jet Fuels and Heating Values

Jet fuels, like other fossil fuels, are produced through fractional distillation of crude oils. The hydrocarbon components in jet fuels are mostly paraffins, having chemical formula C_nH_{2n+2}, with a boiling point of 120–240 °C. Paraffins are more volatile than diesel fuels, and less so than gasoline fuels. The exact composition of jet fuels is quite complex (consisting of over 250 different hydrocarbon species), depending on the jet fuel type and other factors such as seasonal variations, refining process and even storage conditions. Table 6.1 shows the characteristics of some of the jet fuels. Basic properties such as the boiling point and density are certainly of interest, as are flashpoints and melting points. Flashpoint refers to the temperature at one atmospheric pressure that the fuel can be ignited by a spark source, and is determined by standard procedures (e.g. ASTM D-3278). Although it is not an exact, thermal property, it is a useful measure for safe storage of fuel relating to fire and explosion hazards. In Table 6.1, we can see that the JP-5 used by the US Navy has a high flashpoint; this is because ship-board fire safety requirements are more strict than land-based ones.

The most important fuel property for combustor analysis is the fuel heating value and the flame temperature. Let us start with the heating value, sometimes called the heat of combustion. We can consider a constant-pressure control volume that takes the fuel and air inputs and then outputs the combustion products and excess air at a higher temperature, as shown in Figure 6.2. Let us first consider an adiabatic combustor, where the fuel and air enter the control volume at the standard temperature and pressure of 298 K and 1 atm, respectively. Applying the first law of thermodynamics, or the energy balance, for this control volume under steady-state conditions, we simply obtain that the total enthalpy of the reactants, H_o, must be equal that of the exiting products, H_e.

$$H_o = H_e \tag{6.1}$$

How do we calculate the total enthalpy of the reactants and products? For air, we only need the temperature and can look up the air enthalpy (or use the specific heat) at that temperature.

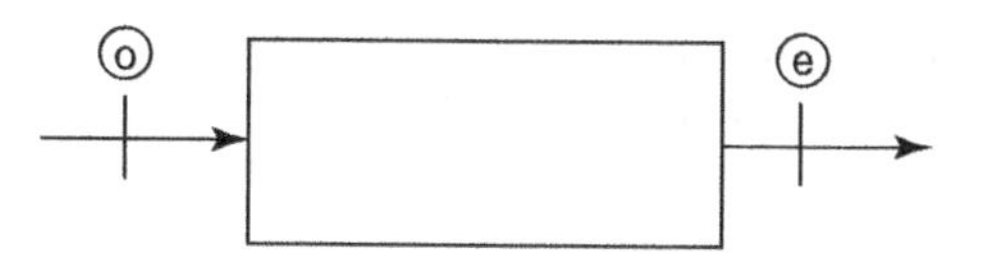

Figure 6.2 A constant-pressure control volume for evaluating combustion parameters.

For chemically reacting flows, we not only have multiple components (species), but these components change as the reaction proceeds. We can either go by the mass or use a molar basis, but it turns out that for chemically reacting systems the molar basis is much more convenient since the chemical species react with one another on a molar basis. For example, for a simple hydrogen–oxygen reaction we can write the chemical reaction as

$$2H_2 + O_2 \rightarrow 2H_2O \tag{6.2}$$

Equation (6.2) states that two moles of hydrogen react with one mole of oxygen to produce two moles of water molecule. Thus, we can find the enthalpy of hydrogen per mole at the reactant temperature, multiply by two moles and add the enthalpy of oxygen. Likewise, on the product side, we look up the enthalpy of water per mole and multiply by two moles. Mathematically, we can write

$$H_o = \sum_i \left(n_i \bar{h}_i\right)_o \tag{6.3a}$$

$$H_e = \sum_i \left(n_i \bar{h}_i\right)_e \tag{6.3b}$$

The subscripts "o" and "e" again refer to the inlet and exit conditions, while "i" is the index for different chemical species. The mole number is denoted by n, while the bar over h indicates that the enthalpy is on a molar basis, typically in kJ/kmole.

The enthalpy itself consists of the sensible enthalpy, which is a function of temperature, and the enthalpy of formation, which represents the energy stored in the chemical structure.

$$\bar{h}_i = \Delta \bar{h}_i(T) + \bar{h}^o_{f,i} \approx C_{p,i}(T - T_{ref}) + \bar{h}^o_{f,i} \tag{6.4}$$

For some common chemical species in combustion systems, the data for the sensible enthalpy ($\Delta \bar{h}(T)$), enthalpy of formation ($\bar{h}^o_f$) and specific heat (C_p) are included in Appendix B.

Using the above calculation methods for enthalpies, we can go back to the energy balance of Eq. (6.2).

$$H_o = \sum_i \left(n_i\left(\Delta \bar{h}_i(T) + \bar{h}^o_{f,i}\right)\right)_o = \sum_i \left(n_i\left(\Delta \bar{h}_i(T) + \bar{h}^o_{f,i}\right)\right)_e = H_e \tag{6.5a}$$

Alternatively, we can approximate the sensible enthalpy change by the specific heats multiplied by the temperature difference from the reference temperature of $T_{ref} = 298$ K.

$$H_o = \sum_i \left(n_i\left(C_{p,i}(T - T_{ref}) + \bar{h}^o_{f,i}\right)\right)_o = \sum_i \left(n_i\left(C_{p,i}(T - T_{ref}) + \bar{h}^o_{f,i}\right)\right)_e = H_e \tag{6.5b}$$

If the reactant enters the combustion chamber at T_{ref}, then the sensible enthalpy term is zero on the reactant side. Thus, we can rewrite Eq. (6.5b) as

$$\sum_i \left(n_i(C_{p,i}(T - T_{ref}))\right)_e = \sum_i \left(n_i \bar{h}^o_{f,i}\right)_o - \sum_i \left(n_i \bar{h}^o_{f,i}\right)_e \tag{6.6}$$

Thus, although we had assumed adiabatic combustion, the temperature at exit will have increased due to combustion. This apparent heat is called the heat of combustion (a more precise definition will be given below), and arises due to the higher enthalpy of formation in the reactants than that in the products. That is, there is more chemical energy in the fuel molecules which is released during combustion.

The enthalpy, or heat, of combustion is defined as the difference in the enthalpy of formation of the reactants and products, on a mole basis of the fuel.

$$q_r \equiv \frac{\left(\sum_i \left(n_i \bar{h}^o_{f,i}\right)_o - \sum_i \left(n_i \bar{h}^o_{f,i}\right)_e\right)}{n_{fuel}} \quad [\text{kJ/kmole-fuel}] \tag{6.7}$$

n_{fuel} is the number of moles of fuel species on the reactant side.

The heating value, h_{PR}, is a similar quantity, except that it is on a mass basis of the fuel.

$$h_{PR} \equiv \frac{\left(\sum_i \left(n_i \bar{h}^o_{f,i}\right)_o - \sum_i \left(n_i \bar{h}^o_{f,i}\right)_e\right)}{n_{fuel} M_{fuel}} \quad [\text{kJ/kg-fuel}] \tag{6.8}$$

M_{fuel} is the molecular weight of the fuel.

Example 6.1 Calculation of the heating value

For the following fuel combustion, find the heating value.

$$C_8H_{18}\text{ (liquid)} + 50(O_2 + 3.76N_2) \rightarrow 8CO_2 + 9H_2O + 37.5O_2 + 188N_2$$

The heating value can be evaluated by using Eq. (6.8). The difference in the enthalpy of formation (the numerator in Eq. (6.8)) is

$$\Delta\bar{h}^o_{f,i} = \sum_i \left(n_i \bar{h}^o_{f,i}\right)_o - \sum_i \left(n_i \bar{h}^o_{f,i}\right)_i = \bar{h}^o_{f,C_8H_{18}} - \left(8\bar{h}^o_{f,CO_2} + 9\bar{h}^o_{f,H_2O}\right)$$

For the enthalpy of formation, we have the following data.

$$\bar{h}^o_{f,C_8H_{18}} = -208\ 447\frac{\text{kJ}}{\text{kmole}};\ \bar{h}^o_{f,CO_2} = -393\ 522\frac{\text{kJ}}{\text{kmole}};\ \bar{h}^o_{f,H_2O(gas)}$$

$$= -241\ 827\frac{\text{kJ}}{\text{kmole}};\ \text{and}\ \bar{h}^o_{f,H_2O(liquid)} = -285\ 838\frac{\text{kJ}}{\text{kmole}}.$$

Using these values along with $M_{C8H18} = 114.23$ kg/kmole, we have

$$\text{LHV (lower heating value)} = 48\ 437.9\ \text{kJ/kg-fuel, using}\ \bar{h}^o_{f,H_2O(gas)}.$$

$$\text{HHV (higher heating value)} = 51\ 905.5\ \text{kJ/kg-fuel, using}\ \bar{h}^o_{f,H_2O(liquid)}.$$

If the water condenses to liquid form on the product side, then some heat (condensation heat) is released resulting in the higher heating value (HHV).

Example 6.2 Calculation of adiabatic flame temperature

For the reaction in Example 6.1, find the adiabatic flame temperature, for initial reactant temperature of $T_{std} = 298.15$ K.

We will use Eq. (6.5a), for this example.

$$\sum_i \left(n_i \left(\Delta \bar{h}_i(T) + \bar{h}^o_{f,i} \right) \right)_o = \sum_i \left(n_i \left(\Delta \bar{h}_i(T) + \bar{h}^o_{f,i} \right) \right)_e$$

On the left-hand side, since the temperature is $T_{std} = 298.15$ K the only enthalpy term that is not zero is the enthalpy of formation of the fuel.

$$\sum_i \left(n_i \left(\Delta \bar{h}_i(T) + \bar{h}^o_{f,i} \right) \right)_o = \bar{h}^o_{f,C_8H_{18}} = -208\,447 \text{ kJ}$$

For the right-hand side, we have

$$\sum_i \left(n_i \left(\Delta \bar{h}_i(T) + \bar{h}^o_{f,i} \right) \right)_e = 8\left(\Delta \bar{h}_{CO_2} + \bar{h}^o_{f,CO_2} \right) + 9(\Delta \bar{h}_{H_2O} + \bar{h}^o_{f,H_2O}) + 37.5\Delta \bar{h}_{O_2} + 188\Delta \bar{h}_{N_2}$$

We have the enthalpy of formation data, from Example 6.1, but we also need the sensible enthalpy ($\Delta \bar{h}_i$) as a function of temperature (see Appendix E).

If we assume $T_e = 900$ K, look up the $\Delta \bar{h}_i$'s at that temperature, and substituting in Eq. (E6.2.3) we get $\sum_i (n_i(\Delta \bar{h}_i(T) + \bar{h}^o_{f,i}))_e = -75\ 5702$ kJ. This does not match the left-hand side (Eq. (6.2.2)).

If we assume $T_e = 1000$ K, we get $\sum_i (n_i(\Delta \bar{h}_i(T) + \bar{h}^o_{f,i}))_e = -62\ 416$ kJ. Interpolating between these two temperatures, we find the adiabatic flame temperature is approximately 962 K.

The temperature on the exit side in Eqs. (6.5) and (6.6) is called the adiabatic temperature, and is also a very useful quantity in combustion analysis.

The adiabatic flame temperature is a function of the mixture ratio. The mixture ratio, f, is defined on a mass basis.

$$f = \frac{\dot{m}_{fuel}}{\dot{m}_{air}} = \frac{n_{fuel} M_{fuel}}{n_{air} M_{air}} \tag{6.9}$$

There is no such thing as an air "molecule" so the averaged molecular weight is used for air by accounting for 21% oxygen (O_2) and 79% nitrogen (N_2), on a mole basis. At a fixed pressure and temperature, the mole composition is equivalent to the volume composition. Thus, one "mole" refers to ($0.21\ O_2 + 0.79\ N_2$), and the molecular weight of air, M_{air}, in Eq. (6.9) is 28.97 kg/kmole. Since whole numbers are preferred in the chemical balance equations, air is more often represented as ($O_2 + 3.76\ N_2$), and its mass would be (4.76 moles) $\times M_{air}$.

An alternate way of defining the mixture ratio is to reference the stoichiometric mixture ratio. A stoichiometric reaction refers to when the fuel and air ratio is such that only combustion products and inerts are produced – that is, there are no excess reactants on the product side. For methane reaction with air, for example, a stoichiometric reaction would be

$$CH_4 + 2(O_2 + 3.76N_2) \rightarrow CO_2 + 2H_2O + 7.52N_2 \tag{6.10}$$

On the product side, there are only combustion products, CO_2 and H_2O, and inert nitrogen, N_2. The mixture ratio for this stoichiometric reaction, f_s, is

$$f_s = \frac{n_{CH4}M_{CH4}}{n_{air}M_{air}} = \frac{(1)(16.04)}{(2)(4.76)(28.97)} = \tag{6.11}$$

Since the mixture ratio tends to be a small number, air-to-fuel ratio, AFR, is also used.

$$\text{AFR} = \frac{1}{f} \tag{6.12}$$

If we use this stoichiometric mixture ratio as a reference, then the equivalence ratio, ϕ, is the ratio of the actual to the stoichiometric mixture ratio.

$$\phi = \frac{f}{f_s} \tag{6.13}$$

Thus, if the equivalence ratio is less than one, there is less fuel than available air for a stoichiometric reaction, and the reaction is said to be fuel-lean, and there will be some excess air on the product side. The other case of $\phi > 1$ is called fuel-rich, and will result in excess fuel on the product side. For fuel-lean combustion, the percentage excess air is defined as

$$\%ExcessAir = \frac{1-\phi}{\phi} \times 100 \tag{6.14}$$

6.3 Fluid Mixing in the Combustor

In a gas-turbine engine, only a small fraction of the air mass is used in combustion of the fuel, and the rest is used as a "momentum" source for thrust generation. In the combustor, this large excess air is useful for dilution and cooling of the combustion gas. The direct combustion of jet fuel with air in the main combustor can result in combustion gas temperatures of approximately 2400 K; therefore, without dilution and cooling, the turbine section cannot withstand such high temperatures. We can estimate the effect of dilution air on the resulting combustor exit, or turbine inlet temperature, based on the energy balance.

$$T_4^o = \frac{\dot{m}_p}{\dot{m}_o} T_f + \left(1 - \frac{\dot{m}_p}{\dot{m}_o}\right) T_3^o \tag{6.15}$$

Example 6.3 Mixture calculations

For the following combustion of some unknown fuel with air, determine *a*, *b*, *c*, *d*, and *e*, and the fuel–air ratio.

$$C_aH_b + cO_2 + dN_2 \rightarrow 8CO_2 + 0.9CO + 8.8O_2 + eH_2O + 82.3N_2$$

Since nitrogen is inert in this reaction, $d = 82.3$.

Also, air has fixed nitrogen to oxygen ratio of $d/c = 3.76$, so that $c = 21.9$.

The remaining constants can be evaluated by counting and equating the number of C, O and H atoms on both sides of the reaction.

$$C : a = 8 + 0.9 \rightarrow a = 8.9$$

$$O : 2(21.9) = 8(2) + 0.9(1) + 8.8(2) + e(1) \rightarrow e = 9.3$$

$$H : b = 2e = 18.6$$

So the fuel is $C_{8.9}H_{18.6}$, with a molecular weight of $8.9(12.011) + 18.6(1.008) =$ 125.65 kg/kmole.

The air in the above reaction consists of 21.9 moles of O_2 and 82.3 moles of N_2, so that the total air mass is $(21.9)(32) + (82.3)(28.014) = 3006$ kg. The fuel–air ratio is then $f = 0.042$.

We can also find the air mass of 3018 kg, using the average molecular weight of air $M_{air} = 28.97$ kg/kmole, with the 0.2% error due to round-off.

Example 6.4 Flame temperatures at various equivalence ratios

For the following reaction and thermodynamic data, we can plot the adiabatic flame temperature as a function of the equivalence ratio for $\gamma > 1$ (Figure E6.4.1).

$$C_8H_{18}\ (\text{liquid}) + 12.5\gamma(O_2 + 3.76N_2) \rightarrow aCO_2 + bH_2O + cO_2 + dN_2$$

For $\gamma > 1$, the reaction is fuel-lean (excess air). For fuel-lean reactions, elemental mass balance for hydrocarbon fuel with air, in general, results in

$$\begin{aligned} &C_\alpha H_\beta + \gamma\left(\alpha + \frac{\beta}{4}\right)(O_2 + 3.76N_2) \rightarrow \\ &\alpha CO_2 + \frac{\beta}{2}H_2O + (\gamma - 1)\left(\alpha + \frac{\beta}{4}\right)O_2 + \gamma\left(\alpha + \frac{\beta}{4}\right)N_2 \end{aligned} \quad \text{(E6.4.1)}$$

We will use the specific heat to calculate the sensible enthalpy change.

$$\Delta\bar{h}_i = C_{p,i}(T - T_o) \quad \text{(E6.4.2)}$$

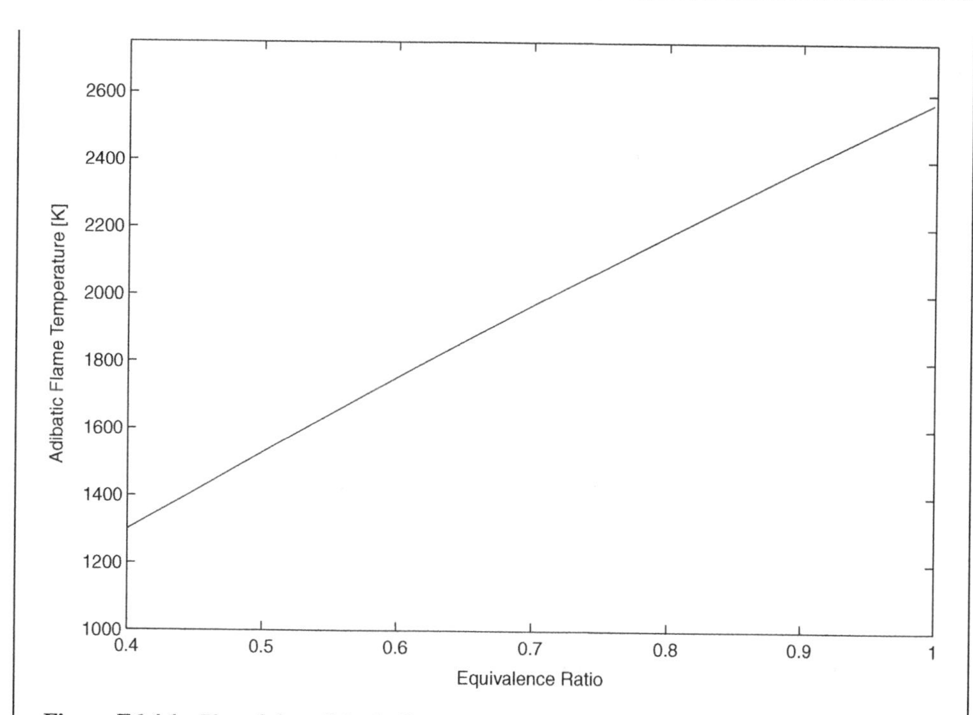

Figure E6.4.1 Plot of the adiabatic flame temperature as a function of the equivalence ratio.

$$C_{p,CO_2} = 50.16\frac{\text{kJ}}{\text{kmolK}}; C_{p,H_2O} = 38.90\frac{\text{kJ}}{\text{kmolK}}; C_{p,O_2} = 33.35\frac{\text{kJ}}{\text{kmolK}}; C_{p,N_2} = 31.50\frac{\text{kJ}}{\text{kmolK}}.$$

The enthalpy of formation for each species is

$$\bar{h}^o_{f,C_8H_{18}} = -208\ 447\frac{\text{kJ}}{\text{kmole}}; \bar{h}^o_{f,CO_2} = -393\ 522\frac{\text{kJ}}{\text{kmole}};\ \text{and}\ \bar{h}^o_{f,H_2O(gas)} = -241\ 827\frac{\text{kJ}}{\text{kmole}}.$$

We can set up a MATLAB® algorithm (MCE64 at the end of this chapter), to calculate and plot the adiabatic flame temperature.

Here, $\dot{m}_p$ and T_f are the mass flow rate for the primary air, used for combustion of fuel, and the adiabatic flame temperature, respectively. Overall energy balance using the heating value is, of course, still valid, and gives the same result.

$$T^o_4 = T^o_3 + \frac{\dot{m}_f h_{PR}}{\dot{m}_o c_p} \tag{6.16}$$

Although Eqs. (6.15) and (6.16) serve as an estimate of the turbine inlet temperature, it assumes perfect mixing of the primary and dilution air in the downstream section of the combustor. In order to achieve thorough mixing, liner holes are used between the primary air and the compressor air at lower temperatures that flow through the annular region between the liner and casing. Although the details of the flow through the liner holes are subject to CFD and testing, the discharge coefficient can be used to estimate the mass flow rates through the liner holes.

$$\dot{m}_H = C_D A \sqrt{2\rho(p_U^o - p_D)} \tag{6.17}$$

C_D = discharge coefficient
A = area of the liner hole
ρ = air density
p_U^o = stagnation pressure upstream of the liner hole
p_D = static pressure downstream of the liner hole

The discharge coefficient is a function of a number of factors, including hole shape (circular or rectangular), hole spacing, swirl and the annular air velocity. A correlation for the discharge coefficient is in terms of two parameters: α, ratio of the hole and annulus mass flow rate, and K, the ratio of the dynamic pressure of the jet through the hole and dynamic pressure of the annulus flow.

$$C_D = \frac{1.25(K-1)}{\sqrt{4K^2 - K(2-\alpha)^2}} \tag{6.18}$$

Note that the input parameters in Eq. (6.18) are related to the sought mass flow rate in Eq. (6.16), so some iterative calculation is needed in order to use Eqs. (6.17) and (6.18).

As noted above, the function of the liner holes is to sequentially add dilution air so that the turbine inlet temperature is brought to acceptable levels. However, mixing between the annulus and combustor flow is not perfect, and some temperature variations may be present at the combustor exit. Any variation in the temperature profile represents possible overshoot of the target temperature, and therefore a uniform temperature distribution is always sought. A measure of the temperature deviation from uniformity is called the pattern factor, and is defined as

$$\text{pattern factor} = \frac{T_{\max} - \bar{T}_4}{\bar{T}_4 - \bar{T}_3} \tag{6.19}$$

The barred quantities indicate mass-averaged temperatures at the combustor inlet (3) and exit (4), while the T_{max} is the maximum observed temperature at the exit. A pattern factor of zero is difficult to achieve, in which case the second-best option is to provide lower temperatures at the turbine blade root (where the mechanical stress is the highest) and the blade tip where turbine cooling is the least accessible.

At the combustor inlet and in injectors, swirls are used to enhance fuel–air mixing. With swirl and attendant flow recirculation, complete combustion in a compact combustor volume would be nearly impossible. Swirls can be induced through swirl vanes in axial flows, or simply diverting the flow and sending it in radial directions, as shown in Figure 6.3.

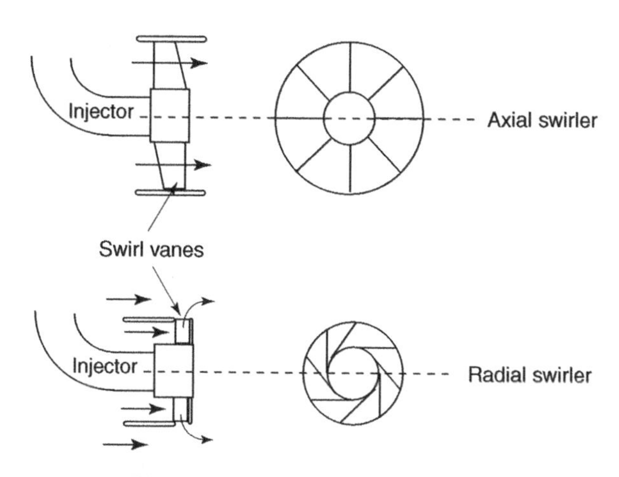

Figure 6.3 Axial and radial swirlers.

The swirl number, S_N is useful in quantifying the tangential momentum added to the flow.

$$S_N = \frac{G_m}{R_o G_t} \tag{6.20}$$

G_m = axial flux of angular momentum = $\int_{Ri}^{Ro} \rho u u_\theta r dr$
G_t = axial thrust = $\int_{Ri}^{Ro} \rho u^2 r dr$
R_o = outer radius of swirl vane (see Figure 6.3)
R_i = inner radius of swirl vane

Although precise, Eq. (6.20) requires detailed velocity data across the swirl vanes. An alternative is to assume that the angle between the axial and tangential velocities is the same as the swirl vane angle, θ. Then, the swirl number can be estimated by

$$S_N = \frac{2}{3}\frac{1-\left(\frac{R_i}{R_o}\right)^3}{1-\left(\frac{R_i}{R_o}\right)^2}\tan\theta \tag{6.21}$$

For $S_N < 0.4$, the swirl is said to be weak and there is no flow recirculation. For $S_N > 0.6$, the tangential momentum is sufficiently strong to spread the flow outward. The resulting low pressure at the centerline draws the flow inward, and a recirculation zone is created near the centerline, as shown in Figure 6.4. It can be seen that such recirculation would be very effective in mixing the injected fuel with the surrounding air. Therefore, swirl vanes are used around a central fuel injector in the primary zone of the combustor.

Fuel injectors deliver liquid fuel into the combustor, in a fine mist of droplets. Typical gas-turbine fuel injectors are shown in Figure 6.5. The schematic shows that swirl is also used in the injector, which spreads the fuel out, so that the bell-shaped liquid sheet can easily be broken into smaller droplets within a short distance. In comparison to a small number of large droplets, a large number of small droplets provides larger surface area for a given volume of

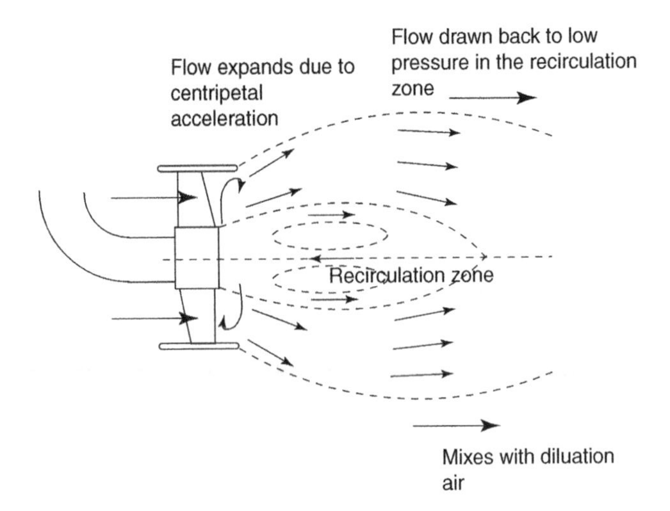

Figure 6.4 Flow recirculation created by strong swirl.

fuel, thus allowing fast evaporation and combustion. Both the drop size and the velocity statistics are desired to characterize the injected fuel, although in a gas-turbine combustor the drop velocity effect is minimal due to the strong swirl in the primary zone. The drop size can be expressed as an average or as a drop size distribution. The relevant drop size for evaporation and combustion processes is the Sauter mean diameter (SMD) or D_{32}.

$$D_{32} = \frac{\sum_i n_i D_i^3}{\sum_i n_i D_i^2} \tag{6.22}$$

Example 6.5 Calculation of the SMD

For the drop size data shown in Table E6.5.1, Eq. (6.22) gives the SMD = 89.54 μm, while the arithmetic mean is 89.54 μm.

Table E6.5.1 An example of a particle size grouped into different size bins.

Particle size range [μm]	Representative particle size, D_i [μm]	ΔD_i [μm]	n_i, Number of particles	$p_i \Delta D_i = \frac{n_i \Delta D_i}{\sum n_i \Delta D_i}$, Size distribution function
0.0–10.0	5.0	10.0	25	0.03151
10.0–20.0	15.0	10.0	127	0.16005
20.0–30.0	25.0	10.0	275	0.34657
30.0–40.0	35.0	10.0	155	0.19534
40.0–50.0	45.0	10.0	75	0.09452
50.0–60.0	55.0	10.0	54	0.06805
60.0–70.0	65.0	10.0	32	0.00403
70.0–85.0	77.5	15.0	15	0.02836
85.0–100.0	92.5	15.0	12	0.02268
100.0–120.0	110.0	20.0	5	0.01260

In Eq. 6.22, n_i is the number of drops in the size bin centered at D_i. SMD has a useful property of having the same volume-to-area ratio in the mean, as the entire group of droplets in the spray, so that it would give an average estimate of the evaporation rate of the fuel spray.

Determination of the drop size, either the SMD or the distribution, is not a simple matter, experimentally or computationally. Some correlations and theories exist, but subject to further interpretations and validations, depending on the precise injection conditions. As noted above, pressure-swirl ("simplex") injectors are almost exclusively used in main combustors, and a correlation for the SMD is (Lefebvre, 1999).

$$\mathrm{SMD} = 4.52\left(\frac{\sigma\mu_L^2}{\rho_A \Delta P_L^2}\right)^{0.25}(t\cos\theta)^{0.25} + 0.39\left(\frac{\sigma\rho_L}{\rho_A \Delta P_L}\right)^{0.25}(t\cos\theta)^{0.75} \tag{6.23}$$

σ = surface tension
μ_L = liquid viscosity
ρ_A = air density
ρ_L = liquid density
ΔP_L = injection pressure difference
t = liquid film thickness at the injector exit
$\cos\theta$ = half-angle of the spray

All of the injection parameters are represented in Eq. (6.23). High injection pressure will result in small SMD, as would low surface tension and viscosity of the liquid. Large spray angle gives low value of $\cos\theta$, so that it also leads to smaller SMD.

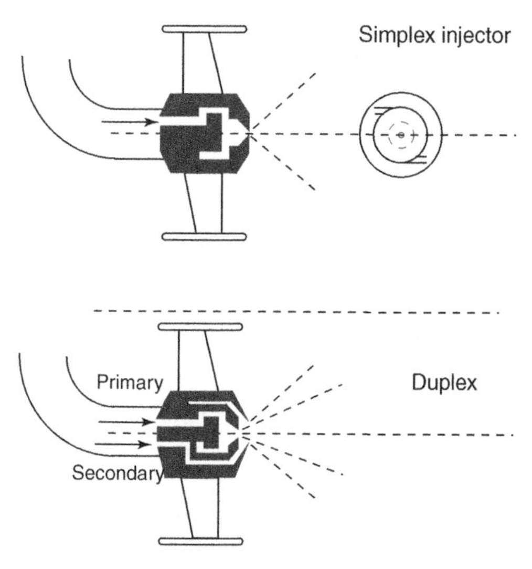

Figure 6.5 Gas-turbine injectors.

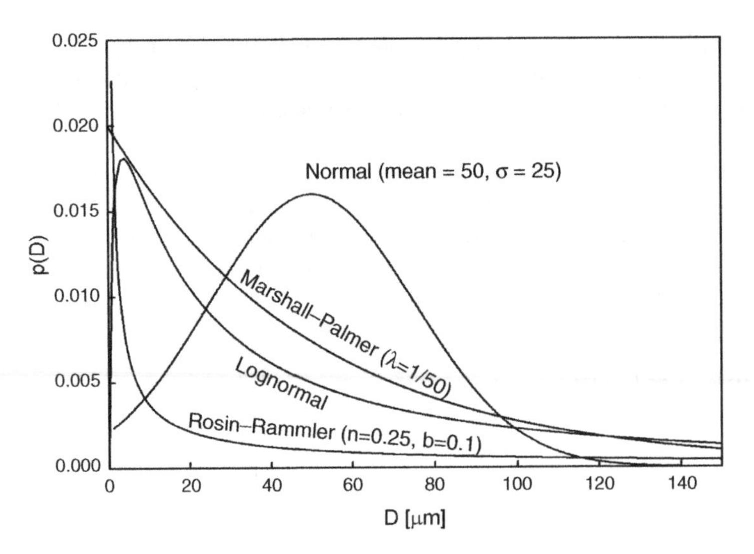

Figure 6.6 Drop size distributions.

Once SMD is estimated, then further details of the spray can be represented by drop size distribution. Drop size distribution gives the expected number of drops in drop size ranges, as shown in Figure 6.6. One can count, for example, based on measured drop size data, the number of drops in a given size bin, then normalize by the total number of drops to determine the drop size distribution. Common mathematical expressions for drop size distributions include Rosin–Rammler and log-normal distributions. For gas-turbine combustion, there is a narrow drop size distribution centered near the design SMD, as the larger drops tend to be the limiting factor in complete combustion. Based on the drop size distribution shown in Figure 6.6, various average drop sizes can be calculated.

$$\text{Rosin – Rammler – Sperling – Bennet Distribution:} \quad p(D) = nbD^{n-1}\exp(-bD^n) \tag{6.24}$$

n and b are adjustable parameters.

$$\text{Log-normal distribution:} \quad p(D) = \frac{1}{\sqrt{2\pi}D\sigma(\ln(D))}\exp\left(-\frac{(\ln(D) - \langle\ln(D)\rangle)^2}{2[\sigma(\ln(D)]^2}\right) \tag{6.25}$$

$\langle\ln(D)\rangle$ and $\sigma(\ln(D))$ are the mean and the standard deviation of the $\ln(D)$, respectively.

$$\beta\text{-distribution:} \quad p(D) = \frac{x^m(\langle D\rangle - D)^n}{\beta(m+1, n+1)} \tag{6.26}$$

β is the beta-function, while m and n are adjustable parameters.

What occurs in a gas-turbine engine main combustor is illustrated in Figure 6.7. The injected fuel spreads out due to the swirl effect, and produces a fine mist of droplets.

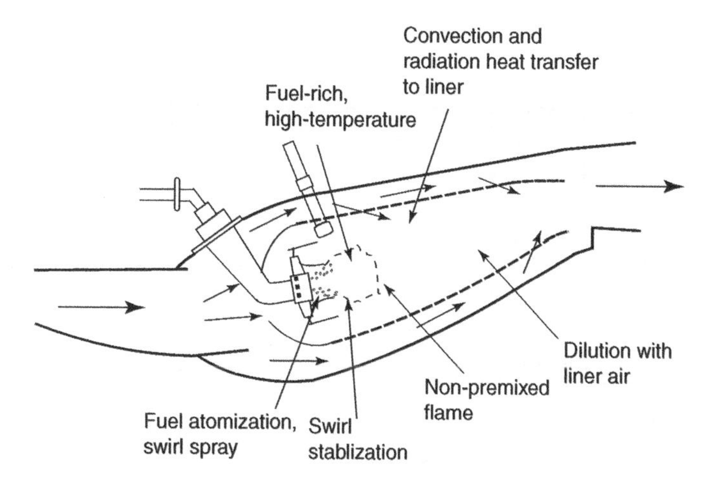

Figure 6.7 Flame structure in the main combustor.

These droplets undergo fairly rapid vaporization in the primary zone. The combustion of the fuel vapor creates a flame ball, which has been shortened due to recirculation of the air. Swirl vanes are designed to cause sufficiently strong swirl to cause air recirculation. The resulting flame is mostly non-premixed flame, where there exists a fuel-rich region inside the flame, and the surrounding air supplies oxygen toward the flame. The heat generated from the flame is convected downstream, to mix with dilution air. Some of the heat is also transferred to the fuel-rich zone, which tends to break down a small portion of the fuel molecules in a process called pyrolysis. The pyrolysis products, hydrocarbons with low molecular weights, then recombine and aggloromerate to form soot particles in the fuel-rich zone. Some of the soot particles may burn out, going through the flame, but some may escape unburned, causing sooty emission. Within the non-premixed flame, the fuel and air mix at the flame in a stoichiometric ratio, and therefore produce the maximum flame temperature, that can be estimated by the adiabatic flame temperature at equivalence ratio of one.

6.4 Afterburners

Detailed schematics of an afterburner are shown in Figure 6.8. The main difference in the afterburner is that fuel is injected upstream of the combustion zone, so that the fuel has enough time to mix with air, and the combustion occurs in a premixed mode. It may be recalled that the burning in the main combustor occurred in non-premixed mode. If the fuel and air are premixed, then the flame will propagate toward the fuel–air mixture to sustain the burning. Thus, the flame stabilizer anchors the flame at a fixed point in the afterburner, and creates a balance of the upstream flame propagation speed and downstream air motion. Similar to swirlers, the flame stabilizers create a small recirculation zone so that the flame is arrested at that point. Typically, the high speed of the air motion will carry the flame downstream, and extinguish the flame without the flame stabilizer. At the flame stabilizer, the flow speed is reduced to nearly zero, so that the flame is stationary. Away from the flame stabilizer, a simple vector diagram (Figure 6.7) explains the balance of the flame

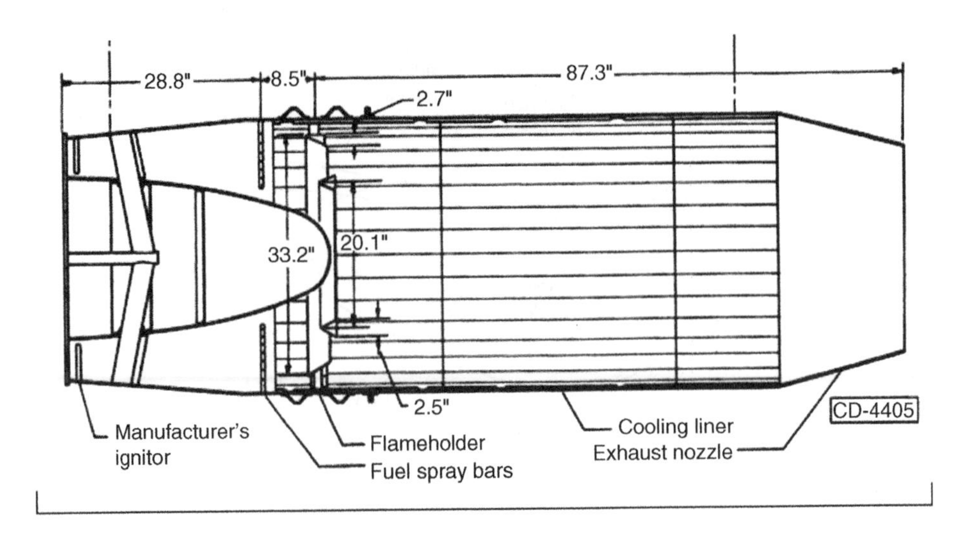

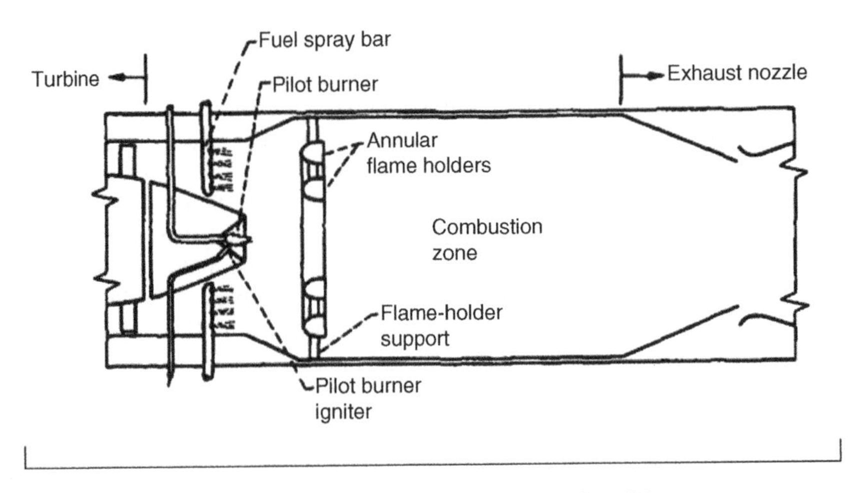

Figure 6.8 Afterburner. Courtesy of NASA.

propagation speed and the air speed.

$$S_T = U_5 \sin \theta_f \tag{6.27}$$

S_T = flame speed
U_5 = flow speed upstream of the flame
θ_f = flame angle

Under laminar flow conditions (low Reynolds numbers), the flame speed is a function only of the fuel and the mixture ratio. For turbulence conditions that exist in the afterburners, the flame speed is greatly enhanced by the turbulent wrinkling of the flame surfaces. The turbulence eddies distort the flame surface so that the net surface area increases as a function of turbulence intensity and length scales. Determination of the flame speed under turbulence conditions is the subject of a vast amount of research, and an exact relationship is not established. As with most problems in turbulence, correlations are used to estimate the turbulent flame speed. An example of a correlation is given below, where the turbulence intensity is parameterized as u'/S_L, where u' is the root-mean-square of the turbulent velocity fluctuations and S_L the laminar flame speed. The laminar flame speed is of the order of 0.05 m/s for hydrocarbon fuels, while the turbulent flame speed can be 10~20 times that depending on the turbulence intensity.

$$S_T = S_L\left[1 + B\left(\frac{u'}{S_L}\right)^2\right]^{0.5} \tag{6.28}$$

S_L = laminar flame speed
B = a constant as a function of mixture properties ~1

The function of the afterburner is to add enthalpy to the flow so that higher kinetic energy can be derived in the nozzle. Due to the density decrease across the flame, there is also an immediate increase in the flow speed. For tilted flames, as occurs in afterburners, there is also bending of the streamlines in a direction opposite to that in the shock waves. Across flames, the density decreases, whereas the opposite is true for shock waves. A simple analysis allows us to estimate the flow acceleration and deflection across flames, as shown in Figure 6.9. For a flame anchored at the stabilizer at an angle θ_f, the velocity component normal to the flame is

$$u_5 = U_5 \sin\theta_f = u_n \tag{6.29}$$

The velocity component parallel to the flame can also be found, and is the same up- and downstream of the flame as the density increase accelerates the flow only in the normal direction.

$$v_5 = U_5 \cos\theta_f = u_n \cot\theta_f = v_6 \tag{6.30}$$

The density ratio under constant-pressure combustion conditions can be related to the temperature ratio across the flame, through the ideal gas equation of state.

$$\frac{\rho_5}{\rho_6} \equiv \alpha = \frac{T_6}{T_5} \tag{6.31}$$

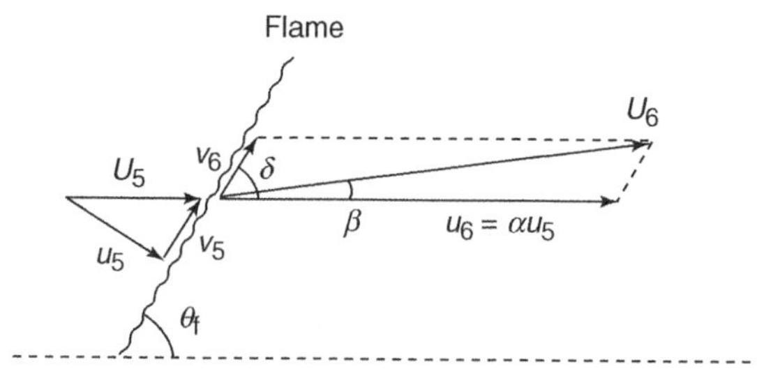

Figure 6.9 Flow deflection across an inclined flame front.

From the mass flux balance across the flame front, we can estimate the velocity downstream of the flame.

$$\rho_5 u_5 = \rho_6 u_6 \rightarrow u_6 = \alpha u_n \tag{6.32}$$

Now, the angle between the downstream velocity components can be calculated.

$$\tan\delta = \frac{v_6}{u_6} = \frac{\cot\theta_f}{\alpha} \tag{6.33}$$

Noting that $\beta + \delta = \pi/2 - \theta_f$, we can find the flow deflection angle, β.

$$\tan\beta = \frac{\alpha - 1}{1 + \alpha\tan^2\theta_f}\tan\theta_f \tag{6.34}$$

In real flames, the inward components cancel one another at the centerline (the flows cannot go through one another), and the result is that the flame will bulge outward and straighten out at a downstream location.

6.5 Combustor Heat Transfer

Due to the high temperature in the combustors, both radiation and convection heat transfers are important. The interior of the combustor imparts heat loading to the liner walls, while the liner wall imparts heat to the air flow in the annulus. The casing itself is not adiabatic, but the liner and annulus heat transfer is of higher importance due to the higher temperatures involved. The convection heat from the combustion gases to the liner wall follows the classical Newton's law of cooling.

$$q''_{conv} = 0.02\frac{k_g}{D_H^{0.2}}\left(\frac{\dot{m}}{A_C\mu_g}\right)(T_g - T_w) \quad [\mathrm{W/m^2}] \tag{6.35}$$

k_g = thermal conductivity of the gas
D_H = *hydraulic diameter of the liner* = $4A_C/P_L$
A_C = cross-sectional area of the combustor inside the liner
P_L = perimeter of the liner
μ_g = viscosity
T_g = gas temperature
T_w = liner wall temperature

As noted above, due to high temperatures in the combustor, the radiant transfer is a large portion of the total heat load.

$$q''_{rad} = \sigma\varepsilon_w(\varepsilon_g T_g^4 - \alpha_g T_w^4) \tag{6.36}$$

$\sigma = 5.67 \times 10^{-8} \dfrac{W}{m^2 K^4}$
ε_w = wall emissivity
ε_g = gas emissivity
α_g = gas absorptivity

Equation (6.36) describes the incident radiation on the liner wall, which is absorbed by the wall (the wall absorptivity is assumed to be equal to the wall emissivity), and the outgoing radiation from the liner wall that is absorbed by the gas. The wall emissivity ranges from 0.85 to 0.95 (0.85 for Nimonic and 0.95 for stainless steel). Also, we can make use of the following correlation.

$$\frac{\alpha_g}{\varepsilon_g} = \left(\frac{T_g}{T_w}\right)^{1.5} \tag{6.37}$$

Then, Eq. (6.36) becomes

$$q_{rad} = \sigma \varepsilon_w \varepsilon_g T_g^{1.5} (T_g^{2.5} - T_w^{2.5}) \tag{6.38}$$

The above two combined heat loading is dissipated to the annular flow between the casing and the liner. In this annular flow, the convection heat transfer is calculated by

$$q''_{conv} = 0.02 \frac{k_a}{D_A^{0.2}} \left(\frac{\dot{m}_{an}}{A_{an}\mu_a}\right)(T_w - T_{an}) \quad [\mathrm{W/m^2}] \tag{6.39}$$

k_a = thermal conductivity of the gas in the annulus
D_A = hydraulic diameter of the annulus = 4 A_{an}/P_{an}
A_{an} = cross-sectional area of the annulus
P_{an} = perimeter of the annulus
μ_a = viscosity of the gas in the annulus
T_{an} = gas temperature in the annulus

6.6 Stagnation Pressure Loss in Combustors

The stagnation pressure can be lost through the combustors and afterburners. One cause is the drag imposed by fuel injectors, swirlers, the liner wall and the flameholder in the afterburners. This effect is due to the complex aerodynamics of the combustor and afterburners, and is usually estimated using empirical relationships or thorough computational simulations. The temperature increase leads to volumetric expansion, which can also lead to reduction in the stagnation pressure. The aerodynamic reduction in the stagnation pressure depends on the flow Mach number in the combustor, and of course on the combustor geometry.

$$\pi_b = \frac{p_4^o}{p_3^o} = 1 - C\frac{\gamma}{2}M_b^2 \tag{6.40}$$

C = empirical constant ($1 < C < 2$)
γ = ratio of specific heats
M_b = average Mach number in the combustor $= (M_3 + M_4)/2$

In modern combustors, the stagnation pressure ratio due to aerodynamic losses typically ranges from 0.92 to 0.96.

From the mass and momentum balance across the combustor, we can see the origin of the stagnation pressure loss due to volumetric expansion.

$$\rho_3 U_3 = \rho_4 U_4 \tag{6.41}$$

$$p_3 + \rho_3 U_3^2 = p_4 + \rho_4 U_4^2 \tag{6.42}$$

The stagnation pressure loss can be calculated as follows.

$$p_3^o - p_4^o = p_3 + \frac{1}{2}\rho_3 U_3^2 - \left(p_4 + \frac{1}{2}\rho_4 U_4^2\right) = p_3 + \rho_3 U_3^2 - \left(p_4 + \rho_4 U_4^2\right) - \frac{1}{2}\left(\rho_3 U_3^2 - \rho_4 U_4^2\right) \tag{6.43}$$

The momentum terms cancel due to Eq. (6.43), and dividing by the dynamic pressure at 3, we have

$$\frac{p_3^o - p_4^o}{\frac{1}{2}\rho_3 U_3^2} = \frac{\rho_4 U_4^2}{\rho_3 U_3^2} - 1 = \frac{\rho_3}{\rho_4} - 1 = \frac{T_4}{T_3} - 1 \tag{6.44}$$

In Eq. (6.44), we use the mass balance (Eq. (6.41)) to go from the second to the third equality, while the ideal gas equation of state is used to go from the third to the final equality. Eq. (6.44) shows that the temperature increase due to combustion can lead to stagnation pressure loss, which could potentially be very high. Fortunately, the static temperature ratio in combustors is not as severe due to the dilution air mixing, and also because the static temperature is already high after the compression. With respect to the stagnation pressure at the combustor inlet, the stagnation pressure drop due to volumetric expansion is typically 0.5 to 1.0%.

$$\frac{p_3^o - p_4^o}{p_3^o} = 0.5 \sim 1\% \tag{6.45}$$

In the afterburners, the fuel injectors and the flame stabilizers present significant flow obstruction by design. The function of the flame stabilizers is to stagnate the flow and create a recirculation zone, for anchoring the flame. The high Mach number, typical of afterburner flows, also increases the drag effect. In the afterburners, these effects lead to afterburner pressure ratios of 0.75 to 0.98, with higher losses at high Mach numbers and larger drag coefficients for the fuel injectors and flameholders.

```
% MCE 64: Calculation of the adiabatic flame temperature
% using specific heats

%Input parameters
To=298.15;

CpCO2=50.16;
CpH2O=38.90;
CpO2=33.35;
CpN2=31.5;

hfCO2=-393522;
hfH2O=-241827;
hfOct=-249952;

mairs=12.5*32+12.5*3.75*28.014;
fs=114.23/mairs;


for i = 1:25;
   g(i) = 1+0.2*(i-1);
   mair=f(i)*(12.5*32+12.5*3.76*28.014);
   phi(i)=114.23/mair/fs;

   nO2=12.5*g(i)-12.5;
   nN2=12.5*3.76*g(i);

   Cp=8*CpCO2+9*CpH2O+nO2*CpO2+nN2*CpN2;
   RHS=hfOct-8*hfCO2-9*hfH2O+Cp*To;
   Tf(i)=RHS/Cp;

end

plot(phi,Tf)
```

6.7 Problems

6.1. Calculate the higher and lower heating values of octane, C_8H_{18}, by using the stoichiometric reaction in air at 298.16 K and 1 atm.

$$C_8H_{18(g)} + 25/2(O_{2(g)} + 3.76N_2) \rightarrow 8CO_{2(g)} + 9H_2O + 25/2(3.76)N_2.$$

6.2. Calculate the lower and upper heating values of the following fuels, during combustion with air: H_2, CH_4, C_3H_8, C_8H_{18}, and $C_{10}H_{22}$.

6.3. Calculate the higher and lower heating values of methyl alcohol, CH_3OH, by using the following reactions at 298.16 K and 1 atm.

$$CH_3OH_{(g)} + 3/2O_{2(g)} \rightarrow CO_{2(g)} + 2H_2O$$

6.4. Balance the following chemical reaction by providing *a*,*b*,*c*, and *d*, and calculate its fuel–oxidizer ratio by mass.

$$C_8H_{18}\ (\text{liquid}) + 3(12.5)O_2 \rightarrow aCO_2 + bH_2O + cO_2$$

a. Calculate the heat of combustion for this fuel–oxidizer.
The enthalpy of formation (h_f^o) for each species is as follows:

Species	*Enthalpy of formation*
O_2	0 kJ/kmol
N_2	0 kJ/kmol
CO_2	−393 522 kJ/kmol
H_2O	−241 827 kJ/kmol
C_8H_{18} (liq.)	−249 952 kJ/kmol.

b. Find its adiabatic flame temperature.

$\Delta h_i = C_{p,i}(T - T_o)$, where	$C_{p,CO2} = 50.16$ kJ/kmol K,
	$C_{p,H2O} = 38.90$ kJ/kmol K,
	$C_{p,O2} = 33.35$ kJ/kmol K,

6.5. One mole of octane is burned with 150% theoretical air, starting from the inlet temperature of 25 °C. Calculate (a) the fuel–air ratio; (b) equivalence ratio, ϕ, and (c) the adiabatic flame temperature.

6.6. A mixture contains 44 kg of CO_2, 112 kg of N_2 and 32 kg of O_2 at a temperature of $T = 287$ K and pressure of 1 atm. and the mixture pressure is $p_m = 1$ bar. Determine:

a. the number of moles of carbon dioxide, nitrogen and oxygen
b. the mole fractions of all the species
c. the partial pressures of all the gases
d. the mixture molecular weight
e. the volume fraction of the constituent gases.

6.7. Plot the adiabatic flame temperature of CH_4, C_3H_8 and C_8H_{18} as a function of the equivalence ratio.

6.8. For hydrocarbon fuel, $C_{8.9}H_{18.6}$, reacting with air $A(O_2 + 3.76N_2)$, plot the fuel–air ratio and equivalence ratio as a function of *A*.

6.9. One mole of octane is burned with 120% theoretical air, Assuming that the octane and air enter the combustion chamber at 25 °C and the excess oxygen and nitrogen in the reaction will not dissociate, calculate

a. the fuel-air ratio
b. the equivalence ratio ϕ
c. the adiabatic flame temperature.

6.10. The swirl number can be estimated by Eq. (6.21). Plot the swirl number as a function of the vane angle θ, for $R_i/R_o = 0.2$, 0.5 and 0.8.

$$S_N = \frac{2}{3}\frac{1-\left(\frac{R_i}{R_o}\right)^3}{1-\left(\frac{R_i}{R_o}\right)^2}\tan\theta \tag{6.21}$$

6.11. For the following fuel droplet size data, plot the drop size distribution, and calculate the mean diameter and SMD.

Particle size range [μm]	Representative particle size, D_i [μm]	ΔD_i [μm]	n_i, Number of particles
0.0–10.0	5.0	10.0	12 500
10.0–20.0	15.0	10.0	12 700
20.0–30.0	25.0	10.0	30 000
30.0–40.0	35.0	10.0	155 000
40.0–50.0	45.0	10.0	750 000
50.0–60.0	55.0	10.0	500 000
60.0–70.0	65.0	10.0	185 000
70.0–85.0	77.5	15.0	150 000
85.0–100.0	92.5	15.0	72 000
100.0–120.0	110.0	20.0	5000

6.12. Plot the SMD as a function of the injection pressure, for water and jet-A fuel at an ambient air density of 2.25 kg/m^3, using Eq. (6.23). $t = 1$ mm, and $\theta = 45°$.

$$SMD = 4.52\left(\frac{\sigma\mu_L^2}{\rho_A \Delta P_L^2}\right)^{0.25}(t\cos\theta)^{0.25} + 0.39\left(\frac{\sigma\rho_L}{\rho_A \Delta P_L}\right)^{0.25}(t\cos\theta)^{0.75} \tag{6.23}$$

6.13. Plot the Rosin–Rammler–Sperling–Bennet and log-normal distributions for various n and b and other parameter values. Find the best fit to the drop size data used in Problem 6.5.

6.14. In a horizontal afterburner, the flow speed is 120 m/s at $T = 1025$ K.

a. Find the required flame speed to hold the flame at an angle of 30° with respect to horizontal.

b. What is the initial deflection angle of the streamline, with respect to horizontal, immediately downstream of the flame?

Bibliography

Lefebvre, A.H. (1999) *Gas Turbine Combustion*, 2nd edn, Taylor and Francis.

7

Gas-Turbine Analysis with Efficiency Terms

7.1 Introduction

In Chapter 3, we applied basic thermodynamics to analyze gas-turbine performance, in which the fundamental assumption of ideal or isentropic flows was used. After discussions of the actual flow processes in Chapters 4–6, we can incorporate some of the real effects through efficiency terms. In addition, we can obtain slightly more accurate results by putting in more realistic gas properties and boundary conditions. The main additions to the current gas-turbine analysis are outlined below. The state designations used are as in Figure 7.1.

1. The diffuser and nozzle flows incur stagnation pressure losses, but are still adiabatic.

$$\pi_d, \pi_n < 1; t_d, t_n = 1. \tag{7.1}$$

2. The specific heat of the gas can vary, by as much as 25%, from upstream of the combustor to downstream, primarily due to temperature differences. This factor is added by using two specific heats, for the cold (upstream of the combustor, "*c*") and hot gases (downstream, "*t*").

$$\begin{aligned} c_{pc} &= \frac{\gamma_c R_c}{\gamma_c - 1}; \\ c_{pt} &= \frac{\gamma_t R_t}{\gamma_t - 1} \end{aligned} \tag{7.2}$$

3. The compressor and turbine flows also incur losses, and are not isentropic. Isentropic efficiency and/or polytropic factor is used to represent the actual process.
4. Allowance is made for the nozzle exit pressure not reaching the ambient pressure.

$$p_7, p_9 \neq p_o \tag{7.3}$$

Aerospace Propulsion, First Edition. T.-W. Lee.
© 2014 John Wiley & Sons, Ltd. Published 2014 by John Wiley & Sons, Ltd.
Companion Website: www.wiley.com/go/aerospaceprop

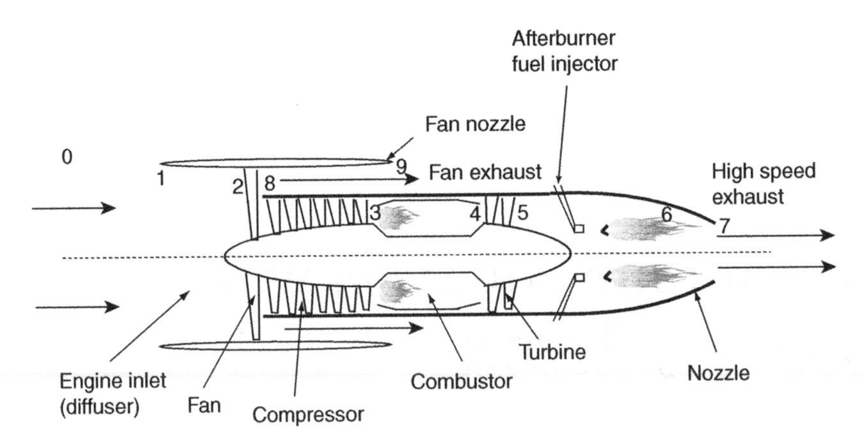

Figure 7.1 Schematic of a turbofan engine.

5. Allowance is made for the combustor efficiency (η_b) and mechanical coupling (between the turbine and compressor/fan) efficiency (η_m) being less than 100%.

$$\eta_b \neq 1 \tag{7.4}$$

$$\dot{W}_c + \dot{W}_{fan} = \eta_m \dot{W}_t \tag{7.5}$$

7.2 Turbofan Engine Analysis with Efficiency Terms

We again start from the thrust equation, but consider the core and fan streams separately, while including the pressure terms.

$$F_C = (\dot{m}_7 U_7 - \dot{m}_C U_o) + A_7(p_7 - p_o) \tag{7.6a}$$

$$F_{fan} = \dot{m}_{fan}(U_9 - U_o) + A_9(p_9 - p_o) \tag{7.6b}$$

The individual specific thrust is obtained by dividing by the appropriate mass flow rate. Division by the mass flow rate also eliminates the cumbersome exit areas in Eq. (7.6).

$$F_{s,C} = \frac{F_C}{\dot{m}_C} = a_o \left(\frac{\dot{m}_7}{\dot{m}_C} \frac{U_7}{a_o} - M_o \right) + \frac{A_7 p_7}{\dot{m}_C} \left(1 - \frac{p_o}{p_7} \right) \tag{7.7}$$

After some algebra, this is equivalent to

$$F_{s,C} = \frac{F_C}{\dot{m}_C} = a_o \left[(1+f)\frac{U_7}{a_o} - M_o + (1+f)\frac{R_t}{R_c}\frac{\frac{T_7}{T_o}}{\frac{U_7}{a_o}} \frac{1 - \frac{p_o}{p_7}}{\gamma_c} \right] \tag{7.8}$$

The specific thrust for the fan stream can also be written in a similar manner.

$$F_{s,fan} = \frac{F_{fan}}{\dot{m}_{fan}} = a_o \left[\frac{U_9}{a_o} - M_o + \frac{\frac{T_9}{T_o}}{\frac{U_9}{a_o}} \frac{1 - \frac{p_o}{p_9}}{\gamma_c} \right] \tag{7.9}$$

Calculations of the thrust now require evaluation of the exit velocities, static temperature and pressure ratios in Eqs. (7.8) and (7.9), which is done in a similar manner as in Chapter 3. For example, for the core exit velocity, we write

$$\left(\frac{U_7}{a_o}\right)^2 = \frac{a_7^2 M_7^2}{a_o^2} = \frac{\gamma_7 R_7 T_7}{\gamma_o R_o T_o} M_7^2 = \frac{\gamma_t R_t T_7}{\gamma_c R_c T_o} M_7^2 \tag{7.10}$$

The exit Mach number is found from the energy balance for the nozzle section.

$$M_7^2 = \frac{2}{\gamma_t - 1} \left[\left(\frac{p_7^o}{p_7}\right)^{\frac{\gamma t - 1}{\gamma t}} - 1 \right] \tag{7.11}$$

The pressure ratio in Eq. (7.11) is computed using all of the pressure ratios.

$$\frac{p_7^o}{p_7} = \frac{p_o}{p_7} \pi_r \pi_d \pi_c \pi_b \pi_t \pi_n \tag{7.12}$$

The static temperature ratio in Eq. (7.8), T_7/T_o, is also needed. The first step is to multiply and divide by the stagnation temperature at 7.

$$\frac{T_7}{T_o} = \frac{\frac{T_7^o}{T_o}}{\frac{T_7^o}{T_7}} = \frac{\frac{T_7^o}{T_7}}{\left(\frac{p_7^o}{p_7}\right)^{\frac{\gamma t - 1}{\gamma t}}} \tag{7.13}$$

The pressure ratio in the denominator is given by Eq. (7.12), so we only need the temperature ratio.

$$\frac{T_7^o}{T_7} = \tau_r \tau_d \tau_c \tau_b \tau_t \tau_n \tag{7.14}$$

Unlike the stagnation pressure ratios, it is reasonable to assume adiabatic processes for diffuser and nozzles, so that these temperature ratios are unity in Eq. (7.14). Defining a new combustion temperature ratio, τ_λ, Eq. (7.14) becomes

$$\frac{T_7^o}{T_7} = \tau_t \tau_\lambda \frac{c_{pc}}{c_{pt}} \tag{7.15}$$

$$\tau_\lambda = \frac{T_4^o c_{pt}}{T_o c_{pc}} = \tau_r \tau_d \tau_c \tau_b \frac{c_{pt}}{c_{pc}} \tag{7.16}$$

So now we have all the elements needed to evaluate Eq. (7.13).

$$\frac{T_7}{T_o} = \frac{\tau_\lambda \tau_t}{\left(\frac{p_7^o}{p_7}\right)^{\frac{\gamma_t - 1}{\gamma_t}}} \frac{c_{pc}}{c_{pt}} \tag{7.17}$$

Using Eqs. (7.17) and (7.11) in Eq. (7.10) gives us the core exit velocity, from which we find the core specific thrust (Eq. (7.8)). Corresponding terms for the fan specific thrust are found in an analogous manner.

$$\left(\frac{U_9}{a_o}\right)^2 = \frac{T_9}{T_o} M_9^2 \tag{7.18}$$

$$M_9^2 = \frac{2}{\gamma_c - 1}\left[\left(\frac{p_9^o}{p_9}\right)^{\frac{\gamma_t - 1}{\gamma_t}} - 1\right] \tag{7.19}$$

The pressure ratio in Eq. (7.19) is again a product of all the pressure ratios in that stream. The fan and fan nozzle pressure are denoted by π_f and π_{fn}, respectively.

$$\frac{p_9^o}{p_9} = \frac{p_o}{p_9} \pi_r \pi_d \pi_f \pi_{fn} \tag{7.20}$$

$$\frac{T_9}{T_o} = \frac{\frac{T_9^o}{T_o}}{\left(\frac{p_9^o}{p_o}\right)^{\frac{\gamma_c - 1}{\gamma_c}}} = \frac{\tau_r \tau_f}{\left(\frac{p_9^o}{p_o}\right)^{\frac{\gamma_c - 1}{\gamma_c}}} \tag{7.21}$$

7.2.1 *Polytropic Factor*

Inclusion of non-ideal effects will alter the pressure and temperature ratios that appear in Eqs. (7.12), (7.14–7.17), (7.20) and (7.21). In particular, the flows are no longer assumed to be isentropic, and are instead approximated as a polytropic process. In the discussion of the gas-turbine components, the deviation from the isentropic factor was treated with the isentropic efficiency. An alternative term, polytropic factor, is also often used. For component "*i*", the polytropic factor, e_i, is defined as the ratio of the isentropic to actual temperature differentials.

$$e_i = \frac{dT_s^o}{dT^o} \tag{7.22}$$

Again, the reference process is the isentropic process, denoted with a subscript "*s*". For an isentropic process, the familiar relationship is

$$pV^\gamma = const. \tag{7.23}$$

In terms of the stagnation pressure and temperature, the isentropic relationship is

$$\frac{(p^o)^{\frac{\gamma-1}{\gamma}}}{T^o} = \text{const.} \tag{7.24}$$

Taking the logarithm and dividing by T°, we get

$$\frac{dT_s^o}{T^o} = \frac{\gamma - 1}{\gamma}\frac{dp^o}{p^o} \tag{7.25}$$

Substituting Eq. (7.22),

$$e_i = \frac{dT_s^o}{dT^o} = \frac{\frac{dT_s^o}{T^o}}{\frac{dT^o}{T^o}} = \frac{\frac{\gamma-1}{\gamma}\frac{dp^o}{p^o}}{\frac{dT^o}{T^o}} \tag{7.26}$$

$$\text{Or,} \quad \frac{dT^o}{T^o} = \frac{\gamma - 1}{\gamma e_i}\frac{dp^o}{p^o} \tag{7.27}$$

If we consider the compressor, and integrate Eq. (7.27) from "2" to "3", we obtain

$$\ln\frac{T_3^o}{T_2^o} = \frac{\gamma_c - 1}{\gamma_c e_c}\ln\frac{p_3^o}{p_2^o} \rightarrow \tau_c = \pi_c^{\frac{\gamma_c-1}{e_c\gamma_c}} \tag{7.28}$$

Thus, the polytropic factor allows us to relate the stagnation temperature with pressure, to a single parameter. Other processes through fans (Eq. (7.29)) and turbines (Eq. (7.30)) can be written in a similar manner, except that for the expansion process the ratio in Eq. (7.26) is reversed and thus the polytropic factor appears in the numerator of the exponent.

$$\tau_f = \pi_f^{\frac{\gamma_c-1}{e_f\gamma_c}} \tag{7.29}$$

$$\tau_t = \pi_t^{\frac{e_t(\gamma_c-1)}{\gamma_c}} \tag{7.30}$$

Example 7.1 Polytropic factor

For compression ratio of 10, compressor inlet stagnation temperature of $T_2^o = 350$ K, $c_p = 1.004$ kJ/(kg · K), and $\gamma = 1.4$, find the isentropic work and polytropic work for $e_c = 0.85$.

For isentropic work, we can use

$$w_{c,s} = h_{3,s}^o - h_2^o = c_p T_2^o\left[\left(\frac{p_3^o}{p_2^o}\right)^{\frac{\gamma-1}{\gamma}} - 1\right] = c_p T_2^o\left[(\pi_c)^{\frac{\gamma-1}{\gamma}} - 1\right] = 326\frac{\text{kJ}}{\text{kg}} \tag{E7.1.1}$$

For $e_c = 0.85$, we can use Eq. (7.28) and substitute in Eq. (E7.1.1).

$$w_c = h_3^o - h_2^o = c_p T_2^o\left[(\pi_c)^{\frac{\gamma-1}{e_c\gamma}} - 1\right] = 410\frac{\text{kJ}}{\text{kg}} \tag{E7.1.2}$$

7.2.2 Diffuser

Now, let us go through the gas-turbine engine and apply the efficiency terms to each of the components. In Chapter 4, the definition of the diffuser efficiency led to the following result.

$$\eta_d = \frac{\tau_r \pi_d^{\frac{\gamma c-1}{\gamma c}} - 1}{\tau_r - 1} \tag{7.31}$$

The temperature ratio, τ_r, is still T_o^o/T_o. The stagnation pressure loss is calculated in terms of the diffuser efficiency, while the stagnation temperature does not change (adiabatic flow).

$$\pi_d = \frac{p_1^o}{p_o^o} = \left[\frac{\eta_d(\tau_r - 1) + 1}{\tau_r}\right]^{\frac{\gamma c-1}{\gamma c}}; \tau_d = \frac{T_1^o}{T_o^o} = 1 \tag{7.32}$$

The above expression is valid for subsonic flows. If the flow is supersonic, then the shock losses need to be accounted for. The stagnation pressure loss across the shock(s), π_s, discussed in Chapter 3 for ramjet analysis, is used.

$$\begin{aligned} \pi_s &= 1 && \text{for } M_o < 1 \\ \pi_s &= 1 - 0.075(M_o - 1)^{1.35} && \text{for } 1 \le M_o < 5 \\ \pi_s &= \frac{800}{M_o^4 + 935} && \text{for } M_o > 5 \end{aligned} \tag{3.72}$$

Thus, for supersonic flows, the total stagnation pressure ratio is the product of Eqs. (3.72) and (7.32).

$$(\pi_d)_{supersonic} = \pi_s \pi_d \tag{7.33}$$

7.2.3 Compressor and Fan

Compressor/fan end conditions can be related either through the isentropic efficiency or the polytropic factor (Eq. (7.28/29)).

$$\tau_c = 1 + \frac{1}{\eta_c}\left[\pi_c^{\frac{\gamma c-1}{\gamma c}} - 1\right] \tag{7.34}$$

$$\tau_c = \pi_c^{\frac{\gamma_c-1}{e_c\gamma_c}} \tag{7.28}$$

$$\tau_f = 1 + \frac{1}{\eta_f}\left[\pi_f^{\frac{\gamma c-1}{\gamma c}} - \right] \tag{7.35}$$

$$\tau_f = \pi_f^{\frac{\gamma_c-1}{e_f\gamma_c}} \tag{7.29}$$

7.2.4 Combustor

The combustor efficiency in modern gas-turbines is very high, close to one. However, we can consider combustor efficiency for the enthalpy balance of the combustor flow.

$$\dot{m}_C c_{pc} T_3^o + \eta_b \dot{m}_f h_{PR} = \dot{m}_4 c_{pt} T_4^o \tag{7.36}$$

Dividing by T_o and rearranging gives us,

$$f = \frac{\dot{m}_f}{\dot{m}_C} = \frac{\tau_\lambda - \tau_r \tau_c}{\frac{\eta_b h_{PR}}{c_{pc} T_o} - \tau_\lambda} \tag{7.37}$$

We can use the fuel–air ratio, f, to compute the TSFC.

$$\text{TSFC} = \frac{\dot{m}_f}{F} = \frac{\frac{\dot{m}_f}{\dot{m}_C}}{\frac{\dot{m}_o}{\dot{m}_C}\frac{F}{\dot{m}_o}} = \frac{f}{(1+\alpha)\frac{F}{\dot{m}_o}} \tag{7.38}$$

7.2.5 Turbine Power Balance

The turbine process is described by the isentropic efficiency or the polytropic factor.

$$\pi_t = \left[1 - \frac{1-\tau_t}{\eta_f}\right]^{\frac{\gamma_c - 1}{\gamma_c}} \tag{7.39}$$

$$\tau_t = \pi_t^{\frac{e_t(\gamma_c - 1)}{\gamma_c}} \tag{7.30}$$

Allowance is made for mechanical coupling loss, between the turbine and compressor/ fan axes.

$$\dot{m}_C c_{pc}(T_3^o - T_2^o) + \dot{m}_{fan} c_{pc}(T_8^o - T_2^o) = \eta_m \dot{m}_4 c_{pt}(T_4^o - T_5^o) \tag{7.40}$$

This provides an updated turbine temperature ratio.

$$\tau_t = 1 - \frac{1}{\eta_m(1+f)}\frac{\tau_r}{\tau_\lambda}\left[\tau_c - 1 + \alpha(\tau_f - 1)\right] \tag{7.41}$$

7.2.6 Nozzle Exit Pressure

The nozzle exit pressures are no longer assumed to be equal to the ambient pressure.

$$\frac{p_o}{p_7} = \frac{\frac{p_o}{p_7^o}}{\frac{p_7}{p_7^o}} = \frac{\frac{p_7^o}{p_7}}{\frac{p_7^o}{p_o}} \tag{7.42}$$

The pressure ratio in the denominator is the product of all the pressure ratios.

$$\frac{p_7^o}{p_o} = \pi_r \pi_d \pi_c \pi_b \pi_t \pi_n \tag{7.43}$$

For choked flow, $p_7 > p_o$ and $M_7 \approx 1$, and the pressure in the numerator of Eq. (7.42) can be approximated as

$$\frac{p_7^o}{p_7} = \left[1 + \frac{\gamma_t - 1}{2} M_7^2\right]^{\frac{\gamma t}{\gamma t - 1}} \approx \left(\frac{\gamma_t + 1}{2}\right)^{\frac{\gamma t}{\gamma t - 1}} \tag{7.44}$$

So Eq. (7.42) becomes

$$\frac{p_o}{p_7} = \frac{\left(\frac{\gamma_t + 1}{2}\right)^{\frac{\gamma t}{\gamma t - 1}}}{\pi_r \pi_d \pi_c \pi_b \pi_t \pi_n} \quad (p_7 > p_o) \tag{7.45}$$

If Eq. (7.45) returns a value $p_o/p_7 > 1$, the flow is not choked and we simply take $p_7 = p_o$. A similar expression can be found for the fan nozzle exit pressure.

$$\frac{p_o}{p_9} = \frac{\left(\frac{\gamma_c + 1}{2}\right)^{\frac{\gamma c}{\gamma c - 1}}}{\pi_r \pi_d \pi_f \pi_{fn}} \quad (p_9 > p_o) \tag{7.46}$$

If $p_o/p_9 > 1$, $p_9 = p_o$.

7.2.7 *Output Parameters*

Now we are ready to use the above results to calculate the output parameters. The thrust for the turbofan is

$$F = (\dot{m}_7 U_7 - \dot{m}_o U_o) + A_7(p_7 - p_o) + \dot{m}_{fan}(U_9 - U_o) + A_9(p_9 - p_o) \tag{7.47}$$

The specific thrust is

$$\frac{F}{\dot{m}_o} = a_o\left(\frac{\dot{m}_7}{\dot{m}_o}\frac{U_7}{a_o} - M_o\right) + \frac{A_7 p_7}{\dot{m}_o}\left(1 - \frac{p_o}{p_7}\right) + \frac{\alpha a_o}{1+\alpha}\left[\left(\frac{U_9}{a_o} - M_o\right) + \frac{A_9 p_9}{\dot{m}_o}\left(1 - \frac{p_o}{p_9}\right)\right] \tag{7.48}$$

For the pressure terms, it can be shown that

$$\frac{A_7 p_7}{\dot{m}_o}\left(1 - \frac{p_o}{p_7}\right) = a_o \frac{\dot{m}_7}{\dot{m}_o}\frac{R_t}{R_c}\frac{\frac{T_7}{T_o}}{\frac{U_7}{a_o}}\frac{1 - \frac{p_o}{p_7}}{\gamma_c} \tag{7.49}$$

Using the similar expression for the fan pressure thrust term and the fact that $\dot{m}_7 = \dot{m}_o + \dot{m}_f = (1+f)\dot{m}_o$, we have

$$F_s = \frac{F}{\dot{m}_o} = \frac{a_o}{1+\alpha}\left[(1+f)\frac{U_7}{a_o} - M_o + (1+f)\frac{R_t}{R_c}\frac{\frac{T_7}{T_o}}{\frac{U_7}{a_o}}\frac{1-\frac{p_o}{p_7}}{\gamma_c}\right] + \frac{\alpha a_o}{1+\alpha}\left[\frac{U_9}{a_o} - M_o + \frac{\frac{T_9}{T_o}}{\frac{U_9}{a_o}}\frac{1-\frac{p_o}{p_9}}{\gamma_c}\right] \tag{7.51}$$

We can incorporate the above results into those of the ideal turbofan from Chapter 3, and summarize in Table 7.1.

Example 7.2 Turbofan calculations

For the following set of parameters, find the specific thrust, TSFC, propulsion, thermal and overall efficiencies.

$M_o = 0.8$, $T_o = 390°\text{R}$, $\alpha = 10$, $h_{PR} = 18\,400$ BTU/lbm, $T_4^o = 3000°\text{R}$, $c_{pc} = 0.24$ BTU/(lbm°R), $c_{pt} = 0.276$ BTU/(lbm°R), $\gamma_c = 1.4$, $\gamma_t = 1.33$, $e_c = 0.9$, $e_f = e_t = 0.89$, $\pi_d = \pi_{fn} = 0.99$, $\eta_b = \eta_m = 0.99$, $\pi_b = 0.96$, $\pi_c = 36$, and $\pi_f = 1.65$. $p_o/p_7 = p_o/p_9 = 0.9$

$$a_o = \sqrt{\gamma_c R_c T_o} = \sqrt{(1.4)(53.36)(32.174)(390)} = 968.2\frac{\text{ft}}{\text{s}}, \text{ where } R_c = \frac{\gamma_c - 1}{\gamma_c}c_{pc}$$

$$= 53.36\frac{\text{ft}\cdot\text{lbf}}{\text{lbm}\cdot°\text{R}}.$$

$$U_o = M_o a_o = 774.6\ \text{ft/s}$$

$$\tau_r = 1 + \frac{\gamma_c - 1}{2}M_o^2 = 1.128;\ \pi_r = \tau_r^{\frac{\gamma_c}{\gamma_c - 1}} = 1.524$$

$$\tau_c = \pi_c^{\frac{\gamma_c - 1}{e_c\gamma_c}} = 3.319 \text{ and } \tau_f = \pi_f^{\frac{\gamma_c - 1}{e_f\gamma_c}} = 1.174$$

$$\tau_\lambda = \frac{c_{pt}T_4^o}{c_{pc}T_o} = 8.846 \rightarrow f = \frac{\tau_\lambda - \tau_r\tau_c}{\frac{h_{PR}\eta_b}{c_{pc}T_o} - \tau_\lambda} = 0.0287$$

$$\tau_t = 1 - \frac{1}{\eta_m(1+f)}\frac{\tau_r}{\tau_\lambda}\left[\tau_c - 1 + \alpha(\tau_f - 1)\right] = 0.516;\ \pi_t = \tau_t^{\frac{\gamma_t}{e_t(\gamma_t - 1)}} = 0.05$$

$$\frac{p_7^o}{p_7} = \frac{p_o}{p_7}\pi_r\pi_d\pi_c\pi_b\pi_t\pi_n = (0.9)(1.5243)(0.99)(36)(0.96)(0.05)(0.99) = 2.329$$

$$M_7 = \sqrt{\frac{2}{\gamma_t - 1}\left[\left(\frac{p_7^o}{p_7}\right)^{\frac{\gamma_t - 1}{\gamma_t}} - 1\right]} = 1.189 \text{ and } \frac{T_7}{T_o} = \frac{\tau_\lambda\tau_t}{\left(\frac{p_7^o}{p_7}\right)^{\frac{\gamma_t - 1}{\gamma_t}}}\frac{c_{pc}}{c_{pt}} = 3.22$$

$$\frac{U_7}{a_o} = M_7 \sqrt{\frac{\gamma_t R_t T_7}{\gamma_c R_c T_o}} = 2.08$$

Similarly for the fan stream, we have

$$\frac{p_9^o}{p_9} = \frac{p_o}{p_9} \pi_r \pi_d \pi_f \pi_{fn} = (0.9)(1.5243)(0.99)(1.65)(0.99) = 2.219$$

$$M_9 = \sqrt{\frac{2}{\gamma_c - 1}\left[\left(\frac{p_9^o}{p_9}\right)^{\frac{\gamma_c-1}{\gamma_c}} - 1\right]} = 1.131 \text{ and } \frac{T_9}{T_o} = \frac{\tau_r \tau_f}{\left(\frac{p_9^o}{p_9}\right)^{\frac{\gamma_c-1}{\gamma_c}}} = 1.055$$

$$\frac{U_9}{a_o} = M_9 \sqrt{\frac{T_9}{T_o}} = 1.161$$

Now we can use the core and fan exit velocities to find the specific thrust.

$$F_s = \frac{F}{\dot{m}_o} = \frac{a_o}{1+\alpha}\left[(1+f)\frac{U_7}{a_o} - M_o + (1+f)\frac{R_t}{R_c}\frac{\frac{T_7}{T_o}}{\frac{U_7}{a_o}}\frac{1 - \frac{p_o}{p_7}}{\gamma_c}\right] + \frac{\alpha a_o}{1+\alpha}\left[\frac{U_9}{a_o} - M_o + \frac{\frac{T_9}{T_o}}{\frac{U_9}{a_o}}\frac{1-\frac{p_o}{p_9}}{\gamma_c}\right]$$
$$= 15.45 \frac{\text{lbf}}{\text{lbm/s}}$$

To bring the specific thrust into the customary units of lbf/(lbm/s), we need to use (divide by) the conversion factor of 1 lbf = 32.174 lbm-f/s^2.

$$\text{TSFC} = \frac{f}{(1+\alpha)F_s} = 0.608 \frac{\text{lbm/hr}}{\text{lbf}}$$

$$\eta_P = \frac{2M_o\left[(1+f)\frac{U_7}{a_o} + \alpha\frac{U_9}{a_o} - (1+\alpha)M_o\right]}{(1+f)\left(\frac{U_7}{a_o}\right)^2 + \alpha\left(\frac{U_9}{a_o}\right)^2 - (1+\alpha)M_o^2} = 0.728$$

$$\eta_T = \frac{a_o^2\left[(1+f)\left(\frac{U_7}{a_o}\right)^2 + \alpha\left(\frac{U_9}{a_o}\right)^2 - (1+\alpha)M_o^2\right]}{2fh_{PR}} = 0.386$$

Again, to have the consistent units in the numerator and denominator, we need to use (multiply) the denominator by the conversion factor of 1 lbf = 32.174 lbm-f/s^2.

$$\eta_o = \eta_P \eta_T = 0.281$$

Example 7.3 Comparison of ideal- and real-cycle calculations

A comparison between the ideal- and real-cycle calculations can be using a MATLAB® program MCE73, at the end of this chapter. Some typical results are shown in Figures E7.3.1 and Figure E7.3.2.

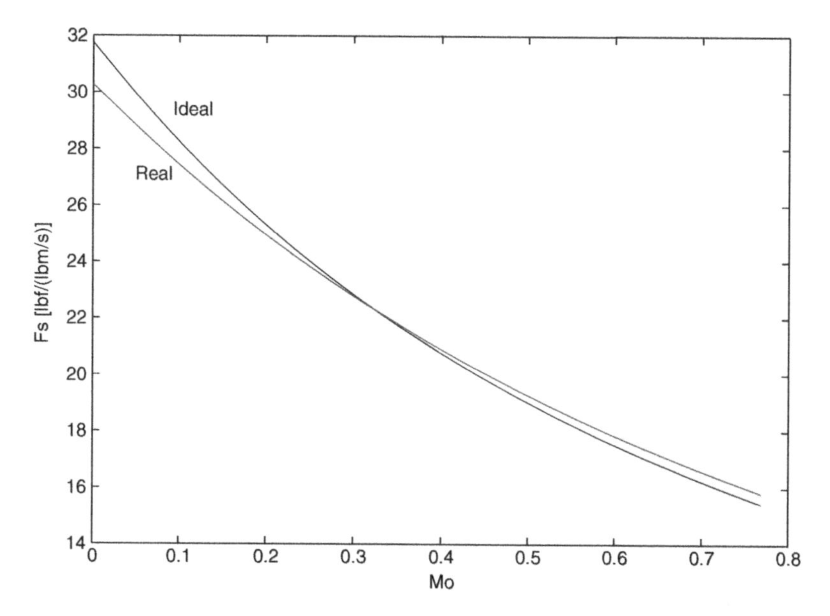

Figure E7.3.1 A comparison of specific thrust, computed using MCE73.

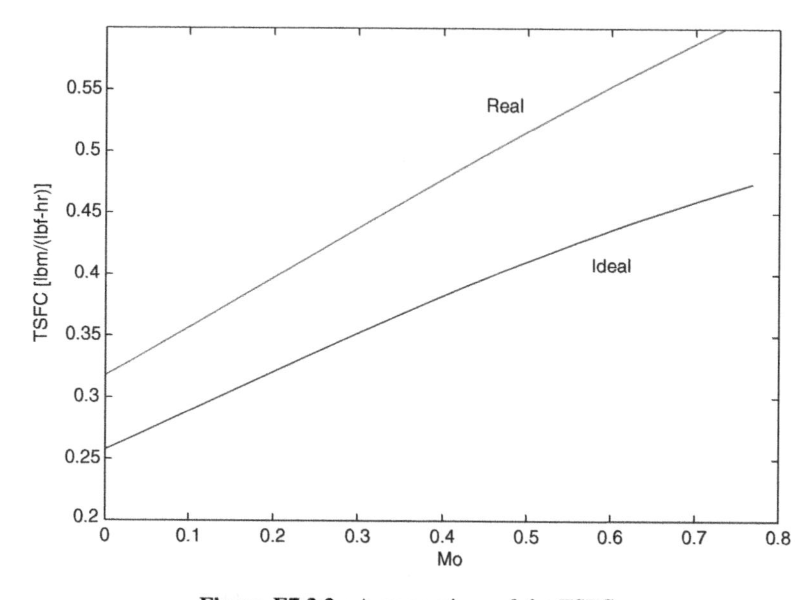

Figure E7.3.2 A comparison of the TSFC.

Example 7.4 Parametric variation of turbo-fan performance with efficiency terms

MCE73 can be used for further analysis of turbofan performance with efficiency terms (see Figure E7.4.1 and Figure E7.4.2).

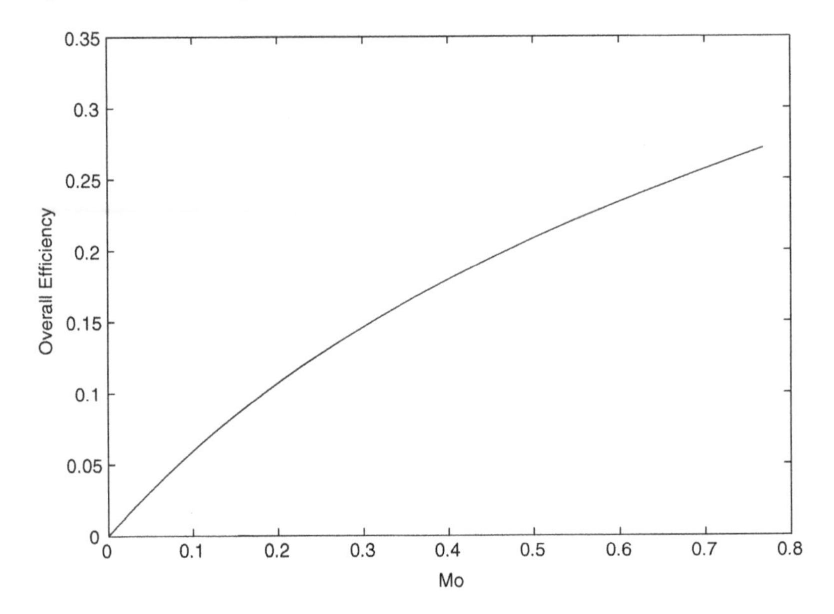

Figure E7.4.1 Plot of the overall efficiency as a function of the Mach number.

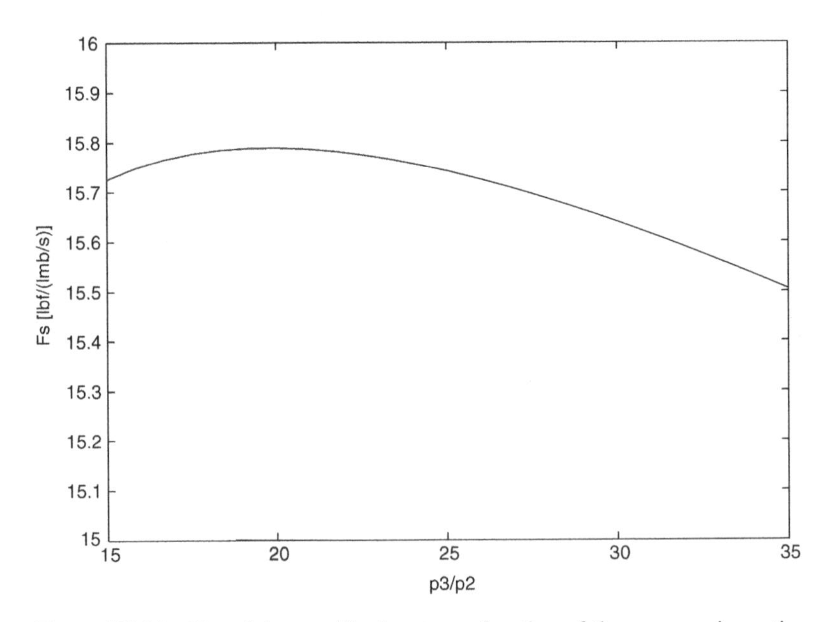

Figure E7.4.2 Plot of the specific thrust as a function of the compression ratio.

Table 7.1 A summary of equations for turbofan analysis with efficiency terms.

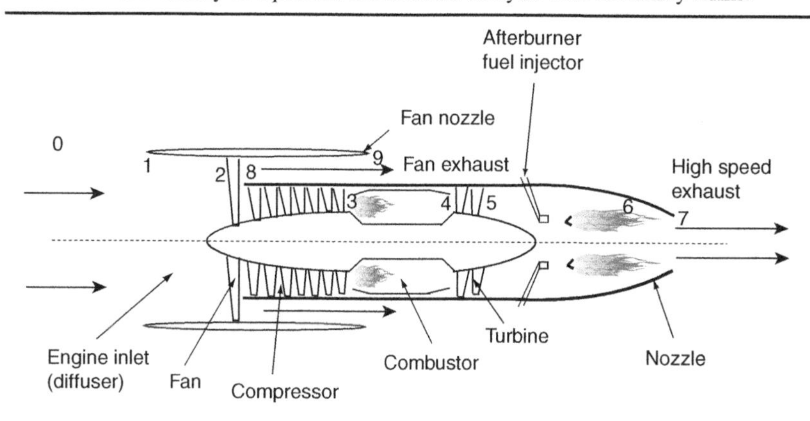

Input parameters: M_o, T_o, γ_c, c_{pc}, γ_t, $c_{pt}h_{PR}$, T_4^o, π_c, π_f(fan pressure ratio), α (bypass ratio), e_c, e_f, e_t, π_d, π_f, π_{fn}, η_b, η_m, π_b, p_o/p_7 and p_o/p_9.

Output parameters: U_7, U_9, F_s, TSFC, η_P, η_T, η_o.

$$R_c = \frac{\gamma_c - 1}{\gamma_c} c_{pc};\ R_t = \frac{\gamma_t - 1}{\gamma_t} c_{pt} \text{ and } a_o = \sqrt{\gamma_c R_c T_o}$$

$$\tau_r = 1 + \frac{\gamma_c - 1}{2} M_o^2;\ \pi_r = \tau_r^{\frac{\gamma_c}{\gamma_c - 1}}$$

$$\tau_c = \pi_c^{\frac{\gamma_c - 1}{e_c \gamma_c}};\ \tau_f = \pi_f^{\frac{\gamma_c - 1}{e_f \gamma_c}}$$

$$\tau_\lambda = \frac{c_{pt} T_4^o}{c_{pc} T_o} \rightarrow f = \frac{\tau_\lambda - \tau_r \tau_c}{\frac{h_{PR}\eta_b}{c_{pc} T_o} - \tau_\lambda}$$

$$\tau_t = 1 - \frac{1}{\eta_m (1 + f)} \frac{\tau_r}{\tau_\lambda} \left[\tau_c - 1 + \alpha(\tau_f - 1)\right];\ \pi_t = \tau_t^{\frac{\gamma_t}{e^t(\gamma_t - 1)}}$$

$$\frac{p_7^o}{p_7} = \frac{p_o}{p_7} \pi_r \pi_d \pi_c \pi_b \pi_t \pi_n \rightarrow M_7 = \sqrt{\frac{2}{\gamma_t - 1}\left[\left(\frac{p_7^o}{p_7}\right)^{\frac{\gamma_t - 1}{\gamma_t}} - 1\right]}$$

$$\frac{T_7}{T_o} = \frac{\tau_\lambda \tau_t}{\left(\frac{p_7^o}{p_7}\right)^{\frac{\gamma_t - 1}{\gamma_t}}} \frac{c_{pc}}{c_{pt}} \rightarrow \frac{U_7}{a_o} = M_7 \sqrt{\frac{\gamma_t R_t T_7}{\gamma_c R_c T_o}}$$

$$\frac{p_9^o}{p_9} = \frac{p_o}{p_9} \pi_r \pi_d \pi_f \pi_{fn} \rightarrow M_9 = \sqrt{\frac{2}{\gamma_t - 1}\left[\left(\frac{p_9^o}{p_9}\right)^{\frac{\gamma_t - 1}{\gamma_t}} - 1\right]}$$

$$\frac{T_9}{T_o} = \frac{\tau_r \tau_f}{\left(\frac{p_9^o}{p_9}\right)^{\frac{\gamma_c - 1}{\gamma_c}}} \rightarrow \frac{U_9}{a_o} = M_9 \sqrt{\frac{T_9}{T_o}}$$

(*continued*)

Table 7.1 *(Continued)*

$$F_s = \frac{F}{\dot{m}_o} = \frac{a_o}{1+\alpha}\left[(1+f)\frac{U_7}{a_o} - M_o + (1+f)\frac{R_t}{R_c}\frac{\frac{T_7}{T_o}\left(1-\frac{p_o}{p_7}\right)}{\frac{U_7}{a_o}\gamma_c}\right] + \frac{\alpha a_o}{1+\alpha}\left[\frac{U_9}{a_o} - M_o + \frac{\frac{T_9}{T_o}\left(1-\frac{p_o}{p_9}\right)}{\frac{U_9}{a_o}\gamma_c}\right]$$

$$TSFC = \frac{f}{(1+\alpha)F_s}$$

$$\eta_P = \frac{2M_o\left[(1+f)\frac{U_7}{a_o} + \alpha\frac{U_9}{a_o} - (1+\alpha)M_o\right]}{(1+f)\left(\frac{U_7}{a_o}\right)^2 + \alpha\left(\frac{U_9}{a_o}\right)^2 - (1+\alpha)M_o^2}$$

$$\eta_T = \frac{a_o^2\left[(1+f)\left(\frac{U_7}{a_o}\right)^2 + \alpha\left(\frac{U_9}{a_o}\right)^2 - (1+\alpha)M_o^2\right]}{2fh_{PR}}$$

$$\eta_o = \eta_P\eta_T$$

7.3 MATLAB® Program

```
% MCE 73: Calculates the turbofan performance based on the following inputs:
% Comparison of ideal and real cycle performance
% Mo, To, gamma, cp, hPR, T4o, pc, pf (fan pressure ratio), alpha (bypass
ratio).
% gc, gt, cpc, cpt, ec, ef, pid, pifn, hb, hm, pib
%pOp7, pOp9
% OUTPUT PARAMETERS: U7, U9, Fs, TSFC, hP, hT, ho.
% Other variations such as turbojets can be calculated, e.g. simply by
setting
% alpha = 0.

% English units are used in this example.
G=32.174; FL=778.16 % conversion factor


% INPUT PARAMETERS
Mo = 0.8; To = 390; po= 629.5;
gc = 1.4; cpc =0.24; gt = 1.33; cpt = 0.276; hPR = 18400;
T4o = 3000;, pc =36; pf = 1.65; alpha = 10;
mo=1125;
```

```
ec = 0.9; ef=0.89; et = 0.89; pid = 0.99; pin = 0.99; pifn = 0.99;
hb=0.99; hm =0.99; pib = 0.96; pop7=0.9; pop9=0.9;

Rc=(gc-1)*cpc*FL/gc; Rt=(gt-1)*cpt*FL/gc;
ao=sqrt(gc*Rc*G*To); gamma = gc; cp = cpc;

%Start of calculations

% Ideal Cycle
%Options to vary X(i) = Mo, To, etc.

N = 25;
Xmax=1; Xmin=0; dX=(Xmax-Xmin)/N;
for i = 1:N;
  X(i) = Xmin+dX*(i-1);
  Mo=X(i);

Uo=Mo*ao

tr=1+(gamma-1)/2*Mo^2; T2o=tr*To

tc=pc^((gamma-1)/gamma)
tf=pf^((gamma-1)/gamma)

tL=T4o/To

tt=1-tr/tL*(tc-1+alpha*(tf-1))

% Exit velocities: U2=(U7/ao)^2; V2=(U9/ao)^2


U2=2/(gamma-1)*tL/(tr*tc)*(tr*tc*tt-1)

U7=sqrt(U2)*ao

V2=2/(gamma-1)*(tr*tf-1);

U9=sqrt(V2)*ao

% Mass flow rates

mc=mo/(1+alpha); mF=mo*alpha/(1+alpha);

% Thrust

F1=mc*(U7-Uo)+mF*(U9-Uo);
F=F1/G

Fs(i) = F/mo;
```

7.4 Problems

7.1. Using the air property data, for $\gamma = c_p/c_v$, plot c_p as a function of temperature for $T = 300$ to 3000 K.

7.2. Show that Eq. (7.7) algebraically leads to the following result.

$$F_{s,C} = \frac{F_C}{\dot{m}_C} = a_o \left[(1+f)\frac{U_7}{a_o} - M_o + (1+f)\frac{R_t}{R_c}\frac{\frac{T_7}{T_o}}{\frac{U_7}{a_o}}\frac{1-\frac{p_o}{p_7}}{\gamma_c} \right] \tag{7.8}$$

7.3. Provide the algebraic steps to the result.

$$F_{s,fan} = \frac{F_{fan}}{\dot{m}_{fan}} = a_o \left[\frac{U_9}{a_o} - M_o + \frac{\frac{T_9}{T_o}}{\frac{U_9}{a_o}}\frac{1-\frac{p_o}{p_9}}{\gamma_c} \right] \tag{7.9}$$

7.4. Non-isentropic processes may be approximated with the polytropic factor, e, or polytropic exponent, $pV^n = \text{const}$. Plot e as function of n, and also the pressure ratio as a function of temperature ratio for $e = 0.8$, 0.9 and 1.0 for $\gamma = 1.4$.

7.5. Plot the stagnation pressure loss as a function of the Mach number, using Eq. (3.72).

$$\pi_s = 1 \quad \text{for } M_o < 1$$

$$\pi_s = 1 - 0.075(M_o - 1)^{1.35} \quad \text{for } 1 \le M_o < 5$$

$$\pi_s = \frac{800}{M_o^4 + 935} \quad \text{for } M_o > 5$$

7.6. Show that the thermal and propulsion efficiencies for turbofans can be written as

$$\eta_P = \frac{2M_o\left[(1+f)\frac{U_7}{a_o} + \alpha\frac{U_9}{a_o} - (1+\alpha)M_o\right]}{(1+f)\left(\frac{U_7}{a_o}\right)^2 + \alpha\left(\frac{U_9}{a_o}\right)^2 - (1+\alpha)M_o^2}$$

$$\eta_T = \frac{a_o^2\left[(1+f)\left(\frac{U_7}{a_o}\right)^2 + \alpha\left(\frac{U_9}{a_o}\right)^2 - (1+\alpha)M_o^2\right]}{2fh_{PR}}$$

7.7. Calculate the propulsion and thermal efficiencies of a turbojet engine during subsonic cruise. The flight Mach number is 0.8, and the ambient temperature is 225 K. The compressor pressure ratio is 15, and the turbine inlet temperature is 1250 K. The efficiencies of the diffuser, compressor, turbine and nozzle are 0.92, 0.85, 0.85 and 0.95. The burner stagnation pressure ratio is 0.97, and the average specific heat during and after combustion is 1.1 kJ/kg-K, and the average molecular weight is 29.

7.8. For a real-cycle turbofan engine, calculate T_5^o for the following operating parameters: $M_o = 0.8$; $T_o = 390°R$; $\gamma_c = 1.4$; $\gamma_t = 1.33$; $c_{pc} = 0.24$ BTU/(lbm-°R); $c_{pt} = 0.276$ BTU/(lbm-°R); $h_{pr} = 18\ 400$ BTU/lbm; $T_4^o = 3000°R$; $\pi_c = 30$; $\pi_f = 2.25$; $\alpha = 8$; $\eta_m = 0.99$; $\pi_{d,\ max} = 0.99$; $\pi_b = 0.96$; $\pi_n = 0.975$; $\pi_{fn} = 0.98$; $\eta_b = 0.99$; $e_c = 0.9$; $e_f = 0.89$; $e_t = 0.89$; $p_o/p_7 = 0.92$, $p_o/p_9 = 0.95$.

7.9. Calculate the specific thrust, TSFC, thermal and propulsion efficiencies for turbofan. $M_o = 0.8$, $T_o = 390°R$, $\alpha = 7.75$, $h_{PR} = 18\ 400$ BTU/lbm, $T_4^o = 3200°R$, $c_{pc} = 0.24$ BTU/(lbm°R), $c_{pt} = 0.276$ BTU/(lbm°R), $\gamma_c = 1.4$, $\gamma_t = 1.33$, $e_c = 0.9$, $e_f = e_t = 0.89$, $\pi_d = \pi_{fn} = 0.99$, $\eta_b = \eta_m = 0.99$, $\pi_b = 0.96$, $\pi_c = 36$, and $\pi_f = 1.65$. $p_o/p_7 = p_o/p_9 = 0.9$.

7.10. For the following turbofan specifications, plot (a) specific thrust as a function of the Mach number for bypass ratio of 7.75, and (b) TSFC as a function of the bypass ratio for $M_o = 0.8$.

$T_o = 390°R$, $h_{PR} = 18\ 400$ BTU/lbm, $T_4^o = 3000°R$, $c_{pc} = 0.24$ BTU/(lbm°R), $c_{pt} = 0.276$ BTU/(lbm°R), $\gamma_c = 1.4$, $\gamma_t = 1.33$, $e_c = 0.9$, $e_f = e_t = 0.89$, $\pi_d = \pi_{fn} = 0.99$, $\eta_b = \eta_m = 0.99$, $\pi_b = 0.96$, $\pi_c = 36$, and $\pi_f = 1.65$. $p_o/p_7 = p_o/p_9 = 0.9$.

7.11. A ramjet is in supersonic flight, under the following conditions. The inlet pressure recovery is $\pi_d = 0.90$. The combustor burns hydrogen with $h_{PR} = 117\ 400$ kJ/kg at a combustor efficiency of $\eta_b = 0.95$. The nozzle expands the gas perfectly, but suffers from a total pressure loss of $\pi_n = 0.92$. Other operational parameters are, $M_o = 2.0$, $a_o = 300$ m/s, $T_4^o = 2200$ K, $c_{pc} = 1004$ J/(kg-K), $c_{pt} = 1243$ J/(kg-K), $\gamma_c = 1.4$, and $\gamma_t = 1.3$.

Calculate (a) fuel-to-air ratio f; (b) nozzle exit Mach number, M_7; and specific thrust Fs, in [N/(kg/s)].

Bibliography

Mattingly, J.D. (2005) *Elements of Gas-Turbine Propulsion*, McGraw Hill.

8

Basics of Rocket Propulsion

8.1 Introduction

Rockets are self-contained propulsion devices, where no intake of air or oxidizer is needed. Conventional rockets make use of materials that contain oxidizer components, or simply carry both the fuel and oxidizer, so that combustion can occur autonomously within the rocket combustion chamber. There are also rockets that do not use chemical fuels at all, but make use of electromagnetic or other novel processes to eject momentum so that propulsive force is achieved. In that regard, the same thrust equation discussed in Chapter 2 is applicable.

$$F = \dot{m}_o[(1+f)U_e - U_o] + (p_e - p_o)A_e \tag{2.10}$$

For rockets, there is no ingestion of mass, and the relevant mass flow rate is the propellant mass flow rate, so that Eq. (2.10) is converted for rockets.

$$F = \dot{m}_p U_e + (p_e - p_o)A_e \tag{8.1}$$

There are novel devices, such as solar sails, where the thrust equation is not directly applicable, but a momentum balance in a different form still gives the thrust for solar sails as well. Rockets are used both within the atmosphere and in outer space. In either case, the pressure term in Eq. (8.1) is typically appreciable, so that it is kept in the evaluation of thrust.

The types of rocket propulsion devices that we will discuss are:

- chemical (solid/liquid/hybrid propellants)
- electric (electrothermal, electrostatic, electromagnetic)
- novel concepts (solar sails, nuclear, etc.).

Figures 8.1–8.3 show the schematics of the above rockets. Solid-propellant rockets have the simplest geometry, involving no internal moving parts, where the propellant is stored in the thrust chamber as shown in Figure 8.1. Upon ignition, the high volume of the combustion product is sent down through the nozzle to generate thrust. Different cross-sectional geometry of the solid propellants provides different surface areas available for

Aerospace Propulsion, First Edition. T.-W. Lee.
© 2014 John Wiley & Sons, Ltd. Published 2014 by John Wiley & Sons, Ltd.
Companion Website: www.wiley.com/go/aerospaceprop

burning, and therefore allows some control over the thrust as a function of time. Once it is ignited, there is little control, other than the solid-propellant "grain" geometry, on the thrust history. Liquid-propellant rockets are more complex in design, as shown in Figure 8.1. Fuel and oxidizer are stored in separate tanks, and turbo-pumps are operated to send these propellant streams into the high-pressure environment of the combustion chamber. Hybrid chemical rockets refer to a system combining solid- and liquid-propellant motors. The United States' space shuttles used a hybrid system of the space shuttle main engine (SSME) and solid rocket boosters (SRB). In that regard, the space shuttle propulsion system is "staged", where the SRBs are jettisoned after their burn-out during the initial boost phase. After the jettison of the body mass of the SRB, the space shuttle is lightened for a more effective boost into high-altitude orbits. Table 8.1 shows the mass and thrust history of the space shuttle propulsion system.

Electric propulsion devices make use of electric energy to generate the propellant mass and energy and/or to propel the propellant to high exit velocities. Arcjet, for example, uses electric energy to generate high-temperate plasmas (electric sparks) to vaporize the solid propellant, as shown in Figure 8.2. The high thermal energy of the vaporized propellant is then used to generate kinetic energy through a gas-dynamic nozzle. This is an example of electrothermal rockets, where the combustion of chemical fuels is replaced by the energy input through electrical means. Electrostatic devices make dual use of the electric energy,

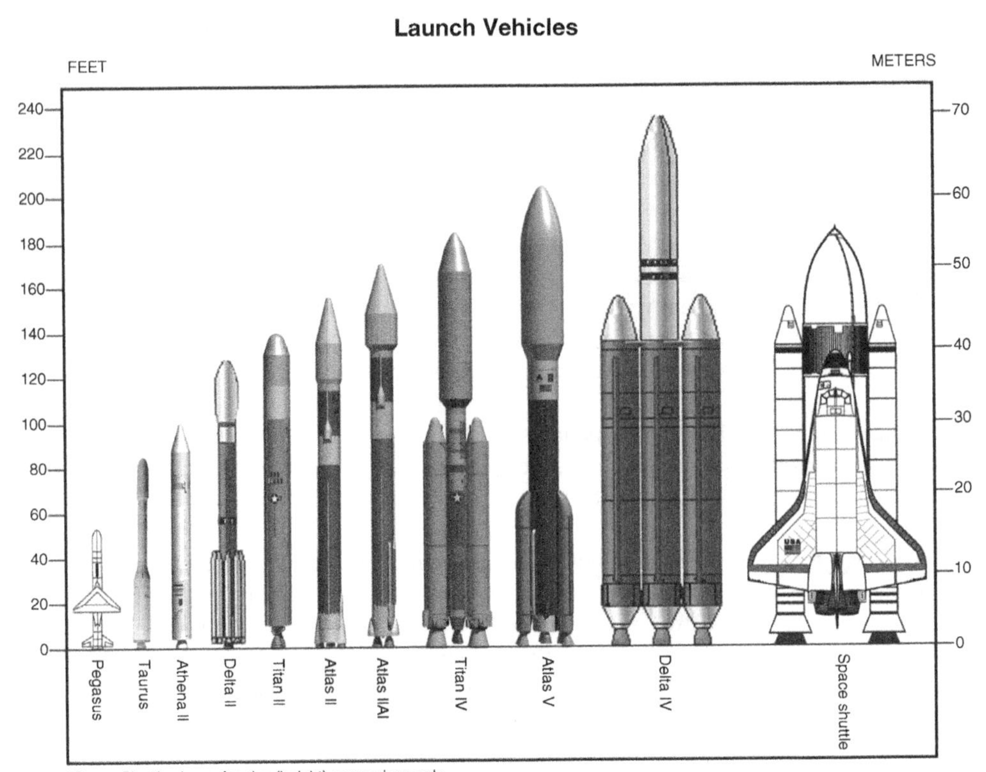

Figure 8.1 Chemical rockets. Courtesy of NASA.

where the high-intensity electric current (electron streams) ablate and ionize the propellant. The ionized propellant is then accelerated through an electric field, as shown in Figure 8.2.

There are some novel concepts, some more seriously considered than others, such as solar sails, nuclear propulsion, gravity-assist and tethers. Figure 8.3 shows some examples, with brief schematics. The solar sails, or other beamed propulsion devices, make use of the "photon pressure" of the light emanating from the sun or other powerful beam sources. Nuclear power can be converted to thermal energy, in various ways for propulsive effects. We will discuss these novel concepts and applications again in a later chapter. Table 8.2 shows the typical range of performance parameters for various rocket propulsion systems.

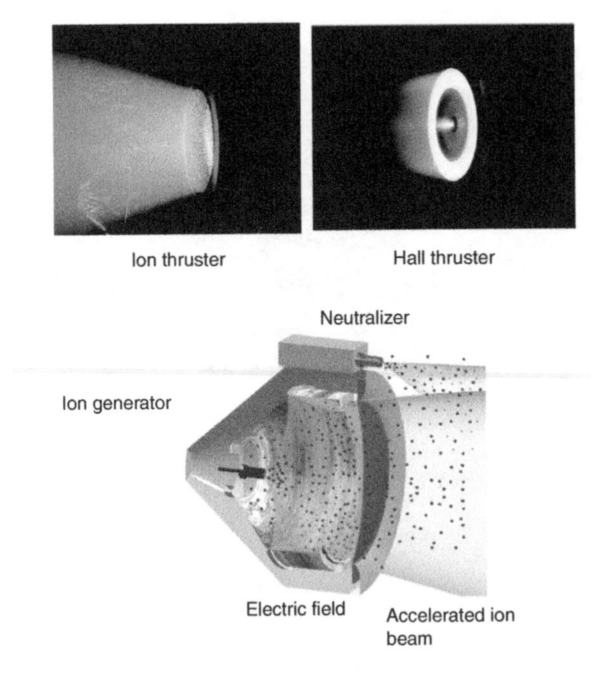

Figure 8.2 Electric propulsion devices. Courtesy of NASA.

Solar sails

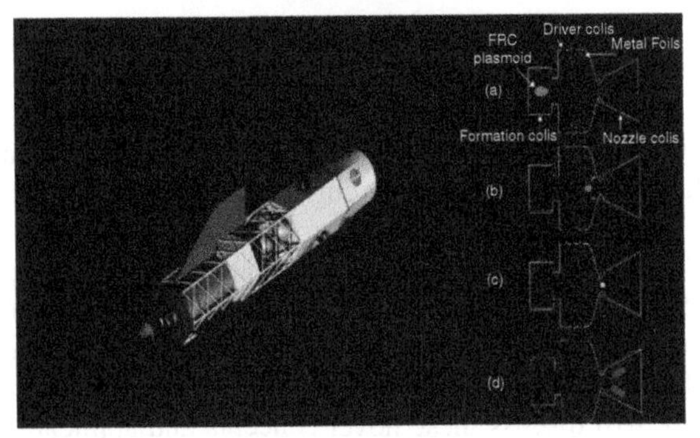

Fusion drive rocket

Figure 8.3 Some examples of novel propulsion systems. Courtesy of NASA.

Table 8.1 Mass and thrust history of the Space Shuttle propulsion system.

	Solid rocket booster (SRB)	Space shuttle main engine (SSME)		
Number	2	3		
Burn time [min]	2	8		
Total thrust [kN]	24 000	6300		
I_{sp} [s]	312	428		
Mass flow rate [kg/s]	8410	1496		
Time [min]	0	2^-	2^+	8
Solid rocket boosters [kg]	1 170 000	164 000	0	0
External fuel tanks [kg]	750 000	571 000	571 000	32 000
Orbiter/*Payload* [kg]	86 400	86 400	86 400	86 400
Total [kg]	2 010 000	820 000	660 000	118 000

Table 8.2 Typical range of performance parameters for various rocket propulsion systems (Source: NASA).

Type	Specific impulse [s]	Propellants
Chemical, low-energy liquid monopropellants	160–190	hydrazine, ethylene oxide, hydrogen peroxide
Chemical, high energy liquid monopropellants	190–230	nitromethane
Chemical, low-energy liquid bipropellants	200–230	perchloryl fluoride-JP4, hydrogen peroxide-JP4
Chemical, medium-energy liquid bipropellants	230–260	ammonia-nitrogen tetroxide
Chemical, high-energy liquid bipropellants	250–270	liquid oxygen-JP4, liquid oxygen-alcohol, hydrazine-chlorine trifluoride
Chemical, very high-energy liquid bipropellants	270–330	liquid oxygen/fluorine-JP4, liquid oxygen-hydrazine
Chemical, super high-energy liquid bipropellants	300–385	fluorine-hydrogen, fluorine-ammonia, ozone-hydrogen
Chemical, low-energy composite solid propellants	170–210	potassium perchlorate Thiokol (asphalt), ammonium perchlorate-rubber, ammonium nitrate-polyester
Chemical, high-energy composite solid propellants	210–250	Potassium perchlorate-polyurethane, ammonium perchlorate-nitropolymer
Chemical, non-metallized double-base solid propellants	170–210	polyglycol adipate (PGA), nitrocellulose (NC), nitroglycerine (NO), hexadiisocryanate (HDI), polythylene glycol (PEG).
Chemical, metallized double-base solid propellants	200–250	boron + oxidant, lithium + oxidant, aluminum + oxidant, magnesium + oxidant
Electrothermal	200–1000	resistojet, arcjet
Electrostatic	1500–5000	ion thruster
Electromagnetic	600–5000	pulsed-plasma, magneto-plasma, Hall-effect thruster

8.2 Basic Rocketry

8.2.1 Specific Impulse

Rockets are characterized by their ability to generate thrust, which is expressed through the specific impulse, I_s. Specific impulse is the total impulse divided by the weight of the expended propellant, and thus represents the effectiveness of the propellant in generating thrust. Total impulse, I_t, is defined as the thrust force integrated over time.

$$I_t = \int_0^t F dt \quad (8.2)$$

If the thrust is constant over time, then the total impulse is simply the thrust times the total elapsed time, Δt.

$$I_t = \int_0^t F dt = F\Delta t \quad (\text{for F} = \text{constant}) \quad (8.3)$$

The specific impulse can thus be written as

$$I_s = \frac{\int_0^t F dt}{g_o \int_0^t \dot{m} dt} = \frac{I_t}{g_o m_p} \quad [1/\text{s}] \quad (8.4)$$

The denominator is the weight of the expended propellant, which can be found by integrating the mass flow rate over time and multiplying that quantity by the gravitational acceleration at the sea level, g_o.

$$g_o = 9.8066\, \text{m/s}^2 = 32.174\, \text{ft/s}^2. \quad (8.5)$$

For constant thrust and propellant mass flow rate, the specific impulse is simply the thrust divided by the "weight flow rate" of the propellant.

$$I_s = \frac{F\Delta t}{g_o \dot{m} \Delta t} = \frac{F}{\dot{w}_p} \quad (8.6)$$

For the thrust calculated by the thrust equation (Eq. (8.1)), we can define an equivalent exhaust velocity, U_{eq}, that absorbs both the momentum and pressure terms.

$$F = \dot{m}_p U_e + (p_e - p_o)A_e \equiv \dot{m}_p U_{eq} \quad (8.7)$$

This can now be related to the specific thrust, under steady-state conditions.

$$U_{eq} = \frac{F}{\dot{m}_p} = U_e + \frac{(p_e - p_o)A_e}{\dot{m}_p} = I_s g_o, \quad \text{or} \quad I_s = \frac{U_{eq}}{g_o} \quad (8.8)$$

Thus, specific impulse is a measure of the rocket's ability to generate thrust, for a given weight of propellant, as expressed by the equivalent exhaust velocity.

8.2.2 Vehicle Acceleration

The importance of the specific impulse and propellant weight is easily seen through consideration of vehicle acceleration by rockets.

$$F = M\frac{dU}{dt} = \dot{m}_p U_{eq} = -\frac{dM}{dt}U_{eq} \tag{8.9}$$

M and U are the vehicle mass and velocity at a given time, respectively. Rearranging Eq. (8.9) leads to

$$MdU = -U_{eq}dM \rightarrow \frac{dM}{M} = -\frac{dU}{U_{eq}} \tag{8.10}$$

Integrating from the initial mass (M_o) at $t=0$ to the mass at burnout (M_b) at $t=t_b$, gives us the vehicle velocity change.

$$\ln\left(\frac{M_b}{M_o}\right) = -\frac{\Delta U}{U_{eq}} \rightarrow \Delta U = U_{eq}\ln\left(\frac{M_o}{M_b}\right) = g_o I_s \ln R_M \tag{8.11}$$

R_M is the "mass ratio" of the initial to the burnout mass. Thus, the larger the specific impulse and the mass ratio, the higher the vehicle speed that can be attained. A large mass ratio is obtained simply by carrying a large amount of propellant, relative to the "empty" vehicle mass, M_b.

We can consider the vehicle speed at an arbitrary time t, using Eq. (8.11), and also add the gravity effect in a purely vertical flight.

$$\Delta U = -U_{eq}\ln\left(\frac{M}{M_o}\right) - gt \tag{8.12}$$

Equation (8.12) can be used to calculate the maximum attainable altitude, for a vertical flight. The altitude at burnout, for example, is found by simply integrating Eq. (8.12). For a static launch ($U=0$ at $t=0$) and constant gravitational acceleration ($g=g_o$),

$$H_b = \int_0^{t_b} Udt = \text{altitude at burnout } (t = t_b) \tag{8.13}$$

Let us consider a constant burning rate, so that the mass of the vehicle in time is

$$\frac{M(t)}{M_o} = 1 - \left(1 - \frac{M_b}{M_o}\right)\frac{t}{t_b} \tag{8.14}$$

Use of Eqs. (8.12) and (8.14) in Eq. (8.13) leads to a lengthy but integrable form.

$$H_b = -U_{eq}t_b\frac{\ln R_M}{R_M - 1} + U_{eq}t_b - \frac{1}{2}g_o t_b^2 \tag{8.15}$$

This is the altitude at burnout, but the vehicle will have residual kinetic energy so that the maximum attainable altitude (H_{max}) is higher and corresponds to the point where the kinetic energy becomes zero. So we consider the energy balance between H_b and H_{max}.

$$M_b \frac{U_b^2}{2} + M_b g_o H_b = M_b g_o H_{\max} \rightarrow H_{\max} = H_b + \frac{U_b^2}{2g_o} \tag{8.16}$$

Combining Eqs. (8.15) and (8.16), we obtain

$$H_{\max} = \frac{\left(U_{eq} \ln R_M\right)^2}{2g_o} - U_{eq} t_b \left(\frac{R_M}{R_M - 1} \ln R_M - 1 \right) \tag{8.17}$$

Thus, it takes large U_{eq} and small t_b, along with a large mass ratio to achieve maximum altitude.

8.2.3 Staging

One way to maximize the mass ratio is to jettison unnecessary mass, and start over: this is referred to as staging. The initial mass of a rocket consists of three main categories: payload (L), vehicle structure (s) (e.g. engine, tanks), and propellant (p).

$$M_o = M_L + M_s + M_p \tag{8.18}$$

The mass at burnout is simply the initial mass minus the propellant mass.

$$M_b = M_L + M_s \tag{8.19}$$

To denote the payload and structural mass with respect to other masses, two parameters are introduced.

$$\text{Payload ratio} = \lambda = \frac{M_L}{M_o - M_L} = \frac{M_L}{M_p + M_s} \tag{8.20}$$

$$\text{Structural mass coefficient} = \varepsilon = \frac{M_s}{M_p + M_s} = \frac{M_b - M_L}{M_o - M_L} \tag{8.21}$$

Using these parameters, it can be shown that the mass ratio can be rewritten as follows.

$$R_M = \frac{M_o}{M_b} = \frac{1 + \lambda}{\varepsilon + \lambda} \tag{8.23}$$

The following example illustrates the benefit of staging.

Example 8.1 Calculation of H_{max}

For $M_o = 1000$ kg, $M_b = 200$ kg, $U_{eq} = 750$ m/s and $t_b = 60$ s, what would the maximum attainable altitude in a vertical trajectory?

The mass ratio is $R_M = M_o/M_b = 5$, so that using Eq. (8.17),

$$H_{\max} = \frac{(750 \times \ln 5)^2}{2(9.81)} - 750(60)\left(\frac{5}{5-1}\ln 5 - 1\right) = 28\ 808\ \text{m}$$

Example 8.2 H_{max} plotted as a function of U_{eq}, t_b, and R_M

We can also use Eq. (8.17) to plot H_{max} as a function of U_{eq}, t_b and R_M (see MATLAB® code MCE82, at the end of this chapter) (Figure E8.2.1).

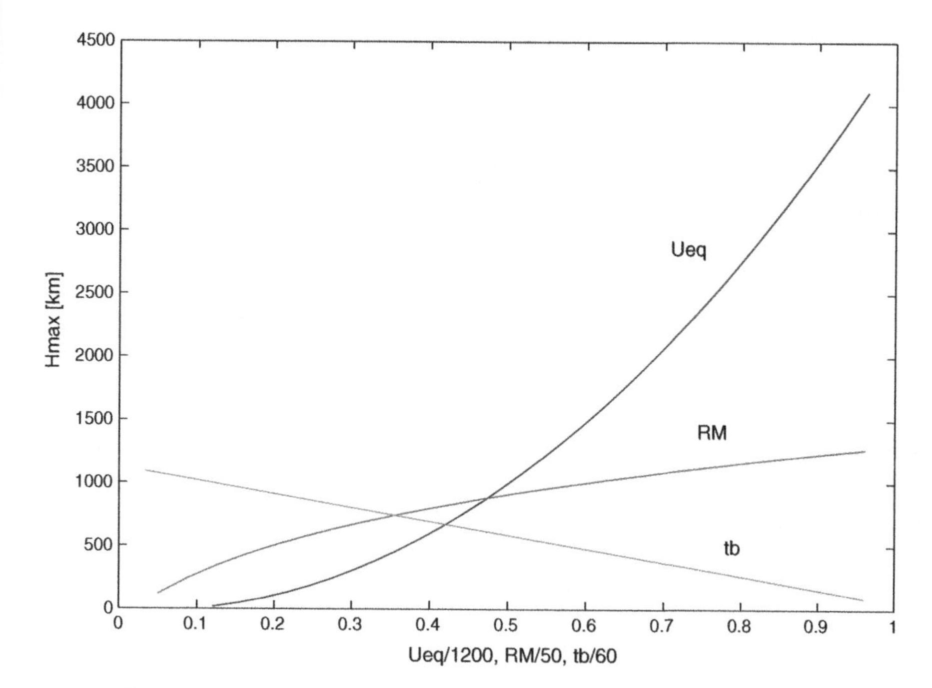

Figure E8.2.1 H_{max} plotted as a function of normalized U_{eq}, t_b and R_M.

Example 8.3 Comparison of single-stage (no staging) and two-stage rockets.

Let us consider the following parameters for rocket propulsion.

$$U_{eq} = 3000\,\text{m/s}$$

$$M_L = 1000\,\text{kg}$$

$$M_o = 15\ 000\ \text{kg}$$

$$M_s = 2000\,\text{kg}$$

For a *single-stage (no staging)* rocket, we would have

$$\varepsilon = \frac{M_s}{M_o - M_L} = 0.143$$

$$\lambda = \frac{M_L}{M_o - M_L} = 0.0714$$

$$R_M = \frac{M_o}{M_s + M_L} = \frac{1+\lambda}{\varepsilon + \lambda} = 5$$

This would produce a vehicle speed of

$$\Delta U = U_{eq} \ln R_M = 4828\,\text{m/s}.$$

For a *two-stage* rocket with similar stages (equal payload ratios and structural coefficients for the stage 1 and 2), we have

$$\lambda_1 = \frac{M_{L1}}{M_{o1} - M_{L1}} = \frac{M_{o2}}{M_o - M_{o2}}$$

$$\lambda_2 = \frac{M_{L2}}{M_{o2} - M_{L2}} = \frac{M_L}{M_{o2} - M_L}$$

We consider the payload of stage 1 as the initial pass of stage 2, since the propellant and structural mass would have been expended at the end of stage 1 operation.

For similar stages with equal payload ratios, we can solve for the unknown M_{o2}.

$$\lambda_1 = \lambda_2 \rightarrow (M_{o2} - M_L)M_{o2} = (M_{o1} - M_{o2})M_L \rightarrow M_{o2} = 3873\,\text{kg}$$

And $\lambda_1 = \lambda_2 = \lambda = 0.348$

We can also equate the structural coefficients.

$$\varepsilon_1 = \varepsilon_2 \rightarrow \frac{M_{s1}}{M_{o1} - M_{o2}} = \frac{M_{s2}}{M_{o2} - M_L}$$

Solving simultaneously with $M_{s1} + M_{s2} = M_s = 2000\,\text{kg}$, we obtain

$$M_{s1} = 1589\,\text{kg},$$

$$M_{s2} = 411\,\text{kg},$$

and $\varepsilon_1 = \varepsilon_2 = \varepsilon = 0.143$

Now we can compute the vehicle speed that would be generated.

$$\Delta U_1 = U_{eq} \ln R_{M1} = U_{eq} \ln \frac{1+\lambda_1}{\varepsilon_1 + \lambda_1} = 3030\,\text{m/s},$$

$$\Delta U_2 = U_{eq} \ln R_{M2} = U_{eq} \ln \frac{1+\lambda_2}{\varepsilon_2 + \lambda_2} = 3030\,\text{m/s},$$

for a total of $\Delta U_{total} = \Delta U_1 + \Delta U_2 = 6060\,\text{m/s}$.

The results of the example above can be generalized for arbitrary number of stages.

$$\Delta U = \sum_{i=1}^{n} \Delta U_i = \sum_{i=1}^{n} \left(U_{eq}\right)_i \ln R_{Mi} \tag{8.24}$$

Let us consider "similar" stages with $(U_{eq})_i = U_{eq} = \text{const.}$, $\varepsilon_i = \varepsilon = \text{const.}$ and $\lambda_i = \lambda = \text{const.}$ Similar stages consist of identical designs with diminishing scales. For such staging,

$$\Delta U = nU_{eq} \ln \frac{1+\lambda}{\varepsilon + \lambda} \tag{8.25}$$

To express this in terms of the overall mass ratio, M_{o1}/M_L, we start by noting that

$$\lambda_i = \frac{M_{o(i+1)}}{M_{oi} - M_{o(i+1)}} \quad \text{or} \quad \frac{M_{oi}}{M_{o(i+1)}} = \frac{1+\lambda_i}{\lambda_i} \tag{8.26}$$

This allows us to write the overall mass ratio in terms of the individual payload ratios.

$$\frac{M_{o1}}{M_L} = \prod_{i=1}^{n} \frac{1+\lambda_i}{\lambda_i} \tag{8.27}$$

For equal payload ratios, this is simply

$$\frac{M_{o1}}{M_L} = \left(\frac{1+\lambda}{\lambda}\right)^n \tag{8.28}$$

Using this result to solve for 1 and inserting into Eq. (8.25), we find

$$\Delta U = nU_{eq} \ln \left\{ \frac{\left(\frac{M_{o1}}{M_L}\right)^{\frac{1}{n}}}{\varepsilon \left[\left(\frac{M_{o1}}{M_L}\right)^{\frac{1}{n}} - 1\right] + 1} \right\} \tag{8.29}$$

We can use Eq. (8.29) to calculate the required mass ratio to achieve rocket acceleration. To achieve the final vehicle speed three times U_{eq} ($\Delta U/U_{eq} = 3$), at least two stages are needed. It turns out there is no value of the overall mass ratio that will achieve this vehicle speed for $n = 1$; that is, staging is a necessity for $\Delta U/U_{eq} = 3$. Higher vehicle speed requires both larger overall mass ratio and number of stages. Also, after $n = 3$, the gain in the overall mass ratio is small, so that staging is preferred with $n = 2$ or $n = 3$ for most rocket mission requirements.

8.2.4 Propulsion and Overall Efficiencies

In Chapter 2, the propulsion and overall efficiencies (η_P and η_o) were defined as

$$\eta_P = \frac{\text{propulsion power}}{\text{engine power output}} \tag{2.19}$$

$$\eta_o = \text{overall efficiency} = \eta_T \eta_P \tag{2.20}$$

Using Eq. (8.7) for thrust, the propulsion efficiency is

$$\eta_P = \frac{FU_o}{FU_o + \frac{\dot{m}_e}{2}(U_e - U_o)^2} \tag{8.30}$$

For the simple case where the pressure thrust is negligible in Eq. (8.7), the propulsion efficiency reduces to

Example 8.4 Staging used in the space shuttle

We can use the data in Table 8.1 to estimate the speed achieved in the space shuttle staging scheme.

$M_{o1} = 2\ 010\ 000\,\text{kg}, M_{b1} = 820\ 000\,\text{kg}, M_{o2} = 660\ 000\,\text{kg}, M_{b2} = 118\ 000\,\text{kg}$, $(I_{sp})_1 = 312\,\text{s}$, and $(I_{sp})_2 = 428\,\text{s}$.

$U_{eq} = g_o I_{sp}$, so that $(\Delta U)_1 = g_o (I_{sp})_1 \ln \frac{M_{o1}}{M_{b1}} = 2740\ \frac{\text{m}}{\text{s}}$ and $(\Delta U)_2 = g_o (I_{sp})_2 \ln \frac{M_{o2}}{M_{b2}} = 7230\ \frac{\text{m}}{\text{s}}$.

The total speed increase is $\Delta U = (\Delta U)_1 + (\Delta U)_2 = 9970\ \frac{\text{m}}{\text{s}}$.

$\frac{\Delta U}{U_{eq}} = 2.8$, which is good enough for boost up to 200 to 300 km orbit.

$$\eta_P = \frac{\dot{m}_e U_e U_o}{\dot{m}_e U_e U_o + \frac{\dot{m}_e}{2}(U_e - U_o)^2} = \frac{2U_e U_o}{U_e^2 + U_o^2} = \frac{2\frac{U_o}{U_e}}{1 + \left(\frac{U_o}{U_e}\right)^2} \tag{8.31}$$

Overall efficiency can also be written.

$$\eta_o = \frac{\dot{m}_e U_e U_o}{\dot{Q} + \frac{\dot{m}_p}{2} U_o^2} \tag{8.32}$$

The denominator is the total rate of energy input, which consists of the heat input rate and rate of kinetic energy input of the propellants. Under steady-state conditions, the propellant mass flow rate is equal to the exit mass flow rate, and we can also write $(KE)_j = \frac{1}{2}\dot{m}_e U_e^2$. Then, Eq. (8.32) becomes

$$\eta_o = \frac{\dot{m}_e r U_e^2}{\dot{Q} + r^2 (KE)_j} = \frac{2r\frac{(KE)_j}{\dot{Q}}}{1 + \frac{(KE)_j}{\dot{Q}} r^2} \tag{8.33}$$

Here, $r = U_o/U_e$. We may also define the internal efficiency, $\eta_i \equiv (KE)_j \dot{Q}$. The internal efficiency is the ratio of the kinetic power of the exhaust jet and heat input, and thus quantifies the ability of the rocket engine to convert thermal energy to kinetic energy useful for thrust production. Then, we have

$$\eta_o = \frac{2r\eta_i}{1 + r^2 \eta_i} \tag{8.34}$$

8.3 MATLAB® Programs

```
% MCE 82: Calculation of Hmax using Equation 8.17.
clear all
%Input parameters
go=9.81;

Umin=750; Umax= 2500; Rmin=5; Rmax=100; tmin=30; tmax =600;
Rm=50;
Ueq=1500;
tb=180;
for i = 1:25;
    U(i) = Umin+ (Umax-Umin)* (i-1)/25;
    R(i) = Rmin+ (Rmax-Rmin)* (i-1)/25;
    t(i) = tmin+ (tmax-tmin)* (i-1)/25;
    HU(i) = (U(i)*log(Rm))^2/(2*go)-U(i)*tb* (Rm*log(Rm)/(Rm-1)-1);
    HR(i) = (Ueq*log(R(i)))^2/(2*go)-Ueq*tb* (R(i)*log(R(i))/(R(i)-1)-1);
    Ht(i) = (Ueq*log(Rm))^2/(2*go)-Ueq*t(i)* (Rm*log(Rm)/(Rm-1)-1);

end

plot(U/Ueq,HU,R/Rm,HR,t/tb,Ht)
```

8.4 Problems

8.1. A rocket engine is required to generate a thrust of 10 MN at sea level with the exit pressure matched to the ambient pressure. The combustion chamber pressure and temperature are 250 atm and 3000 K, respectively. Find the specific impulse, required propellant flow rate, throat and exit diameters. R = 0.320 kJ/kgK, and $\gamma = 1.2$.

8.2. In a rocket nozzle operating in a vacuum, the exit conditions are $U_e = 11\ 264$ ft/s and $p_o = 629.5$ lbf/ft^2, and $\dot{m}_p = 2305$ lbm/s.

a. What is the nominal thrust?
b. If this thrust is augmented by a vectored jet at 30° from the x-direction, at $U_e = 7503$ ft/s and $\dot{m}_p = 305$ lbm/s, what is the resultant thrust in x- and y-directions? The exit pressure stays the same.

8.3. Estimate the nozzle exit area for a rocket booster delivering a sea-level thrust of 125 000 lbf and vacuum thrust of 175 000 lbf.

8.4. Plot rocket thrust as a function of the mass flow rate, exhaust velocity and altitude, for a reasonable range of each of these parameters.
Use $A_e = 4.87$ m^2. $p_e = 5$ atm.

8.5. Derive the expressions for the propulsion efficiency of rockets and turbojets, neglecting the pressure force terms in both.

8.6. Plot the propulsion and overall efficiency of rockets as a function of the velocity ratio. Use $\eta_i = 0.8$.

8.7. Find the thrust for a rocket booster with the following mass flow rate history, and constant $U_{eq} = 1{,}200$ m/s. Pressure effects are negligible.

$$\dot{m}_p = 1500t \quad [\text{kg/s}] \quad \text{for } t = 0 \text{ to } 0.5 \text{ s}$$

$$\dot{m}_p = 750 + 1200t \quad \text{for } t = 0.5 \text{ to } 5 \text{ s}$$

8.8. Plot the required initial mass of a rocket as a function of the payload mass to achieve $\Delta U = 1000$ m/s, for $I_{sp} = 200$, 250, 300 and 350 s.

8.9. Plot the maximum achievable altitude, for a vertical launch as a function of the burn-time, for mass ratio for $R_M = 10$, 100, 500 and 1000.

8.10. Plot the maximum achievable altitude, for a vertical launch as a function of the mass ratio for $I_{sp} = 200$, 250, 300 and 350 s.

8.11. Maximum allowable acceleration during a vertical flight is 5g for a payload of 500 kg in a rocket system. Maximum propellant mass is 1000 kg, while $\varepsilon = 0.1$ and $I_{sp} = 250$ s. What are the minimum burn time and maximum altitude achieved?

8.12. For a rocket consisting of two similar stages ($\varepsilon_1 = \varepsilon_2$; $\lambda_1 = \lambda_2$), find the maximum altitude achieved. Assume $g = g_o$.

$$(I_{sp})_1 = 280 \text{ s}; (I_{sp})_2 = 310 \text{ s}$$

Burn-out time: $(t_b)_1 = 60$ s; $(t_b)_2 = 120$ s
Payload mass = M_L = 1000 kg; total structural mass = M_s = 3000 kg; Total initial mass = $M_o = 25\ 000$ kg.

8.13. Find the configuration and the final velocity change that can be achieved with a rocket consisting of two equivalent stages.

$$I_{eq} = 300\ \text{s}$$

$$M_L = 500\ \text{kg}$$

$$M_o = 12\ 000\ \text{kg}$$

$$M_s = 1800\ \text{kg}$$

8.14. a. Plot the final velocity change as a function of the overall mass ratio, M_{o1}/M_L, for $n = 1, 2, 3, 4$ and 5. $\varepsilon = 0.05$. Use similar stages and $I_{sp} = 300$ s.
b. Plot the final velocity change a function of the overall mass ratio, for $n = 3$ and $\varepsilon = 0.02$, 0.05 and 0.075.

8.15. Plot M_o/M_L vs. n (number of similar stages) for $\Delta U/U_{eq} = 2$, 3 and 6. Use $\varepsilon = 0.1$. Similar stages: $U_{eq} = \text{const.}$, $\lambda = \text{const.}$

$$\Delta U = nU_{eq} \ln \left\{ \frac{\left(\frac{M_{o1}}{M_L}\right)^{\frac{1}{n}}}{\varepsilon\left[\left(\frac{M_{o1}}{M_L}\right)^{\frac{1}{n}} - 1\right] + 1} \right\}$$

8.16. For the following I_{sp} available for stages 1, 2, 3 and 4, write a MATLAB® program to plot and find the minimum M_o/M_L to generate $\Delta U = 7500$ m/s. Use $\varepsilon = 0.08$ for all stages.
Stage 1: $I_{sp} = 315$ s; All other stages: $I_{sp} = 425$ s.

8.17. For the same I_{sp}'s as in Problem 8.11, find the optimum λ that maximizes the ΔU. Use $\varepsilon = 0.08$ for all stages.

8.18. Compare a liquid- (A) and solid-propellant (B) booster systems by calculating the ratio of velocity changes, ΔU. Both have initial mass of 500 000 kg, and payload mass of 10 000 kg. System A has specific impulse of 300 s, while B has $I_{sp} = 210$ s. However, A has structural mass that is 1.33 times that of B, which has structural mass coefficient of $\varepsilon = 0.075$.

Bibliography

Hill, P. and Petersen, C. (1992) *Mechanics and Thermodynamics of Propulsion*, 2nd edn, Addison-Wesley Publishing Company.

9

Rocket Propulsion and Mission Analysis

9.1 Introduction

Rockets are used within and outside the atmosphere. Missiles of various mission lengths are propelled by rockets in atmospheric or trans-atmospheric flights. Exo-atmospheric missions may involve launch to various earth orbits, lunar and planetary missions. Inter-stellar or galactic flights may also be considered, but with current propulsion systems the flight duration becomes excessive.

Let us start from the flight conditions of the earth. The earth atmosphere has been discussed in Chapter 1, and some atmospheric data is included in Appendix A. Here, we will look at the gravitational field. Newton's law of gravitation states that the gravitational force is directed along the line the connects the mass of two bodies, and has the magnitude of

$$F_g = \frac{Gm_1m_2}{r^2} \tag{9.1}$$

$$G = \text{gravitational constant} = 6.673 \times 10^{-11}\frac{\text{m}^3}{\text{kg}\cdot\text{s}^2}$$

For the Earth's mass (m_E), the earth gravitational constant is

$$\begin{aligned}\mu &= \text{Earth's gravitational constant} = Gm_E \\ &= 6.673 \times 10^{-11}\frac{\text{m}^3}{\text{kg}\cdot\text{s}^2} \times 5.794 \times 10^{12}\text{kg} = 3.987 \times 10^{14}\frac{\text{m}^3}{\text{s}^2}\end{aligned} \tag{9.2}$$

This gives the acceleration at the Earth's surface of

$$g_o = \frac{\mu}{r_E^2} = 9.806\frac{\text{m}}{\text{s}^2} \tag{9.3}$$

Aerospace Propulsion, First Edition. T.-W. Lee.
© 2014 John Wiley & Sons, Ltd. Published 2014 by John Wiley & Sons, Ltd.
Companion Website: www.wiley.com/go/aerospaceprop

Thus, an object of mass m at the Earth surface is subject to a gravitational force, or weight, of

$$F = mg_o \tag{9.4}$$

At higher altitude (H), the gravitational acceleration becomes inversely proportional to the square of the distance.

$$g = \frac{\mu}{r^2} = g_o\left(\frac{r_E}{r}\right)^2 = g_o\left(\frac{r_E}{r_E + H}\right)^2 \tag{9.5}$$

The gravity force is associated with the potential energy of any object in the Earth's gravitational field.

$$PE = m\int g dr \approx mg_o(r_E + H) \tag{9.6}$$

Thus, for an object to escape the Earth's gravitational field, a kinetic energy that can overcome this potential energy is required. (Later, we will set this up as a conservation of mechanical energy.)

$$\frac{1}{2}mU_o^2 = m\int g dr \tag{9.7}$$

For Earth surface launch ($H = 0$), the escape velocity is

$$U_{esc} = r_E\sqrt{\frac{2g_o}{r_E}} = \sqrt{\frac{2\mu}{r_E}} = 11,179\frac{\text{m}}{\text{s}} \tag{9.8}$$

For a circular orbit, the centrifugal force is equal to the gravitational force, so that the spacecraft is subject to a zero net force, or maintains a constant altitude.

$$m\frac{U_s^2}{r} = mg = mg_o\left(\frac{r_E}{r_E + H}\right)^2 \rightarrow U_s = r_E\sqrt{\frac{g_o}{r_E + H}} = \text{orbiting speed} \tag{9.9}$$

Example 9.1 Orbiting speed

Equation (9.9) gives the orbiting speed for a $H = 200$ km orbit is $U_s = 7.78$ km/s = 17 514 mph.

Example 9.2 Lunar gravity and escape velocity

Using Eq. (9.8) with the data from Table 9.1, we have

$$U_{esc} = r_E\sqrt{\frac{2g_o}{r_E}} = \sqrt{\frac{2\mu}{r_E}} = 2.375\frac{\text{km}}{\text{s}} = 5344 \text{ mph}$$

Table 9.1 Gravity parameters for the solar system.

	Mass [kg]	Radius [m]	g [m/s^2]	μ [km^2/s^2]	g/g_{Earth}
Earth	5.98×10^{24}	6.38×10^6	9.81	398600.4	1
Sun	1.99×10^{30}	6.96×10^8	274.13	1.327×10^{11}	27.95
Moon	7.36×10^{22}	1.74×10^6	1.62	4902.8	0.17
Mercury	3.18×10^{23}	2.43×10^6	3.59	22032.1	0.37
Venus	4.88×10^{24}	6.06×10^6	8.87	324858.8	0.90
Mars	6.42×10^{23}	3.37×10^6	3.77	42828.3	0.38
Jupiter	1.90×10^{27}	6.99×10^7	25.95	1.267×10^8	2.65
Saturn	5.68×10^{26}	5.85×10^7	11.08	3.794×10^7	1.13
Uranus	8.68×10^{25}	2.33×10^7	10.67	5.792×10^6	1.09
Neptune	1.03×10^{26}	2.21×10^7	14.07	6.873×10^6	1.43
Pluto	1.40×10^{22}	1.50×10^6	0.42	1020.9	0.04

9.2 Trajectory Calculations

For trajectory calculations in a gravitational field, we can start from the vector form of the Newton's second law.

$$m\frac{d^2\vec{r}}{dt^2} = m\vec{g} \rightarrow \frac{d^2\vec{r}}{dt^2} = -\frac{\mu}{r^2}\vec{e}_r \tag{9.10}$$

Equation (9.10) simply states that the gravitation force is directed toward to center of the attracting mass, as denoted by the gravitational acceleration vector or the unit vector in the same direction. Cross-product of two vectors pointing in the same direction is zero, as in Eq. (9.11).

$$\vec{r} \times \frac{d^2\vec{r}}{dt^2} = \vec{r} \times \left(-\frac{\mu}{r^2}\vec{e}_r\right) = 0 \tag{9.11}$$

But, the left-hand side can be rewritten as

$$\vec{r} \times \frac{d^2\vec{r}}{dt^2} = \frac{d}{dt}\left(\vec{r} \times \frac{d\vec{r}}{dt}\right) = \frac{d}{dt}\left(\vec{r} \times \vec{U}\right) = \frac{d\vec{H}}{dt} = 0 \tag{9.12}$$

Here, vector U represents the velocity, $\frac{d\vec{r}}{dt} = \vec{U}$, so that the expression inside the parenthesis is the angular momentum, and Eq. (9.12) is the conservation of angular momentum, H.

$$\vec{H} = const. \quad \text{in a gravitational field} \tag{9.13}$$

If we consider a spacecraft velocity as sketched in Figure 9.1, then the angular momentum is

$$H = rU\cos\gamma = rU_\varphi = r^2\frac{d\varphi}{dt} \tag{9.14}$$

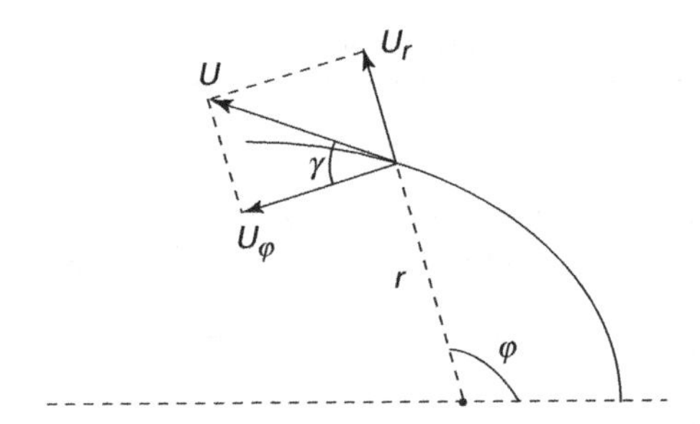

Figure 9.1 Velocity vector in an orbit.

The area swept by the spacecraft trajectory then is

$$\Delta A \approx \frac{1}{2} r^2 \Delta\varphi \tag{9.15}$$

For infinitesimal time interval dt, the above expression becomes exact, and we have

$$\frac{dA}{dt} = \frac{1}{2} r^2 \frac{d\varphi}{dt} = \frac{H}{2} = \text{const.} \tag{9.16}$$

This is the proof of the Kepler's second law, which states that the radius vector of an orbiting body in a gravitational field sweeps equal areas in equal times.

Next, we can write the conservation of mechanical energy by performing dot products of both sides of Eq. (9.10).

$$\frac{d\vec{r}}{dt} \cdot \frac{d^2\vec{r}}{dt^2} = \frac{d\vec{r}}{dt} \cdot \left(-\frac{\mu}{r^2} \vec{e}_r \right) \tag{9.17}$$

$$\frac{1}{2}\frac{d}{dt}\left(\frac{d\vec{r}}{dt} \cdot \frac{d\vec{r}}{dt} \right) = -\frac{\mu}{r^3} \frac{1}{2} \frac{d}{dt} (\vec{r} \cdot \vec{r}) = \frac{d}{dt}\left(\frac{\mu}{r} \right) \tag{9.18}$$

$$\frac{d}{dt}\left(\frac{1}{2} U^2 - \frac{\mu}{r} \right) = 0 \rightarrow \frac{1}{2} U^2 - \frac{\mu}{r} = E = \text{const.} \tag{9.19}$$

In Eq. (9.19), E is the total specific mechanical energy, which is constant.

Now, we can examine the exact shape of the trajectory in a gravitational field. The velocity vector can be decomposed into radial and tangential components.

$$\frac{d\vec{r}}{dt} = \frac{dr}{dt} \vec{e}_r + r \frac{d\varphi}{dt} \vec{e}_\varphi \tag{9.20}$$

We can differentiate once more to find the acceleration in this coordinate frame.

$$\frac{d^2\vec{r}}{dt^2} = \left[\frac{d^2 r}{dt^2} - r \left(\frac{d\varphi}{dt} \right)^2 \right] \vec{e}_r + \left[r \frac{d^2\varphi}{dt^2} + 2 \frac{dr}{dt} \frac{d\varphi}{dt} \right] \vec{e}_\varphi \tag{9.21}$$

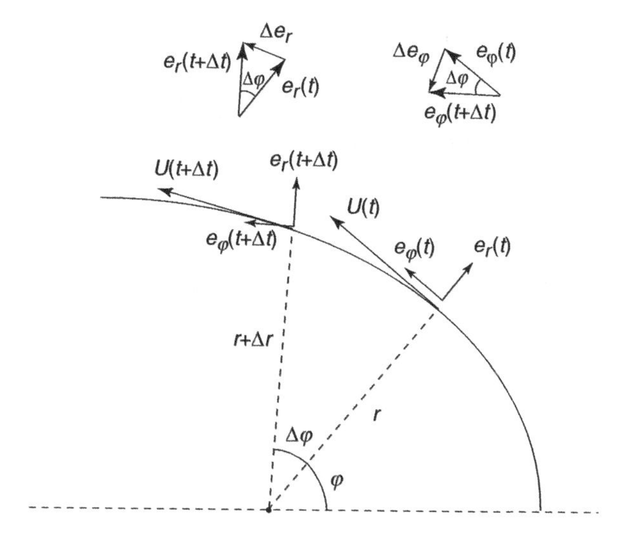

Figure 9.2 Velocity vectors in a radial coordinate frame.

Additional terms for acceleration arise since the unit normal vectors are no longer constant, but change direction as a function of time, as shown in Figure 9.2.

$$\begin{aligned}\frac{d\vec{e}_r}{dt} &= \frac{d\varphi}{dt}\vec{e}_\varphi \\ \frac{d\vec{e}_\varphi}{dt} &= -\frac{d\varphi}{dt}\vec{e}_r\end{aligned} \tag{9.22}$$

Comparison of Eq. (9.20) with (9.10) gives us the dynamical equation for trajectories in a gravitational field.

$$\frac{d^2r}{dt^2} - r\left(\frac{d\varphi}{dt}\right)^2 = -\frac{\mu}{r^2} \tag{9.23}$$

$$r\frac{d^2\varphi}{dt^2} + 2\frac{dr}{dt}\frac{d\varphi}{dt} = 0 \tag{9.24}$$

Zero acceleration in the tangential direction, as denoted in Eq. (9.24), implies conservation of angular momentum, already verified in Eq. (9.12).

$$r\frac{d^2\varphi}{dt^2} + 2\frac{dr}{dt}\frac{d\varphi}{dt} = \frac{d}{dt}\left(r^2\frac{d\varphi}{dt}\right) = 0 \rightarrow r^2\frac{d\varphi}{dt} = r\left(r\frac{d\varphi}{dt}\right) = rU_\varphi = const. \tag{9.25}$$

So we use Eq. (9.23) to find the trajectory, but to obtain an explicit trajectory, $r = r(\varphi)$, we need to make the following substitutions.

$$\frac{d}{dt} = \frac{d\varphi}{dt}\frac{d}{d\varphi} = \frac{H}{r^2}\frac{d}{d\varphi} \tag{9.26}$$

$$r = \frac{1}{u} \tag{9.27}$$

These two changes of variables give us

$$\frac{dr}{dt} = Hu^2 \frac{d}{dt}\left(\frac{1}{u}\right) = -H\frac{du}{d\varphi} \tag{9.28}$$

$$\frac{d^2r}{dt^2} = -Hu^2 \frac{d^2u}{d\varphi^2} \tag{9.29}$$

Substitution of the above two terms in Eq. (9.23) brings us to the trajectory equation.

$$\frac{d^2u}{d\varphi^2} + u = \frac{\mu}{H^2} \tag{9.30}$$

The solution of this second-order ODE is a mathematical exercise, and for now it suffices to write the solution.

$$r = \frac{H^2/\mu}{1 + \left(\sqrt{\frac{2EH^2}{\mu^2}} + 1\right)\cos\theta} \tag{9.31}$$

Here, the new tangential angle is simply, $\theta = \varphi - \omega$, where ω is some reference angle.

Comparison of Eq. (9.31) to so-called conic sections reveals that trajectories under gravitational field follow curves represented by conic sections. Conic sections include circles, ellipses, hyperbolas and parabolas, and have the property as shown in Figure 9.3.

$$\frac{r}{d} = \text{const.} \tag{9.32a}$$

r = distance to F (focus)
d = distance to line *l*.

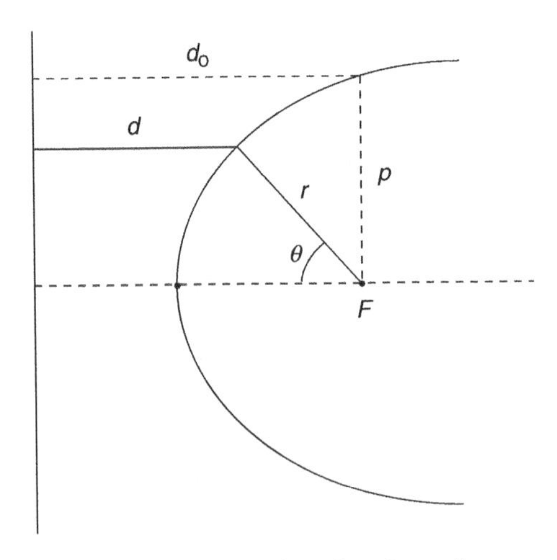

Figure 9.3 Properties of conic sections.

The ratio in Eq. (9.32) is called "eccentricity (e)", and we can use this relationship to find a general equation for conic sections, again with the aid of Figure 9.2.

$$\frac{r}{d} = e = \frac{p}{d_o} \tag{9.32b}$$

$$\frac{p}{e} = \frac{r}{e} + r\cos\theta \rightarrow r = \frac{p}{1 + e\cos\theta} \tag{9.33}$$

Comparison of (9.31) with (9.33) tells us that the trajectory of an object under gravitational field follows conic sections (Kepler's first law), with

$$e = \sqrt{\frac{2EH^2}{\mu^2} + 1} \tag{9.34}$$

$$p = \frac{H^2}{\mu} \tag{9.35}$$

The trajectories have specific forms depending on the eccentricity.

$$e = 0: \quad \text{circle } (r = p) \tag{9.36a}$$

$$0 < e < 1: \quad \text{ellipse} \tag{9.36b}$$

$$e = 1: \quad \text{parabola} \tag{9.36c}$$

$$e > 1: \quad \text{hyperbola} \tag{9.36d}$$

Let us look at each of these trajectories, starting from the ellipse. The circular orbit turns out to be a special case of elliptical ones. As shown in Figure 9.4, an ellipse is defined as a trace of

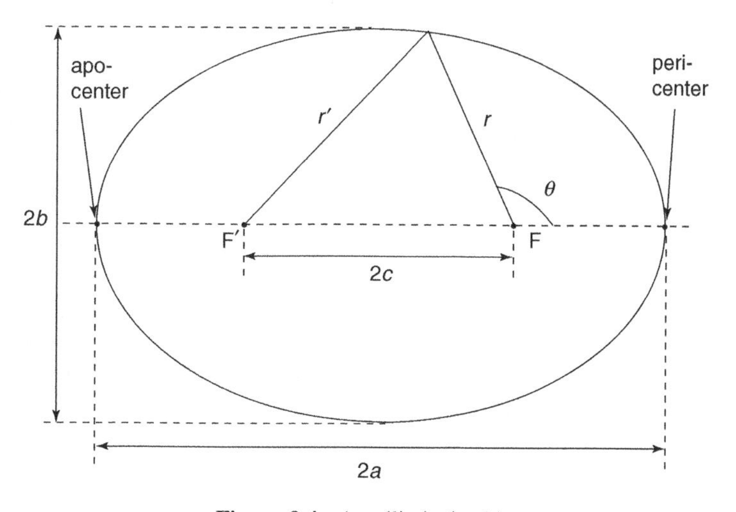

Figure 9.4 An elliptical orbit.

points for which $r + r' = \text{constant}$. The focus of the elliptical orbit, F, is located at the center of the larger (central) mass, and the orbital motion takes place on a fixed plane, the same as the one that the ellipse is on. Also, from our previous observations, the mechanical energy and angular momentum are conserved. Using Eq. (9.33), the minimum distance to the focus (called the pericenter distance), F, occurs at $\theta = 0$.

$$r_p = \frac{p}{1+e} = \text{pericenter distance} \tag{9.37}$$

Similarly, the maximum distance from F occurs at $\theta = \pi$, and is called the apocenter distance.

$$r_a = \frac{p}{1-e} = \text{apocenter distance} \tag{9.38}$$

Then, the semi-major axis, a, is calculated, as shown in Figure 9.4.

$$a = \frac{r_p + r_a}{2} = \frac{p}{1-e^2} = \text{semi-major axis} \tag{9.39}$$

From Eqs. (9.38) and (9.39), the eccentricity, e, is

$$e = \frac{r_a - r_p}{r_a + r_p} = \frac{c}{a} \tag{9.40}$$

When $c = 0$, then the eccentricity is zero in Eq. (9.40), and the two foci, F and F′, merge to become a single center of a circle. This is the condition for the circular orbit ($e = 0$).

The parabolic orbit ($e = 1$) is at the limit of elliptical and hyperbolic orbits, and thus represents a threshold between the two orbits. It corresponds to the limit where the closed orbit (elliptic) becomes an open orbit, so that escape from the gravitational field is possible. However, $u \to 0$, as $r \to \infty$ for parabolic orbits, so the kinetic energy is the minimum needed to escape from the gravitational potential. At $\theta = 0$, $r_p = \frac{p}{2}$ in Eq. (9.33), so that the equation for the parabolic orbit is

$$r = \frac{2r_p}{1 + \cos\theta} \tag{9.41}$$

For $e > 1$, the denominator of Eq. (9.33) changes sign at $e = 1/\cos\theta$, which results in two branches of the hyperbolic orbits. We are only interested in the one near the real focus, F, as shown in Figure 9.5. The turning angle, δ, which is the angle that an object is turned by the central mass at F, can be calculated as

$$\sin\frac{\delta}{2} = \frac{1}{e} \tag{9.42}$$

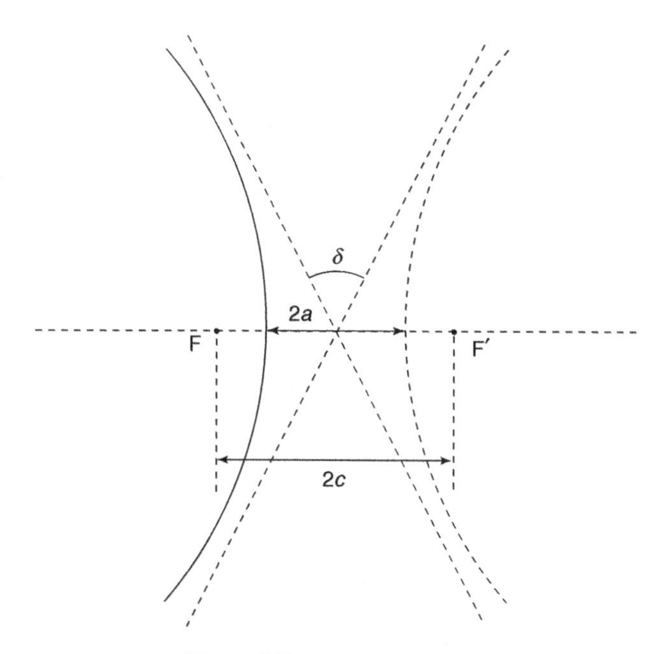

Figure 9.5 A hyperbolic orbit.

From Eqs. (9.34) and (9.36), we can identify the physical parameters that determine the orbital shape.

$$e < 1 \text{ (ellipse)} \rightarrow E < 0, U_o < \sqrt{\frac{2\mu}{r_o}} \tag{9.43a}$$

$$\mathrm{e} = 1 \text{ (parabola)} \rightarrow \mathrm{E} = 0, U_o = \sqrt{\frac{2\mu}{r_o}} \tag{9.43b}$$

$$e > 1 \text{ (hyperbola)} \rightarrow \mathrm{E} > 0, U_o > \sqrt{\frac{2\mu}{r_o}} \tag{9.43c}$$

Equation (9.43b) is the same as Eq. (9.8) obtained earlier, where U_o represents the velocity at $r = r_o$.

The circular orbit occurs when $e = 0$, which from Eq. (9.34) is equivalent to

$$\frac{2EH^2}{\mu^2} + 1 = 0 \tag{9.44}$$

The mechanical energy, E, and the angular momentum can be computed using the parameters in Figure 9.1.

$$\frac{2}{\mu}\left(\frac{U_o^2}{2} - \frac{\mu}{r_o}\right)(r_o U_o \cos\gamma_o)^2 + 1 = 0 \tag{9.45}$$

This condition for circular orbit can be rewritten as

$$\sin^2\gamma_o + \cos^2\gamma_o\left(1 - \frac{r_o U_o^2}{\mu}\right)^2 = 0 \tag{9.46}$$

Since all of the terms are squared in Eq. (9.46), the only way the sum can equal to zero is

$$\gamma_o = 0 \text{ and } U_o = U_{circular} = \sqrt{\frac{\mu}{r_o}} \tag{9.47}$$

These are the conditions for a circular orbit, where the velocity vector is normal to the radius vector ($\gamma_o = 0$).

The time to complete a closed orbit (ellipse or circle) can be found from the areal velocity, from Eq. (9.16).

$$\text{T} = \text{period of the orbit} = (\text{area})/(\text{areal velocity}) = \frac{\pi ab}{\frac{dA}{dt}} = \frac{2\pi a\sqrt{1-e^2}}{H} \tag{9.48}$$

Using $H = \sqrt{\mu a(1-e^2)}$ from Eqs. (9.35) and (9.39),

$$T = 2\pi\sqrt{\frac{a^3}{\mu}} \tag{9.49}$$

For a circular orbit, the speed is constant, as given by Eq. (9.47). What about the orbiting speed in elliptical orbits? We know that the areal velocity, angular momentum and mechanical energy are all constant in an elliptical orbit, from the discussions above. Constant angular momentum means

$$rU = H = \text{const.} \rightarrow r_1 U_1 = r_2 U_2 = H \tag{9.50}$$

$$\text{For example,} \quad r_a U_a = r_p U_p \tag{9.51}$$

Similarly, constant mechanical energy is written as

$$\frac{1}{2}U_1^2 - \frac{\mu}{r_1} = \frac{1}{2}U_2^2 - \frac{\mu}{r_2} = E \tag{9.52}$$

$$\text{Or,} \quad \frac{1}{2}\left(U_1^2 - U_2^2\right) = \mu\left(\frac{1}{r_1} - \frac{1}{r_2}\right) \tag{9.53}$$

From Eq. (9.50), we have $U_1 = H/r_1$ and $U_2 = H/r_2$. Substituting these into Eq. (9.53), and rearranging gives the following result.

$$\frac{H^2}{2} = \frac{\mu}{\frac{1}{r_1} + \frac{1}{r_2}} = \frac{r_1 r_2 \mu}{r_1 + r_2} \tag{9.54}$$

We can also add the mechanical energy at two points, and the sum will be $2E$.

$$\frac{1}{2}U_1^2 - \frac{\mu}{r_1} + \frac{1}{2}U_2^2 - \frac{\mu}{r_2} = 2E \tag{9.55}$$

Again, using $U_1 = H/r_1$ and $U_2 = H/r_2$, we have

$$\frac{H^2}{2}\left(\frac{1}{r_1^2} + \frac{1}{r_1^2}\right) - \mu\left(\frac{1}{r_1} + \frac{1}{r_2}\right) \tag{9.56}$$

Noting that

$$\left(\frac{1}{r_1^2} + \frac{1}{r_1^2}\right) = \left(\frac{1}{r_1} + \frac{1}{r_2}\right)^2 - \frac{2}{r_1 r_2} \tag{9.57}$$

we have, after also using Eq. (9.54) and a little algebra,

$$E = \frac{\mu}{r_1 + r_2} = \frac{\mu}{2a} \tag{9.58}$$

The second part of the equality arises from the fact that $r + r' = \text{constant}$ in an ellipse, and Eq. (9.39).

Thus, we can use Eq. (9.58) to calculate the orbiting speed at any point in an elliptical orbit, simply using the semi-major axis of the ellipse, a.

$$\frac{1}{2}U^2 - \frac{\mu}{r} = E = \frac{\mu}{2a} \tag{9.59}$$

$$U = \sqrt{\frac{2\mu}{r} - \frac{\mu}{a}} \tag{9.60}$$

9.3 Rocket Maneuvers

From the discussion in the previous section, we can see that the orbit is completely determined by the speed of the object in a given gravitational field. Therefore, by using rocket propulsion to alter the spacecraft speed, the trajectories can be controlled. For example, to go from a circular orbit to an elliptical orbit, the following velocity change is required.

$$\Delta U = U_p - U \tag{9.61}$$

U_p = spacecraft velocity at the periapsis of the elliptical orbit
U = spacecraft velocity at any point in the circular orbit

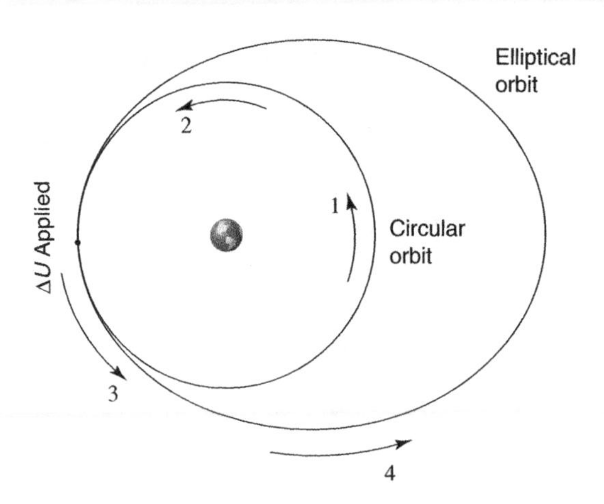

Figure 9.6 Circular to elliptical orbit.

Periapsis is the point in the elliptical orbit corresponding to the pericenter distance (minimum distance to the central body). Circular to elliptical orbital change implies that the starting point of the elliptical orbit will necessarily be at the periapsis, as shown in Figure 9.6. The velocities in Eq. (9.61) can be calculated using Eqs. (9.47) and (9.60).

Example 9.3 Change from circular to elliptical orbit

Let us start from a circular orbit at 300 km altitude. From Eq. (9.47), the orbiting speed at this altitude is

$$U = \sqrt{\frac{\mu}{r}} = \sqrt{\frac{398\ 600.4\dfrac{\text{km}^3}{\text{s}^2}}{(6378.1 + 300)\text{km}}} = 7.73\frac{\text{km}}{\text{s}}$$

In order to transition to a 300 km × 3000 km elliptical orbit, first we need the semi-major axis (Eq. (9.39))

$$a = \frac{r_a + r_p}{2} = \frac{(6378.1 + 3000) + (6278.1 + 300)}{2} = 8028\ \text{km}$$

Thus, if the velocity is increased at periapsis (at some point in the circular orbit) according to Eq. (9.60), then the orbit will be changed to the elliptical orbit of above semi-major axis.

$$U = \sqrt{\frac{2\mu}{r} - \frac{\mu}{a}} = \sqrt{\frac{2(398600.4)}{6678.1} - \frac{398600.4}{8028}} = 8.35\frac{\text{km}}{\text{s}}$$

Example 9.4 Repositioning

The geosynchronous orbit is useful since it maintain a constant position above the Earth at all times. At times, the position of geosynchronous satellites may need to adjusted, and this maneuver is called repositioning. A reposition can be done by sending the spacecraft into a temporary elliptical orbit. First, the geosynchronous orbit for the Earth has the following characteristics.

$$T = 86\ 164.09\ \text{s}$$

$$r = 42\ 164.17\ \text{km} = \text{constant}$$

$$U = 3.0747\ \text{km/s}$$

Let us consider a 2° adjustment in the direction opposite to the satellite motion, that is, a 2° lag from the current motion is desired. The 2° lag corresponds to

$$\Delta T = \frac{2}{360}(86\ \ 164.09) = 478.689\ \text{s}$$

Thus, the temporary elliptical orbit needs to have a period of $T_e = 86\ \ 164.09 + 478.689 = 86\ \ 642.78\ \text{s}$. Using Eq. (9.49), the corresponding semi-major axis is

$$a = \sqrt[3]{\frac{T^2\mu}{4\pi^2}} = 42\ \ 320\ \text{km}$$

$$U = \sqrt{\frac{2\mu}{r} - \frac{\mu}{a}} = \sqrt{\frac{2(398600.4)}{42164} - \frac{398600.4}{42320}} = 3.0803\frac{\text{km}}{\text{s}}$$

Thus, if the velocity is increased according to Eq. (9.60) to $U = 3.0803$ km/s from the circular orbit and the satellite allowed to complete exactly one elliptical orbit, then the satellite will return to the same position lagging 2° from the original orbit. At that point, the satellite needs to be decelerated back to 3.0747 km/s, in order to return to the geosynchronous orbit.

9.3.1 Coplanar Orbit Change

In general, an orbit in a given plane can be changed to any other closed orbits, as shown in Figure 9.7. The resulting circle/ellipse will be a function of the velocity and the gravitational field. Using the cosine law, the velocity required for a coplanar orbit change is

$$\Delta U = U_i^2 - U_f^2 - 2U_iU_f\cos\alpha \tag{9.62}$$

From Eq. (9.62), it is evident that the least amount of energy is required for coplanar orbit change when $\alpha = 0$, that is, when the two orbits are tangent.

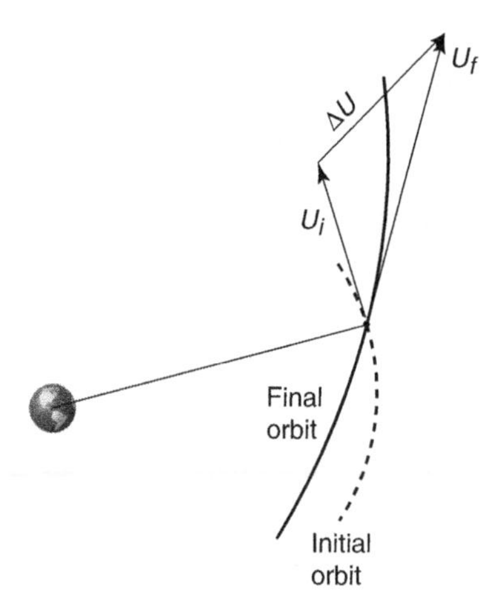

Figure 9.7 Coplanar orbit change.

9.3.2 Hohmann Transfer

Energy-efficient transfer between non-intersection circular orbits can be achieved using the Hohmann transfer, which connects the periapsis to the apoapsis of an elliptical orbit called the transfer ellipse, as shown in Figure 9.8. Apoapsis is the point in the elliptical orbit corresponding to the apocenter distance (maximum distance to the central body). The radii are then related as follows

$$r_{pt} = r_i$$
$$r_{at} = r_f \quad (9.63)$$

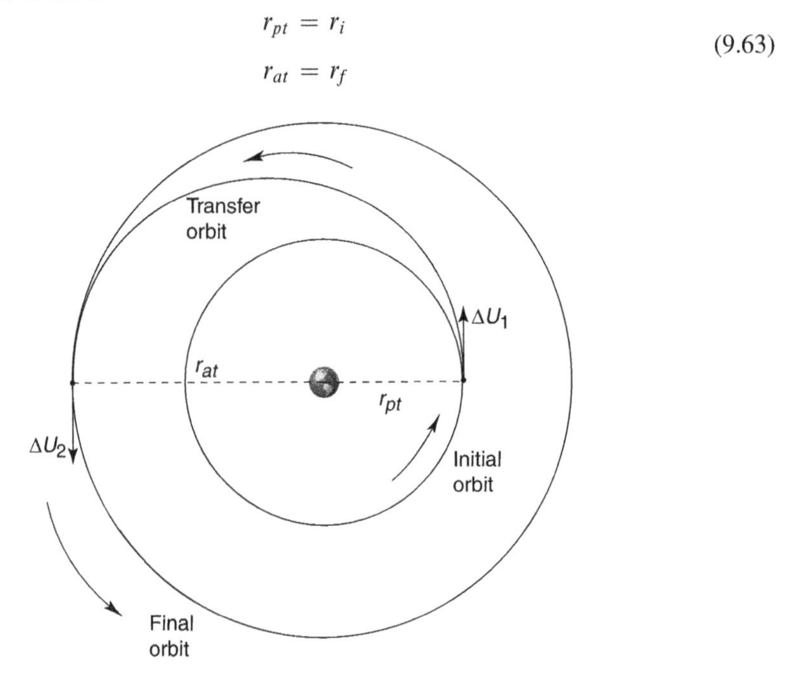

Figure 9.8 Hohmann transfer.

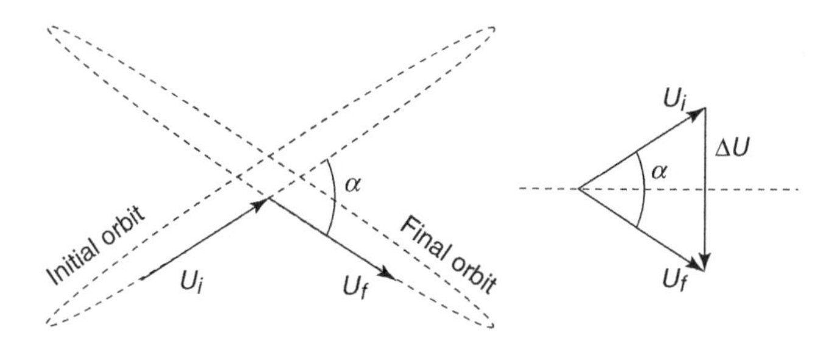

Figure 9.9 Plane change.

Two velocity changes are needed: one to go from the lower-radius circular orbit to the transfer ellipse, and the other to stabilize to the larger-radius circular orbit.

$$\begin{aligned} \Delta U_1 &= U_{pt} - U_i \\ \Delta U_2 &= U_{ap} - U_f \end{aligned} \tag{9.64}$$

The velocities in Eq. (9.64) can be computed using Eqs. (9.47) and (9.60).

9.3.3 Plane Change

Plane change refers to altering the inclination angle of the orbit, while keeping intact the orbital shape itself (i.e. eccentricity, semi-major axis or radius are unaffected). This kind of maneuver may be required to change the satellite orbit from, say, equatorial to inclined orbits to access different parts of the globe. The required velocity in the normal direction of the orbital plane, as shown in Figure 9.9, is computed using the cosine law.

$$\Delta U = 2U_i \sin\frac{\alpha}{2} \tag{9.65}$$

Maneuvers can be optimized by combining multiple orbit changes, for example, plane change and orbit transfer. The reason is clear from Figure 9.10, where the hypotenuse of the triangle is always less than the sum of the two other sides.

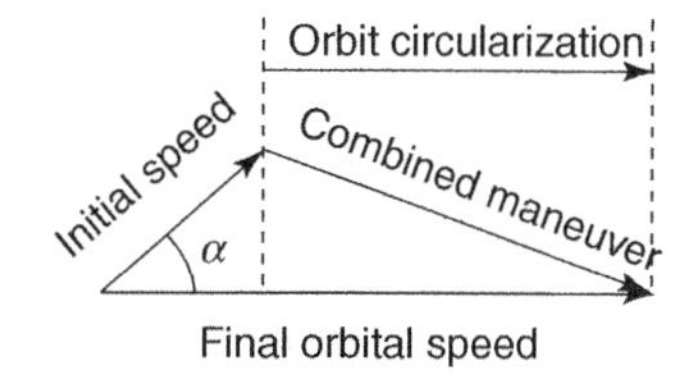

Figure 9.10 Combined maneuvers.

Example 9.5 Hohmann transfer

Let us consider a Hohmann transfer from an 8000 km radius circular orbit to the geosynchronous orbit around the Earth. Initially, the low circular orbit has the speed of

$$U = \sqrt{\frac{\mu}{r}} = \sqrt{\frac{398600.4}{8000}} = 7.059\frac{\text{km}}{\text{s}}$$

The transfer orbit has the semi-major axis of

$$a = \frac{8000 + 42\ \ 164.17}{2} = 25\ \ 082.1\ \text{km}$$

Acceleration at the periapsis to a velocity of 9.152 km/s would send the spacecraft into the transfer orbit. At the apoapsis, the orbiting speed would be

$$U = \sqrt{\frac{2\mu}{r} - \frac{\mu}{a}} = \sqrt{\frac{2(398600.4)}{42164} - \frac{398600.4}{25082.1}} = 3.015\frac{\text{km}}{\text{s}}$$

Then, at that point deceleration to $U = 3.0747$ km/s would stabilize to the geosynchronous orbit.

9.3.4 *Attitude Adjustments*

For satellite or spacecraft attitude control, thrust with intermittent on-off operation is required for rotation in three axes. For convenience, the three axes are referred to as pitch, roll and yaw, although there is no formal way to define these angles as in an aircraft. Figure 9.11 shows typical thruster arrangements for attitude control. From basic dynamics,

$$\theta = \frac{1}{2}\frac{T}{I}t_b^2 = \text{angle of the spacecraft[rad]} \tag{9.66}$$

$$\omega = \frac{T}{I}t_b = \text{angular speed of the spacecraft[rad/s]} \tag{9.67}$$

$$T = nFL = \text{constant torque} \tag{9.68}$$

I = moment of inertia
t_b = burn-time
n = number of thrusters
F = thrust for each thruster
L = moment arm, distance from the center of mass of the spacecraft to the thruster

Equation (9.68) shows that a large moment arm can accomplish the attitude control with a small amount of thrust. The corresponding expenditure in propellant weight is

$$w_p = \frac{nFt_b}{I_{sp}} \tag{9.69}$$

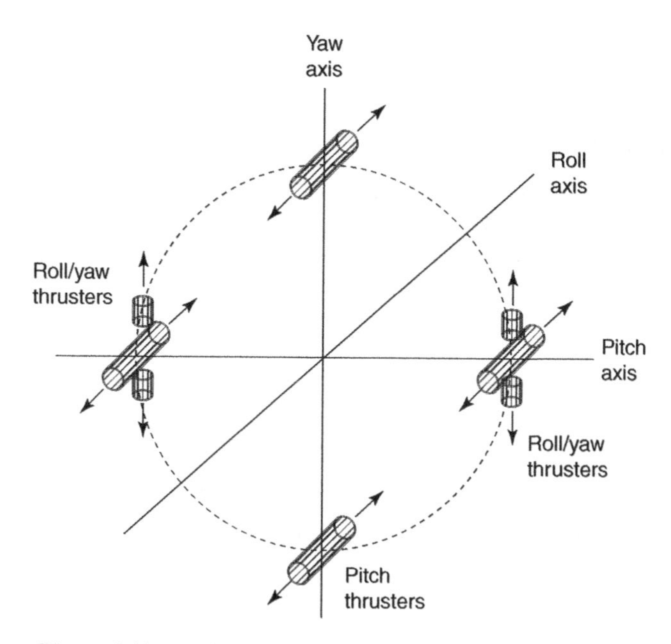

Figure 9.11 Typical thruster arrangement for attitude control.

Equations (9.66) and (9.67) tell us that the angle and the angular speed continue to increase during firing, and for attitude control reverse thrust needs to be applied to stop the rotation. Typically, an attitude control maneuver (one-axis maneuver) consists of acceleration, coasting and braking phase. With no other external forces, the braking torque equals the initial rotation torque to bring the spacecraft to a constant attitude angle. During coasting, the angle increases at a constant angular speed, achieved at the end of firing.

$$\theta = \omega t_c = \frac{nFL}{I} t_b t_c = \text{angle during coasting} \tag{9.70}$$

t_c = coast time

The angle traversed during braking is the same as the angle covered during acceleration, so that the total angular change for the maneuver, θ_m, is

$$\theta_m = \frac{nFL}{I}\left(t_b^2 + t_b t_c\right) \tag{9.71}$$

Total maneuver time is simply the sum of the time of all the phases.

$$t_m = 2t_b + t_c \tag{9.72}$$

9.4 Missile Pursuit Algorithms and Thrust Requirements

Missiles are another common application of rocket thrust. For stationary targets, thrust requirements are only required for the initial boost, after which the missiles will undergo coast, and gravity-driven fall or re-entry phase, outside of small trajectory adjustments. For moving or accelerating targets (executing evasive maneuvers), active control of thrust using various pursuit algorithms is needed.

9.4.1 Velocity Pursuit

The velocity pursuit algorithm is based on the idea that the missile heading should always be on the target. If the missile has a higher speed than the target, then this algorithm will theoretically result in an intercept. From Figure 9.12, the kinematics of the velocity pursuit is as follows:

$$\dot{r} = U_t \cos\phi - U_m \tag{9.73}$$

$$r\dot{\phi} = -U_t \sin\phi \tag{9.74}$$

U_t = target velocity
U_m = missile velocity
ϕ = angle between the target and missile velocity vectors

Integrating Eqs. (9.73) and (9.74) gives, with subscript "o" denoting the initial conditions,

$$r = r_o \frac{(1+\cos\phi_o)^{\frac{U_m}{U_t}}}{(\sin\phi_o)^{\frac{U_m}{U_t}}} \frac{(\sin\phi)^{\frac{U_m}{U_t}-1}}{(1+\cos\phi)^{\frac{U_m}{U_t}}} \tag{9.75}$$

r_o, ϕ_o = initial distance and heading

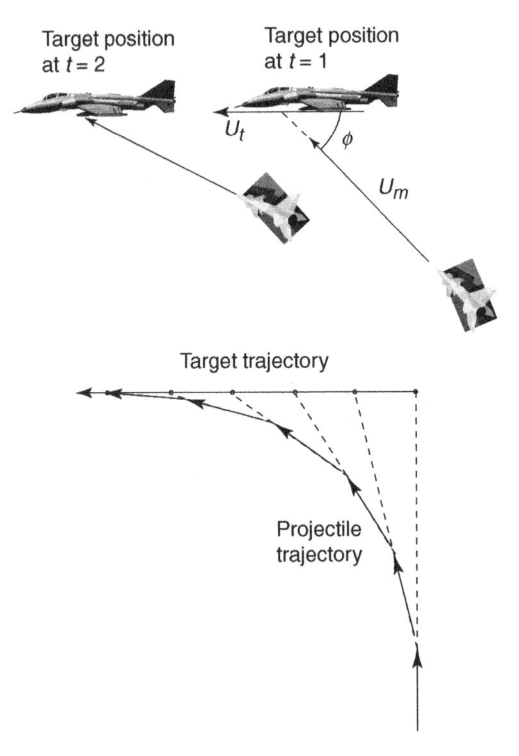

Figure 9.12 Velocity pursuit.

The intercept ($r=0$) occurs at $\phi=0$ or π. Since $\phi=\pi$ (head-on) is an unstable pursuit mode, intercept is sought at $\phi=0$ (tail hit). From Eq. (9.74), the required lateral speed becomes high at small r, which means the missile needs to have a very large lateral acceleration near the intercept.

9.4.2 *Proportional Navigation*

Proportional navigation maintains a nearly constant bearing to the target, in order to achieve intercept, as shown in Figure 9.13. In the ideal case of constant target speed, a constant heading required for the missile can be computed. The target's heading and speed, as well as the missile's own heading as a function of time, are required data to perform proportional navigation. This method can be implemented through

$$\dot{\gamma} = c\dot{\phi} \tag{9.76}$$

Here, γ is the direction of the missile's velocity vector and ϕ the bearing of the target, which are measured with respect to some reference angle, and c an adjustable constant. From Figure 9.13, the dynamical equations can be developed.

$$r\dot{\phi} = U_m \sin(\phi - \gamma) - U_t \sin\phi \tag{9.77}$$

$$\dot{r} = -U_m \cos(\phi - \gamma) + U_t \cos\phi \tag{9.78}$$

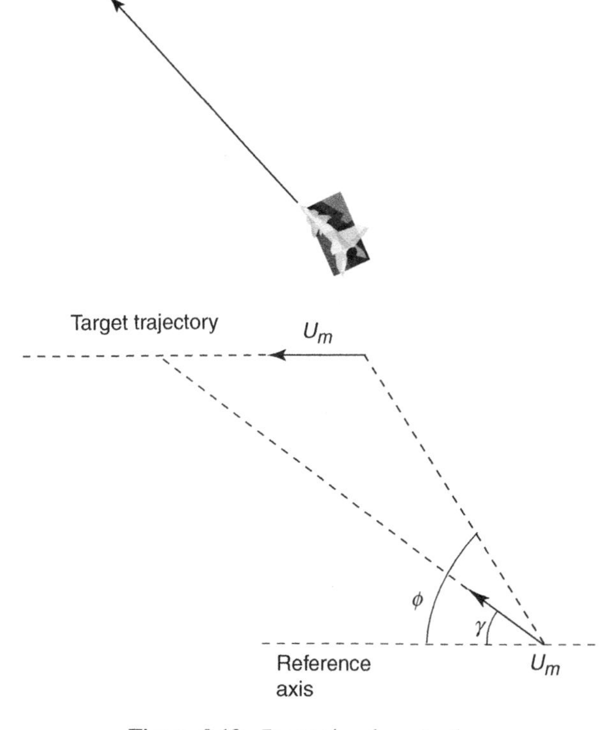

Figure 9.13 Proportional navigation.

Equations (Equations (9.76)–(9.78) can be numerically integrated to simulate the intercept scenarios. For analytical purposes, we can also examine small deviations (Δ) of the angles, from the ideal collision triangle, denoted by subscript "o".

$$\phi = \phi_o + \Delta\phi \tag{9.79}$$

$$\gamma = \gamma_o + \Delta\gamma \tag{9.80}$$

Substitutions of Equations (9.79) and (9.80) in (9.77 and (9.78)), and further analysis yields the following requirement for the required missile acceleration (α_m) relative to the target acceleration (α_t), both in the lateral direction

$$\left|\frac{\alpha_m}{\alpha_t}\right| = \frac{c_{PN}}{c_{PN} - 1} \tag{9.81}$$

$$c_{PN} = \frac{cU_m \cos\phi_o}{r} \tag{9.82}$$

9.4.3 Command-to-Line-of-Sight (CLOS)

CLOS and the so-called beam-riding guidance use the concept that the missile should follow the straight line between the launcher and the target at all times, as shown in Figure 9.14. From Figure 9.14, the following dynamical equations can be written.

$$r_m\dot{\phi}_m = -U_m \sin(\phi_m - \gamma_m) \tag{9.83}$$

$$\dot{r}_m = U_m \cos(\phi_m - \gamma_m) \tag{9.84}$$

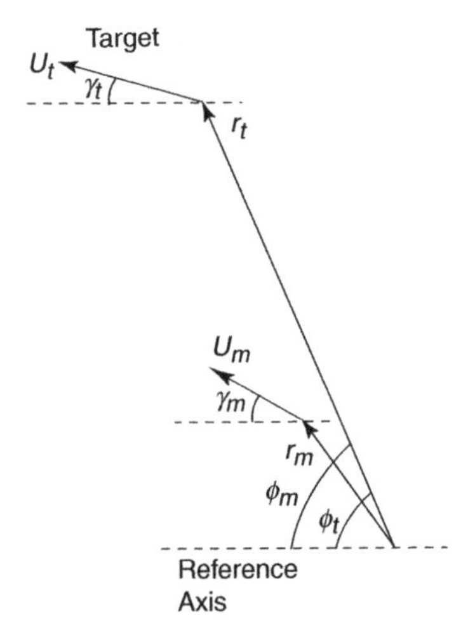

Figure 9.14 Command-to-line-of-sight guidance.

Taking the time derivatives and combining these equations yields the following missile equation.

$$\dot{\gamma}_m = \left(2 - \frac{\dot{U}_m}{U_m}\frac{r_m}{\dot{r}_m}\right)\dot{\phi}_t + \frac{r_m}{\dot{r}_m}\left[\frac{\dot{r}_t}{r_t}\dot{\gamma}_t - 2\dot{\phi}_t\left(\frac{\dot{r}_t}{r_t} - \frac{\dot{U}_t}{U_t}\right)\right] \tag{9.85}$$

The first term represents the missile acceleration required to keep up with the rotation of the line-of-sight, while the second term involves the requirement to adapt to the target maneuver (e.g. evasive tactics).

9.5 Problems

9.1. Plot the gravitational acceleration of the Earth, Venus, Mars, the Moon and Jupiter, as a function of the distance from the surface.

9.2. Calculate (a) orbiting speed at 100 km altitude; and (b) the escape velocity from the Earth, the Sun, the Moon, Venus, Mars, and Jupiter.

9.3. State and provide proofs of Kepler's first and second laws, concerning the motion of planets in a gravitational field.

9.4. Write a MATLAB® program for plotting the traces of an ellipse, with arbitrary semi-major and eccentricity, a and e.

9.5. Write a MATLAB® program for plotting the traces of parabolic and hyperbolic orbits.

9.6. Show that the condition for a circular orbit is $\sin^2\gamma_o + \cos^2\gamma_o\left(1 - \frac{r_o U_o^2}{\mu}\right)^2 = 0$, as given by Eq. (9.46).

9.7. Show that the period for a closed orbit is $T = 2\pi\sqrt{\frac{a^3}{\mu}}$.

9.8. Prove that the orbiting speed at an arbitrary point in an elliptical orbit is $U = \sqrt{\frac{2\mu}{r} - \frac{\mu}{a}}$. Plot U as a function of r, for orbits around the Earth, the Moon, Mars and the Sun.

9.9. For an orbit booster with an $I_{sp} = 300$ s and initial mass of 350 kg, calculate the propellant mass needed to move the spacecraft from a 350 km circular orbit to a 350 km × 3500 km elliptical orbit.

9.10. Estimate the launch speed needed to reach the geosynchronous Earth orbit (GEO) from the Earth's surface. Note that the launch speed is in no way equal to the orbital speed, and you need to consider all the energy changes to go from the Earth's surface to GEO.

9.11. Find the energy required per unit mass to send a payload from the Earth's surface to a circular orbit of 500 km altitude.

9.12. Find the radius of the GEO in km. GEO has the orbital period, which is the same as the time for one Earth rotation (24 hours).

9.13. Find the velocity requirements to go from a 160 km altitude circular orbit, through a Hohmann transfer to a 42 200 km altitude circular orbit.

9.14. Consider the Hohmann transfer to send a probe from Earth to Mars orbits, both of which may be assumed as circular orbits of 1 AU and 1.524 AU, respectively, where 1 AU = 149 597 871 km.

a. What is the semi-major axis and eccentricity of the transfer orbit?
b. What is the speed of the probe needed to reach Mars from the Earth orbit?
c. What is the time of flight for the probe to reach Mars?
d. What is the relative angle between the Earth and Mars, that the probe should be launched at?

9.15. Find the total velocity change needed to send a satellite from an orbit of 552 km altitude to GEO, using Hohmann transfer.

9.16. Calculate the velocity change requirements to provide a 5° lag in a geosynchronous orbit around the Earth.

9.17. Calculate the velocity change requirements for a Hohmann transfer from a 12 000 km radius circular orbit to the geosynchronous orbit around the Earth, and then back to 12 000 km circular orbit.

9.18. a. Plot the trajectories in the Sun's gravitational field, using the equation of motion below, for all of the planets in the solar system. Do not include the inclination angle in the plot, that is, plot in the same plane. Look up some data on the planets' orbits, for example, pericenter distance, eccentricity, time to orbit, and so on, to calculate E and H. Some of them can be found on http://solarsystem.nasa.gov/planets/charchart.cfm

b. Tabulate E, H, semi-major axis, semi-minor axis, eccentricity, inclination angle, pericenter distance, apocenter distance, time to orbit, aerial velocity and other orbit parameters for all the planets.

$$r = \frac{H^2/\mu}{1 + \left(\sqrt{\frac{2EH^2}{\mu^2} + 1}\right)\cos\theta}$$

c. Add (a) separate plot(s) of some of the other possible trajectories around the solar system, including parabolic and hyperbolic ones.

9.19. Calculate $\pm\Delta U$ required for various incremental orbital changes, starting from 200 km circular orbit to orbit distances around the Earth, using Hohmann transfer. Include at least five examples.

9.20. Plot the trajectories of various co-planar orbit changes around the Earth, starting from 200 km circular orbit. Use increments of $\Delta U = 0.1 U_{esc}$ and $\Delta\alpha = \pi/36$. Include at least six combinations of ΔU and $\Delta\alpha$, and at least one orbit change to hyperbolic.

9.21. Plot the trajectory of a missile with constant speed of $U_m = 500$ m/s, following a velocity pursuit toward a target at $U_t = 415$ m/s. $r_o = 10$ km and $\phi_o = 30°$.

9.22. a. Use MATLAB® to numerically integrate the following missile equations, and plot the target and missile trajectories for $U_m = 3U_t$. The target velocity is horizontal at a constant $U_t = 427$ m/s. $r_o = 5$ km $\phi_o = \pi/3$.

$\dot{r} = U_t \cos\phi - U_m$
$r\dot{\phi} = -U_t \sin\phi$
U_t = target velocity
U_m = missile velocity
ϕ = angle between the target and missile velocity vectors

b. Compare with

$$r = r_o \frac{(1+\cos\phi_o)^{\frac{U_m}{U_t}}}{(\sin\phi_o)^{\frac{U_m}{U_t}}} \frac{(\sin\phi)^{\frac{U_m}{U_t}-1}}{(1+\cos\phi)^{\frac{U_m}{U_t}}}$$

r_o, ϕ_o = initial distance and heading

c. Plot the lateral velocity and acceleration as a function of r.

9.23. a. Verify that for proportional navigation, the following missile equations hold.

$$r\dot{\phi} = U_m \sin(\phi - \gamma) - U_t \sin\phi$$

$$\dot{r} = -U_m \cos(\phi - \gamma) + U_t \cos\phi$$

b. Numerically integrate and plot trajectories, lateral velocities and accelerations for some reasonable initial configurations and parameters.

9.24. Verify that for CLOS pursuit, the following equations hold.

$$r_m\dot{\phi}_m = -U_m \sin(\phi_m - \gamma_m)$$

$$\dot{r}_m = U_m \cos(\phi_m - \gamma_m)$$

Bibliography

Brown, C.D. (1996) *Spacecraft Propulsion, AIAA Education Series.*

Cornelisse, J.W., Schoyer, H.F.R., and Wakker, K.F. (1979) *Rocket Propulsion and Spaceflight Dynamics*, Pitman.

10

Chemical Rockets

10.1 Rocket Thrust

Chemical rockets can be classified into liquid-, solid- and hybrid-propellant rockets (using a combination of liquid and solid propellants). The thrust chamber in chemical rockets consists of the combustion chamber and the nozzle. The combustion chamber is where the propellant is either injected (liquid propellants) or stored (solid propellants) and burned. The enthalpy generated through the combustion process is converted to the kinetic energy through the nozzle. We have seen in Chapter 8 that the thrust of a rocket is given by

$$F = \dot{m}_p U_e + (p_e - p_o)A_e \equiv \dot{m}_p U_{eq} \tag{8.7}$$

Maximum exhaust velocity is achieved when $p_e = p_o$ at the nozzle exit. However, in most instances this is not the case, and the ambient pressure will affect the rocket thrust (it will increase with decreasing ambient pressure at higher altitudes).

10.1.1 Ideal Rocket Thrust

Let us start by considering an ideal rocket process as shown in Figure 10.1, where the nozzle expansion is isentropic. First, the combustion of the propellants will raise the stagnation enthalpy in the combustion chamber by the heat of combustion, q_r.

$$h_1^o + q_r = h_2^o \tag{10.1}$$

This is the enthalpy available for conversion to kinetic energy in the nozzle.

$$h_2^o = h_e^o = h_e + \frac{U_e^2}{2} \tag{10.2}$$

$$U_e \approx \sqrt{2c_p(T_2^o - T_e)} = \sqrt{2c_p T_2^o \left[1 - \left(\frac{p_e}{p_2^o}\right)^{\frac{\gamma-1}{\gamma}}\right]} \tag{10.3}$$

Aerospace Propulsion, First Edition. T.-W. Lee.
© 2014 John Wiley & Sons, Ltd. Published 2014 by John Wiley & Sons, Ltd.
Companion Website: www.wiley.com/go/aerospaceprop

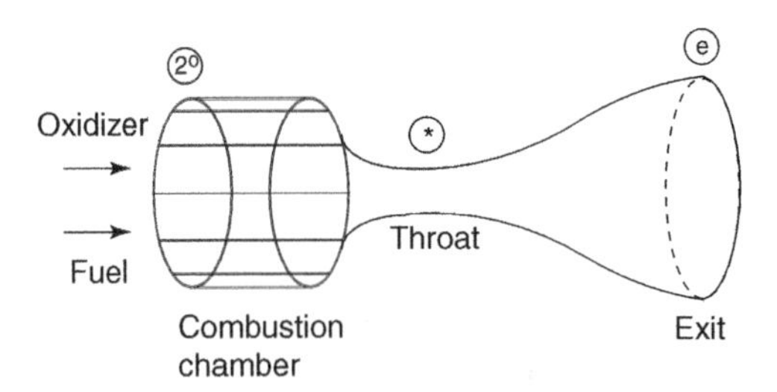

Figure 10.1 Ideal rocket process.

Using $c_p = \frac{\gamma R}{\gamma-1} = \frac{\gamma \bar{R}}{(\gamma-1)\bar{M}}$, where $\bar{R}$ and $\bar{M}$ are the universal gas constant and the molecular weight of the mixture in the combustion chamber, respectively, we have

$$U_e = \sqrt{\frac{2\gamma\bar{R}\left(T_1^o + \frac{q_r}{c_p}\right)}{(\gamma-1)\bar{M}}\left[1 - \left(\frac{p_e}{p_2^o}\right)^{\frac{\gamma-1}{\gamma}}\right]} \tag{10.4}$$

From Eqs. (10.3) and (10.4), we can see how various thermal parameters affect the exhaust velocity. High specific heat, c_p, large heat of combustion, q_r, and large combustion chamber pressure, p_2^o, increase the exhaust velocity, while small molecular weight and exhaust pressure, c_p, also increase it. Eq. (10.4) also shows that the theoretical maximum exhaust velocity occurs when $p_e = 0$ (which is, of course, impossible since the nozzle length would have to be infinite), where $U_{e,\max} = \sqrt{\frac{2\gamma\bar{R}T_2^o}{(\gamma-1)\bar{M}}}$.

10.1.2 Thrust Coefficient and Characteristic Velocity

Now we can examine the effectiveness of the actual combustion chamber and the nozzle. The thrust terms can be separated as follows.

$$F = \dot{m}_e c^* C_F \tag{10.5}$$

Example 10.1 Plot $U_e/U_{e,max}$ as a function of the p_e/p_2^o for $\gamma = 1.15, 1.25,$ and 1.35 for molecular weight of 10.0 kg/kmole.

From Eq. (10.4),

$$\frac{U_e}{U_{e,\max}} = \sqrt{1 - \left(\frac{p_e}{p_2^o}\right)^{\frac{\gamma-1}{\gamma}}}$$

The plot is shown in Figure E10.1 (MATLAB® program: MCE101). We can see that for $p_e/p_2^o > 10$, the gain in the velocity ratio, $U_e/U_{e,max}$, is diminishingly small.

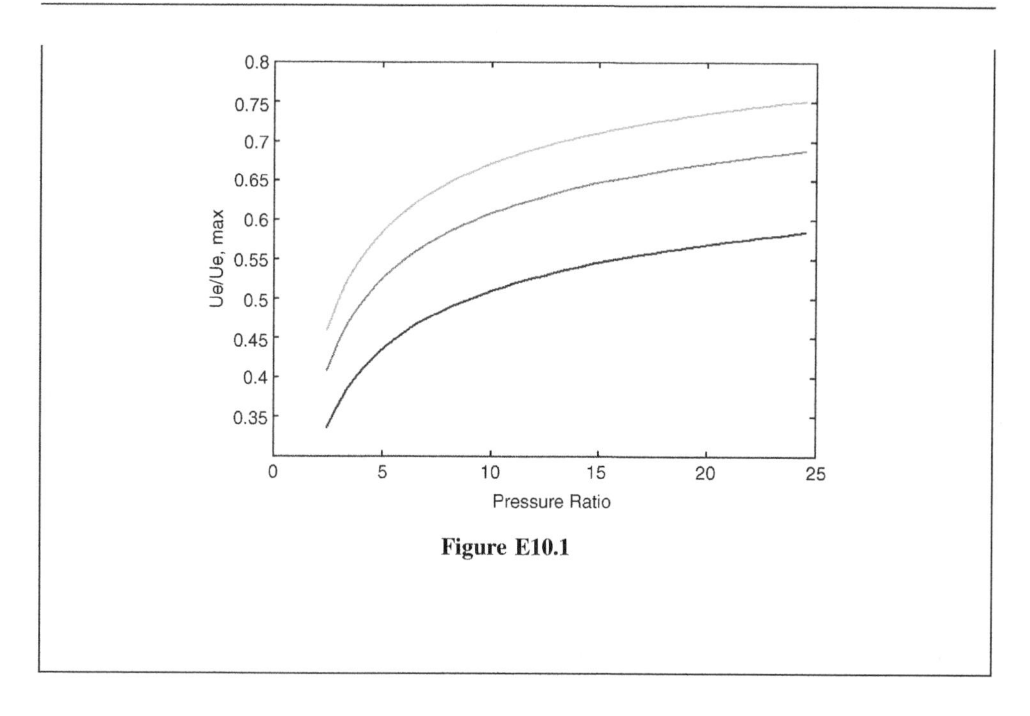

Figure E10.1

$\dot{m}_e = \dot{m}_p$ = exit mass flow rate
c^* = characteristic velocity (combustion chamber effectiveness)
C_F = thrust coefficient (nozzle effectiveness)

The characteristic velocity is defined as the force developed at the throat area ($A*$) by the combustion chamber stagnation pressure, divided by the propellant mass flow rate (equal to the exit mass flow rate). The characteristic velocity can simply be evaluated by the expression for the mass flow rate under choked conditions.

$$\dot{m}_e = \frac{A^* p_2^o}{\sqrt{RT_2^o}} \sqrt{\gamma \left(\frac{2}{\gamma + 1}\right)^{\frac{\gamma+1}{\gamma-1}}} \tag{10.6}$$

A simple rearrangement gives

$$c* = \frac{A^* p_2^o}{\dot{m}_e} = \sqrt{\frac{RT_2^o}{\gamma} \left(\frac{\gamma + 1}{2}\right)^{\frac{\gamma+1}{\gamma-1}}} \tag{10.7}$$

Since the characteristic velocity represents the stagnation pressure and temperature developed in the combustion chamber, normalized by the propellant mass flow rate, it is a measure of effectiveness of the combustion chamber or the injection/combustion processes. In practice, the combustion chamber stagnation pressure is measured as a function of the propellant mass flow rate, and compared with the theoretical value on the right-hand side of Eq. (10.7), to rate the actual to theoretical performance.

Thrust coefficient is defined as the thrust divided by the force exerted on the throat area ($A*$) by the combustion chamber stagnation pressure. Using Eqs. (10.3) and (10.6), we can write

$$C_F = \frac{F}{A^* p_2^o} = \sqrt{\frac{2\gamma}{\gamma - 1}\left(\frac{2}{\gamma + 1}\right)^{\frac{\gamma+1}{\gamma-1}}\left[1 - \left(\frac{p_e}{p_2^o}\right)^{\frac{\gamma-1}{\gamma}}\right]} + \left(\frac{p_e}{p_2^o} - \frac{p_o}{p_2^o}\right)\frac{A_e}{A^*} \tag{10.8}$$

Since the denominator contains the stagnation pressure available for subsequent nozzle expansion process, the thrust coefficient is a measure of effectiveness of the nozzle expansion. In practice, the rocket thrust or the exhaust velocity is measured as a function of the combustion stagnation pressure, and compared with the theoretical value on the right-hand side of Eq. (10.8), to rate the actual to theoretical performance. Taking Eqs. (10.7) and (10.8) together, we have

$$F = \dot{m}_e \frac{A^* p_2^o}{\dot{m}_e} \frac{F}{A^* p_2^o} = \dot{m}_e c^* C_F \tag{10.9}$$

Comparing with Eq. (8.7) ($F = \dot{m}_p U_{eq}$), it is evident that the equivalent exhaust velocity $U_{eq} = c{*}C_F$.

10.2 Liquid Propellant Rocket Engines

Liquid propellant rocket engines mainly consist of the thrust chamber (the combustion chamber and the nozzle) and propellant pumps. Other components are also necessary to operate and control the engine, such as the propellant control and possibly a thrust-vector system. The main component is the thrust chamber, where the propellants are injected, mixed, burned and the expanded down the nozzle. Due to the high thrust chamber pressure, propellant pumps may be driven by turbines, which in turn may be powered by pre-burners (small combustors), heat or gas exchanger.

The space shuttle main engine (SSME), although decommissioned as of 2010, is a good example of liquid-propellant rocket engine, and is shown in Figure 10.2. The SSME uses liquid hydrogen and liquid oxygen (LOX), as the fuel and oxidizer, although hydrogen is actually burned in the gas phase in the main combustion chamber (MCC). The maximum power, at 109% of the "rated power level (RPL)", is 512 900 lbf per engine in vacuum and 418 000 lbf at sea level. Note that at higher altitudes the pressure force contributes to the thrust. These thrust levels translate to specific impulses of 452 and 366 seconds, again in vacuum and at sea level respectively. The thrust can be throttled at 4700 lbf increments, from the minimum power level (MPL) of 316 100 lbf (67% of RPL), through 470 800 lbf (RPL), up to 512 900 lbf, the (full power level (FPL). This throttling is accomplished by metering the flow of propellants (LOX, liquid H_2) into the pre-burner. The mixture ratio going into the pre-burner is extremely fuel-rich, so that the combustion in the pre-burner generates sufficient energy to power the turbine, but also converts the liquid hydrogen to gas-phase hydrogen. The pre-burner power output determines the turbine and propellant pump power, so that the propellant mass flow rate into the MCC can be controlled in this sequence. The operation of the pre-burner thus drives high-pressure propellant feed, but

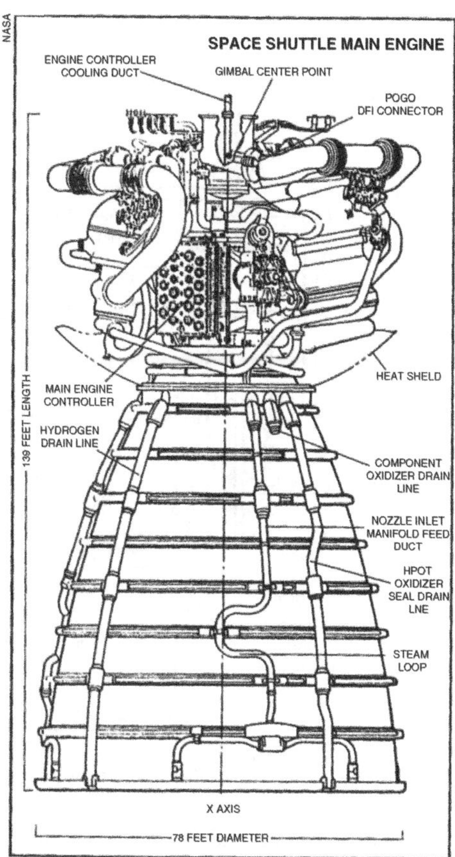

Figure 10.2 Space shuttle main engine. Courtesy of NASA.

also allows for burning of the fuel hydrogen in gas phase (the oxygen is injected as liquid) at the mixture ratio of 6.032 : 1 (oxygen: hydrogen). The flow diagram for the SSME is shown in Figure 10.3.

As shown in Figure 10.3, the oxygen flow from the tank has three branches, where the main branch carries about 89% of the oxygen directly into the MCC through the injector head. A small portion of the flow is sent to the heat exchanger coil, where the vaporized oxygen is looped back to pressurize the tank. The third branch takes about 11% of the oxygen into two pre-burners, one each for the fuel and oxidizer staged-combustion turbine-pump assemblies. In contrast, a large portion (76%) of the hydrogen flow is sent into the pre-burners. The exhaust from the pre-burner is joined in the fuel feedline by the remaining hydrogen that has been circulated in the hot-gas manifold for cooling.

The MCC operating pressure is 2994 psia. The thrust chamber dimensions are 168 inches in length, and 96 inches in diameter, with the nozzle expansion ratio of 69 : 1 with respect to the throat area. Each thrust chamber weighs 7775 lbs. In addition to the MCC, there are the orbital maneuver system (OMS) and the reaction control system (RCS) that are used for thrust in the space shuttle. As the name suggest, the OMS provides boost for orbital maneuvers, while the

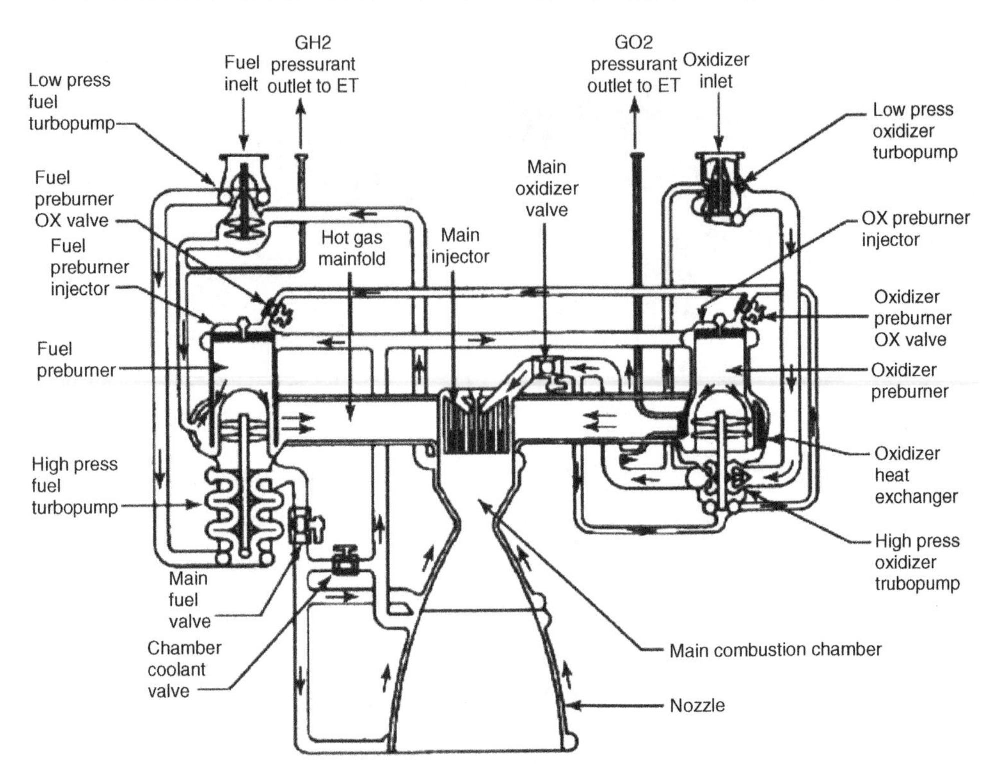

Figure 10.3 The flow diagram for SSME. Courtesy of NASA.

RCS is used for attitude control (yaw, pitch and roll). These systems use different combinations of propellants as discussed in the next section (see Table 10.2). Some of the main specifications of the space shuttle OMS are shown as an example in Table 10.1.

10.2.1 Liquid Propellants and Their Chemistry

Table 10.2 lists bi-propellant combinations in some of the US launch vehicles. Some properties of the propellants, along with those of the oxidizer-fuel combinations, are shown in Table 10.3 and 10.4, respectively.

Table 10.1 Specifications of the space shuttle OMS.

Function	Orbit insertion/exit, circularization
Mass/length/diameter	118 kg (dry)/1.96 m/1.17 m (max.)
Mounting/thrust vector control	Gimbaled for ±7° yaw, and ±6° pitch
Oxidizer/fuel/mixture ratio	Nitrogen tetroxide/MMH/1.65
Thrust/specific impulse	26.7 kN (vacuum)/316 s (vacuum)
Nozzle expansion ratio	55 : 1
Combustion chamber pressure	8.62 atm
Burn time	500 starts/15 hour/100 missions; 1250 s (longest firing), 150–250 s (de-orbit firing)

Table 10.2 Propellants used in some US launch vehicles.

Rocket	Stage	Engines (number deployed)	Propellant	Specific impulse [s] (sea level/vacuum)
Atlas/Centaur	0	Rocketdyne YLR89-NA7 (2)	LOX/RP-1	259/292
	1	Rocketdyne YLR105-NA7	LOX/RP-1	220/309
	2	P&W RL-10A-3-3 (2)	LOX/LH2	444 (vacuum)
Titan II	1	Aerojet LR-87-AJ-5 (2)	NTO/Aerozine 50	259/285
	2	Aerojet LR-91-AJ-5	NTO/Aerozine 50	312 (vacuum)
Saturn V	1	Rocketdyne F-1 (5)	LOX/RP-1	265/304
	2	Rocketdyne J-2 (5)	LOX/LH2	424 (vacuum)
	3	Rocketdyne J-2	LOX/LH2	424 (vacuum)
Space Shuttle	0	Thiokol SRB (2)	PBAN (Solid)	242/268
	1	Rocketdyne SSME (3)	LOX/LH2	363/453
	OMS	Aerojet OMS (2)	NTO/MMH	313 (vacuum)
	RCS	Kaiser Marquardt R-40 and R-1E	NTO/MMH	280 (vacuum)
Delta II	0	Castor 4A (9)	HTPB (Solid)	238/266
	1	Rocketdyne RS-27	LOX/RP-1	264/295
	2	Aerojet AJ10-118K	NTO/Aerozine 50	320 (vacuum)

Some of the representative properties of propellants, such as the specific impulse and characteristic velocity, are calculated at some fixed reference chamber pressures and nozzle expansion ratios, for comparison purposes. Therefore, actual performance can vary, depending on the rocket motor designs.

Some liquid propellants such as oxygen, hydrogen, fluorine and oxygen difluoride are in gas phase under standard conditions. In order to increase the volume density, these propellants are stored under cryogenic conditions, and pumped into the combustion

Table 10.3 Properties of commonly used liquid propellants.

Propellant (F = fuel, O = oxidizer)	Chemical formula	Enthalpy of formation [kcal/mole]	Density [g/cm^3]
Hydrogen (F)	H_2	−1.89	0.071
Hydrazine (F)	N_2H_4	12.05	1.004
UDMH (F)	$C_2H_8N_2$	11.30	0.784
RP-1 (F)	C_xH_y	10.38	0.806
Fluorine (O)	F_2	−3.03	1.507
Tetrafluorohydrazine (O)	N_2F_4	−3.90	1.140
Chlorine trifluoride (O)	ClF_3	−45.3	1.809
Chlorine pentafluoride (O)	ClF_5	−60	1.750
Perchloryl fluoride (O)	ClO_3F	−10.42	1.414
Oxygen difluoride (O)	OF_2	3.90	1.521
Oxygen (O)	O_2	−3.08	1.144
Hydrogen peroxide (O)	H_2O_2	−44.85	1.442
Nitrogen tetroxide (O)	N_2O_4	−4.676	1.434
Nitric acid (O)	HNO_3	−41.46	1.504

Table 10.4 Maximum specific impulse of various bi-propellant combinations. I_{sp} [s].

	H_2	N_2H_4	UDMH	CH_2
F_2	412	365	348	328
N_2F_4	364	335	321	304
ClF_3	321	295	281	260
ClF_5	343	311	298	276
ClO_3F	344	295	290	281
OF_2	412	346	352	351
O_2	391	313	310	300
H_2O_2	322	287	284	278
N_2O_4	341	291	285	276
HNO_3	320	279	272	263

chamber in liquid or gas phase. These propellants are referred to as cryogenic propellants. For example, oxygen has a boiling point of −183 °C, below which it becomes a light blue, odorless liquid. Hydrogen has an even lower boiling point of −253 °C, with a transparent, odorless property in liquid phase.

The nitrogen hydride group of propellants include hydrazine, monomethyl hydrazine (MMH) and unsymmetrical dimethyl hydrazine (UDMH), and are characterized by combinations of nitrogen and hydrogen in their chemical structures. When used with oxidizers, such as nitrogen tetroxide or nitric acid, they undergo hypergolic reactions, meaning that these mixtures will spontaneously ignite without any external means. Thus, it simplifies the ignition mechanism; however, hypergols tend to be extremely toxic, so there are some handling and storage issues. Blended fuel such as Aerozine 50 involves a mixture of 50% hydrazine and 50% UDMH, to compensate for the high freezing point of hydrazine while retaining its performance. Nitric acid, used as the oxidizer, is a highly corrosive liquid, so the alternative is to use the so-called inhibited red-fuming nitric acid (IRFNA). IRFNA consists of HNO_3 + 14% N_2O_4 + 2% H_2O + 0.6% HF (corrosive inhibitor). In general, in addition to the oxidizer and fuel components, additives are used to reduce corrosive effects, improve ignitability and also to stabilize combustion.

Hybrid propellants refer to a combination where either the fuel or the oxidizer component is in a solid or gel form, so that liquid injection into the solid propellant feeds the combustion process. For example, a solid propellant known as HTPB can be used with a nitrous oxide oxidizer.

Monopropellants contain both oxidizer and fuel components within their chemical structure, and are capable of sustaining combustion by themselves once ignited by catalytic or thermal means. Their main advantage is in simpler storage and pumping requirements, as only a single feed into the combustion chamber is needed. They are typically used in small-thrust applications such as for orbit/attitude maneuvers or in powering turbopumps for the main propellants. Table 10.5 shows typical performance of monopropellants.

The same principles of combustion used for gas-turbine combustion can be applied to rocket propellant chemistry. However, there are some major differences, such as the wider range of oxidizers and fuels used in rockets. In addition, due to the high temperature and pressure of the rocket combustion, the equilibrium products may contain appreciable amounts of radicals. Thus, we will start our discussion of rocket propellant chemistry with a brief review of equilibrium chemistry.

Table 10.5 Monopropellants.

Chemical	Chemical formula	c* [ft/s]	I_{sp} [s]	Adiabatic flame temperature [°F]
Nitromethane	CH_3NO_2	5026	244	4002
Nitroglycerine	$C_3H_5N_3O_9$	4942	244	5496
Ethyl nitrate	$C_2H_5NO_3$	4659	224	3039
Hydrazine	N_2H_4	3952	230	2050
Tetranitromethane	$C(NO_2)_4$	3702	180	3446
Hydrogen peroxide	H_2O_2	3418	165	1839
Ethylene oxide	C_2H_4O	3980	189	1760
n-propyl nitrate	$C_3H_7NO_3$	4265	201	2587

Let us start by considering the hydrogen–oxygen reaction. Typically, the equilibrium chemistry will yield the combustion product of water, and any excess oxygen.

$$aH_2 + bO_2 \leftrightarrow aH_2O + \left(b - \frac{a}{2}\right)O_2 \quad (b > a/2) \tag{10.10}$$

For this reaction, we can assume that the reaction proceeds until all reactants are converted to products on the right-hand side, and thus simply use the conservation of atomic mass to find the mole numbers on the product side. However, at high temperature and pressures, some radical species may exist in appreciable quantities.

$$aH_2 + bO_2 \leftrightarrow n_{H_2O}H_2O + n_{H_2}H_2 + n_{O_2}O_2 + n_HH + n_OO + n_{OH}OH \tag{10.11}$$

Thus, the number of unknown mole numbers on the product side is much larger than the number of equations available from the conservation of atomic mass. In addition, the amounts of dissociated atomic (H, O) and radical species (OH) need to be determined. So how do we find the molar concentrations of all the product species, that may exist in the combustion chamber? The determination of the molar concentrations is necessary for calculations of heat of combustion (i.e. the heating value of the propellant combination), the adiabatic flame temperature and also the mixture specific heats. This question can be addressed using the chemical equilibrium principles.

10.2.2 Chemical Equilibrium

From thermodynamics, chemical equilibrium is achieved when the Gibbs free energy is at a minimum at the given temperature and pressure, so that the condition for equilibrium is

$$dG_{T,p} = 0 \tag{10.12}$$

For a mixture consisting of ideal gases, the Gibbs free energy can be calculated as follows:

$$G = \sum_i n_i\bar{g}_i \tag{10.13}$$

n_i = mole number of species i.

The Gibbs free energy for an ideal gas species, i, can also be calculated.

$$\bar{g}_i = \bar{g}_i^o(T) + \bar{R}T \ln\left(\frac{p_i}{p_{REF}}\right) \tag{10.14}$$

$\bar{g}_i^o(T)$ = Gibbs function at temperature T and standard pressure, p_{REF}.
$\bar{R}$ = universal gas constant
p_{REF} = standard pressure = 1 atm
p_i = partial pressure of species, i.

Equation (10.13) now can be rewritten with Eq. (10.14) inserted in the summation sign.

$$G = \sum_i n_i\left[\bar{g}_i^o(T) + \bar{R}T \ln\left(\frac{p_i}{p_{REF}}\right)\right] \tag{10.15}$$

So that $dG_{T,p} = 0$ (Eq. (10.12)) becomes

$$\sum_i dn_i\left[\bar{g}_i^o(T) + \bar{R}T \ln\left(\frac{p_i}{p_{REF}}\right)\right] + \sum_i n_i d\left[\bar{g}_i^o(T) + \bar{R}T \ln\left(\frac{p_i}{p_{REF}}\right)\right] = 0 \tag{10.16}$$

We can verify that the second term in Eq. (10.16) is zero, $d(lnp_i) = dp_i/p_i$ and $\Sigma dp_i = 0$, since all the changes in the partial pressures must sum to zero, due to the constant pressure condition. Thus, the condition for chemical equilibrium of an ideal-gas mixture is

$$dG = \sum_i dn_i\left[\bar{g}_i^o(T) + \bar{R}T \ln\left(\frac{p_i}{p_{REF}}\right)\right] = 0 \tag{10.17}$$

Let us consider an arbitrary chemical reaction.

$$a\mathrm{A} + b\mathrm{B} \leftrightarrow c\mathrm{C} + d\mathrm{D} \tag{10.18}$$

Any change in the mole numbers of the species will follow the above stoichiometry.

$$dn_A = -\varepsilon a; dn_B = -\varepsilon b; dn_C = -\varepsilon c; dn_D = -\varepsilon d \tag{10.19}$$

ε = proportionality constant

Substituting Eq. (10.19) into (10.17) gives us

$$\begin{aligned} -a\left[\bar{g}_A^o(T) + \bar{R}T\ln\left(\frac{p_A}{p_{REF}}\right)\right] - b\left[\bar{g}_B^o(T) + \bar{R}T\ln\left(\frac{p_B}{p_{REF}}\right)\right] + \\ c\left[\bar{g}_C^o(T) + \bar{R}T\ln\left(\frac{p_C}{p_{REF}}\right)\right] + d\left[\bar{g}_D^o(T) + \bar{R}T\ln\left(\frac{p_D}{p_{REF}}\right)\right] = 0 \end{aligned} \tag{10.20}$$

Collecting the Gibbs energy terms,

$$-\left(c\bar{g}_C^o(T)+d\bar{g}_D^o(T)-a\bar{g}_A^o(T)-b\bar{g}_B^o(T)\right)=\bar{R}T\ln\left[\frac{\left(\frac{p_C}{p_{REF}}\right)^c\left(\frac{p_D}{p_{REF}}\right)^d}{\left(\frac{p_A}{p_{REF}}\right)^a\left(\frac{p_B}{p_{REF}}\right)^b}\right] \tag{10.21}$$

The set of terms on the left-hand side of Eq. (10.21) is called Gibbs function change.

$$\Delta G^o\equiv\left(c\bar{g}_C^o(T)+d\bar{g}_D^o(T)-a\bar{g}_A^o(T)-b\bar{g}_B^o(T)\right) \tag{10.22}$$

The expression involving the pressure ratios (inside the logarithmic sign) is defined as the equilibrium constant.

$$K_p=\frac{\left(\frac{p_C}{p_{REF}}\right)^c\left(\frac{p_D}{p_{REF}}\right)^d}{\left(\frac{p_A}{p_{REF}}\right)^a\left(\frac{p_B}{p_{REF}}\right)^b} \tag{10.23}$$

Using the mole fraction, X_i, and the fact that $p_i=X_ip$,

$$K_p=\frac{X_C^cX_D^d}{X_A^aX_B^b}\left(\frac{p}{p_{REF}}\right)^{c+d-a-b} \tag{10.24}$$

The equilibrium constant can be calculated using the tabulated values of $\bar{g}_i^o(T)$, sometimes written as $\bar{g}_{f,i}^o(T)$, and furnishes us with additional equations to solve for the equilibrium mole numbers, n_i. Combining Eqs. (10.21) and (10.24), we have

$$\frac{X_C^cX_D^d}{X_A^aX_B^b}\left(\frac{p}{p_{REF}}\right)^{c+d-a-b}=\exp\left[-\frac{\Delta G^o}{\bar{R}T}\right] \tag{10.25}$$

Thus, in summary, equilibrium concentrations or mole numbers of gas species in a mixture can be found from three sets of equations: conservation of atomic mass, equilibrium constants (Eq. (10.25)) and the fact that the sum of all the mole fractions is equal to 1.

$$\sum_i X_i=1,\text{where } X_i=\frac{n_i}{\sum_i n_i}=\text{mole fraction} \tag{10.26}$$

Example 10.2 Oxygen dissociation

Let us start from a simple example of using the equilibrium constants. What are the equilibrium concentrations of O_2 and O at $p=1$ and 0.1 atm, for $T=3000$ K? The oxygen dissociation reaction is: $O_2 \leftrightarrow 2O$

The Gibbs function change, according to Eq. (10.22), is

$$\Delta G^o = 2\bar{g}^o_O(T) - \bar{g}^o_{O_2}(T) = 2(52\,554) - 1(0) = 109\,108\frac{\text{kJ}}{\text{kmole}}$$

Then, we can use Eqs. (10.24) and (10.25) to find

$$K_p = \exp\left[-\frac{\Delta G^o}{\bar{R}T}\right] = \exp\left[-\frac{109\,108}{8314(3000)}\right] = 0.0126 = \frac{X_O^2}{X_{O_2}^1}\left(\frac{p}{p_{REF}}\right)^{2-1}$$

$$= \frac{\left(\frac{p_O}{p}\right)^2}{\frac{p_{O_2}}{p}}\frac{p}{p_{REF}} = \frac{p_O^2}{p_{O_2}}\frac{1}{p_{REF}}$$

Since $p=p_O+p_{O2}$, we have

$$K_p = \frac{p_O^2}{p-p_O}\frac{1}{p_{REF}} \rightarrow p_O^2 + K_p p_{REF} p_O - K_p p_{REF} p = 0 \qquad \text{(E10.2.1)}$$

Using $p_{REF}=p=1$ atm, we can solve the above quadratic equation, whose non-negative solution is $p_O = 0.1061$ atm and $p_{O_2} = p - p_O = 0.8939$ atm.

Now that we have set up Eq. (10.2.1), we can solve for the equilibrium concentrations at $T=3000$ K at any pressure. At $p=0.1$ atm, for example, $X_O = 0.2975$ and $X_{O_2} = 1 - X_O = 0.7025$.

Example 10.3 Carbon dioxide dissociation

Let us consider an example involving two elements, C and O.

$$CO_2 \leftrightarrow CO + \frac{1}{2}O_2$$

We would like to find the equilibrium concentrations or mole numbers for CO_2, CO and O_2, following the reaction below at $T=2500$ K.

$$CO_2 \leftrightarrow CO + \frac{1}{2}O_2 \qquad \text{(E10.3.1)}$$

For calculation of the Gibbs function change, $a=1$, $c=1$ and $d=1/2$. At $T=2500$ K, we can look up the data and insert into Eq. (10.22).

$$\Delta G^o = \frac{1}{2}\bar{g}^o_{O_2}(T) + (1)\bar{g}^o_{CO}(T) - (1)\bar{g}^o_{CO_2}(T) - = 0.5(0) + (1)(-327\,245)$$
$$- 1(396\,152) = 68\,907\frac{\text{kJ}}{\text{kmole}}$$
$$K_p = \exp\left[-\frac{\Delta G^o}{\bar{R}T}\right] = \exp\left[-\frac{68\,907}{8314(2500)}\right] = 0.03635 = \frac{X_{CO}X^{0.5}_{O_2}}{X_{CO_2}}\left(\frac{p}{p_{REF}}\right)^{0.5} \tag{E10.3.2}$$

Since there are three unknown mole fractions, we need two additional equations. We can use the ratio of carbon to oxygen atoms, which must remain the same on both sides of the reaction.

$$\frac{\text{C}}{\text{O}} = \frac{1}{2} = \frac{X_{CO} + X_{CO_2}}{X_{CO} + 2X_{O_2} + 2X_{CO_2}} \tag{E10.3.3}$$

$$\text{Also}, X_{CO_2} + X_{CO} + X_{O_2} = 1 \tag{E10.3.4}$$

Solving Eqs. (E10.3.2)–(E10.3.4), simultaneously we obtain an equation for oxygen mass fraction.

$$2\left(\frac{p}{p_{REF}}\right)^{0.5} X^{1.5}_{O_2} - K_p(1 - 3X_{O_2}) = 0 \tag{E10.3.5}$$

Equation (E10.3.5) can be solved using a non-linear equation solver or iteratively. At $T=2500$ K, and $p=1$ atm, the solution is

$$X_{O_2} = 0.1790; X_{CO} = 0.3581; \text{and}\, X_{CO_2} = 0.4629.$$

Example 10.4 Hydrocarbon reaction

More complex reactions involving hydrocarbons can be considered using the same approach.

$$C_aH_b + c(O_2 + 3.76N_2) \leftrightarrow$$
$$n_1CO_2 + n_2CO + n_3H_2O + n_4H_2 + n_5O_2 + n_6OH + n_7N_2 + n_8H + n_9O + n_{10}NO \tag{E10.4.1}$$

In addition to the combustion products, this reaction includes various dissociation products. The dissociation reactions are:

$$CO_2 \leftrightarrow CO + \frac{1}{2}O_2 \tag{E10.4.2}$$

$$H_2O \leftrightarrow H_2 + \frac{1}{2}O_2 \tag{E10.4.3}$$

$$H_2O \leftrightarrow OH + \frac{1}{2}H_2 \tag{E10.4.4}$$

$$\frac{1}{2}H_2 \leftrightarrow H \tag{E10.4.5}$$

$$\frac{1}{2}O_2 \leftrightarrow O \tag{E10.4.6}$$

$$\frac{1}{2}N_2 + \frac{1}{2}O_2 \leftrightarrow NO \tag{E10.4.7}$$

For a given hydrocarbon fuel, for which a and b, are known, the reaction has 11 unknowns. We can use the equilibrium constants for the dissociation reactions, and elemental mass balances, to set up a system of 11 equations.

$$\text{First, } \sum_i X_i = \sum_i \frac{n_i}{N} = 1, \text{ where } N = \sum_i n_i \tag{E10.4.8}$$

Elemental mass balances:

$$C: \quad n_1 + n_2 = a \tag{E10.4.9}$$

$$H: \quad n_3 + n_4 + \frac{1}{2}(n_6 + n_8) = b \tag{E10.4.10}$$

$$O \quad n_1 + \frac{1}{2}(n_2 + n_3 + n_6 + n_9 + n_{10}) + n_5 = c \tag{E10.4.11}$$

$$N: \quad n_7 + \frac{1}{2}n_{10} = 3.76c \tag{E10.4.12}$$

For each of the dissociation reactions, we can write the equilibrium constant.

$$(E10.4.2) \rightarrow K_i = \frac{X_2 X_5^{0.5}}{X_1}\left(\frac{p}{p_{REF}}\right)^{0.5} \tag{E10.4.13}$$

$$(E10.4.3) \rightarrow K_j = \frac{X_4 X_5^{0.5}}{X_3}\left(\frac{p}{p_{REF}}\right)^{0.5} \tag{E10.4.14}$$

$$(E10.4.4) \rightarrow K_k = \frac{X_6 X_4^{0.5}}{X_3}\left(\frac{p}{p_{REF}}\right)^{0.5} \tag{E10.4.15}$$

$$(E10.4.5) \rightarrow K_l = \frac{X_4^{0.5}}{X_8}\left(\frac{p}{p_{REF}}\right)^{0.5} \tag{E10.4.16}$$

$$(E10.4.6) \rightarrow K_m = \frac{X_5^{0.5}}{X_9}\left(\frac{p}{p_{REF}}\right)^{0.5} \tag{E10.4.17}$$

$$(E10.4.7) \rightarrow K_n = \frac{X_5^{0.5} X_7^{0.5}}{X_{10}} \tag{E10.4.18}$$

Equations (E10.4.8)–(E10.4.18) furnish the 11 equations to solve for 11 X_i's. A MATLAB® non-linear equation solver, dedicated software, or web-based solver (e.g. http://cearun.grc.nasa.gov/) can be used for chemical equilibrium calculations.

Once the product mole numbers or fractions have been obtained, then the adiabatic flame temperatures and heat of combustion can be found in the same manner, as discussed in Chapter 6.

$$\sum_i \left(n_i\left(C_{p,i}(T - T_{ref})\right)\right)_e = \sum_i \left(n_i \bar{h}_{f,i}^o\right)_o - \sum_i \left(n_i \bar{h}_{f,i}^o\right)_e \tag{6.6}$$

T = adiabatic flame temperature

$$q_r \equiv \frac{\left(\sum_i \left(n_i \bar{h}_{f,i}^o\right)_o - \sum_i \left(n_i \bar{h}_{f,i}^o\right)_e\right)}{n_{fuel}} = \text{heat of combustion}[\text{kJ/kmole-fuel}] \tag{6.7}$$

n_{fuel} = the number of moles of fuel species on the reactant side

In addition, average properties of the combustion gas product mixture, such as the specific heat, can be evaluated.

$$(C_p)_{mixture} = \frac{\sum_i n_i C_{p,i}}{\sum_i n_i} = \sum_i X_i C_{p,i} \tag{10.27}$$

Similar to the hydrocarbon and hydrogen reactions shown in previous examples, performance of monopropellants can also be evaluated using the chemical equilibrium method. Monopropellants, once injected into the combustion chamber, are typically ignited by a catalytic bed. Hydrazine (N_2H_4), for example, undergoes the decomposition reactions, when in contact with catalytic materials such as ruthenium and iridium. These catalytic materials are embedded into "support" structures made of aluminum oxide. The support structure is designed to ensure maximum contact surface area, and can involve monolithic or wire-gage shape. The hydrazine decomposition reactions are

$$3N_2H_4 \leftrightarrow 4NH_3 + N_2 \qquad \Delta H_r = -36.36\,\text{kcal} \tag{10.28a}$$

$$4NH_3 \leftrightarrow 2N_2 + 6H_2 \qquad \Delta H_r = 199.57\,\text{kcal} \tag{10.28b}$$

The first reaction is endothermic, while the second reaction produces heat.

The decomposition also produces hydrogen, and thus hydrazine is also used as the fuel component in bi-propellant combinations, as shown in Table 10.4. Tables 10.6–10.9 show the typical performance of propellants such as hydrazine paired with various oxidizers.

10.2.3 Liquid Propellants Combustion Chambers

Due to the large mass flow rate requirement, a large number of injectors are used, often occupying the entire top surface of combustion chamber. Also, for rapid atomization and mixing of the fuel and oxidizer streams, various impingement or internal mixing schemes are used, as shown in Figure 10.4. Two impinging liquid jets result in a fan-shaped liquid sheet at

Table 10.6 Theoretical performance of hydrazine with various oxidizers.

Oxidizer	F_2	N_2F_4	ClF_3	ClF_5	ClO_3F	OF_2	O_2	H_2O_2	N_2O_4	HNO_3
O/F ratio [%]	2.37	3.25	2.70	2.71	1.50	1.50	0.92	2.03	1.33	1.50
T_c^o [K]	4727	4481	3901	4157	3467	4047	3406	2927	3247	3021
MW_c [g/mole]	19.49	21.28	23.08	22.25	21.88	18.69	19.52	19.39	20.86	20.87
T_e^o [K]	2777	2289	1832	2069	1921	2435	1974	1533	1703	1530
MW_e [g/mole]	21.51	22.67	24.35	23.74	23.09	20.50	20.54	19.77	21.53	21.29
$c*$ [ft/s]	7257	6745	5995	6324	5888	6863	6203	5753	5839	5623
ρ [g/cm^3]	1.314	1.105	1.507	1.458	1.327	1.263	1.065	1.261	1.217	1.254
I_{sp} [s]	364.4	334.7	294.6	312	295.3	345.9	312.9	286.9	291.1	279.1

Similar data can be computed for other fuels in liquid bi-propellant combinations.

Table 10.7 Theoretical performance of hydrogen with various oxidizers.

Oxidizer	F_2	N_2F_4	ClF_3	ClF_5	ClO_3F	OF_2	O_2	H_2O_2	N_2O_4	HNO_3
O/F ratio [%]	8.09	11.50	11.50	11.50	6.14	5.67	4.00	7.33	5.25	6.14
T_c^o [K]	3988	3814	3434	3705	3003	3547	2977	2419	2640	2474
MW_c [g/mole]	12.10	14.53	16.40	15.83	12.73	10.74	10.00	11.70	11.28	12.02
T_e^o [K]	1881	1665	1390	1605	1288	1622	1355	1050	1106	1043
MW_e [g/mole]	12.82	15.10	16.78	16.44	12.85	11.09	10.08	11.71	11.30	12.04
$c*$ [ft/s]	8376	7453	6634	7039	7065	8407	7975	6599	7005	6567
ρ [g/cm^3]	0.468	0.517	0.616	0.605	0.403	0.375	0.284	0.435	0.353	0.393
I_{sp} [s]	411.8	363.5	320.7	343.0	344.0	412.2	391.1	322.4	340.7	319.7

Table 10.8 Theoretical performance of RP-1 with various oxidizers.

Oxidizer	F_2	N_2F_4	ClF_3	ClF_5	ClO_3F	OF_2	O_2	H_2O_2	N_2O_4	HNO_3
O/F ratio [%]	2.70	3.55	3.35	3.00	4.26	3.85	2.70	6.69	4.00	4.88
T_c^o [K]	4430	4132	3594	3798	3720	4716	3686	3006	3438	3147
MW_c [g/mole]	23.83	25.02	28.93	27.26	26.27	20.69	23.66	22.00	25.53	25.72
T_e^o [K]	2883	2334	1951	2120	2221	2670	2457	1745	2016	1838
MW_e [g/mole]	25.79	26.17	30.03	28.46	28.47	22.62	25.79	22.67	26.86	26.73
$c*$ [ft/s]	6434	6046	5195	5512	5565	7028	5879	5509	5459	5213
ρ [g/cm^3]	1.282	1.080	1.495	1.422	1.453	1.339	1.067	1.341	1.295	1.353
I_{sp} [s]	327.5	303.5	259.8	276.3	280.6	351.9	300.1	278.2	275.7	263.4

Table 10.9 Theoretical performance of UDMH with various oxidizers.

Oxidizer	F_2	N_2F_4	ClF_3	ClF_5	ClO_3F	OF_2	O_2	H_2O_2	N_2O_4	HNO_3
O/F ratio [%]	2.45	3.17	3.00	2.85	2.70	2.69	1.70	4.26	2.57	3.00
T_c^o [K]	4464	4226	3799	4003	3657	4493	3608	3008	3415	3147
MW_c [g/mole]	21.18	22.57	25.60	24.31	24.16	19.81	21.54	21.10	23.55	23.62
T_e^o [K]	2866	2301	1990	2170	2114	2705	2280	1731	1966	1746
MW_e [g/mole]	23.18	23.85	26.98	25.80	25.89	21.99	23.14	21.77	24.72	24.38
$c*$ [ft/s]	6850	6429	5663	5989	5755	7006	6091	5622	5658	5426
ρ [g/cm^3]	1.190	1.028	1.381	1.325	1.288	1.214	0.976	1.244	1.170	1.223
I_{sp} [s]	347.9	321.0	281.4	298.3	289.6	352.2	309.7	283.7	285.2	272.4

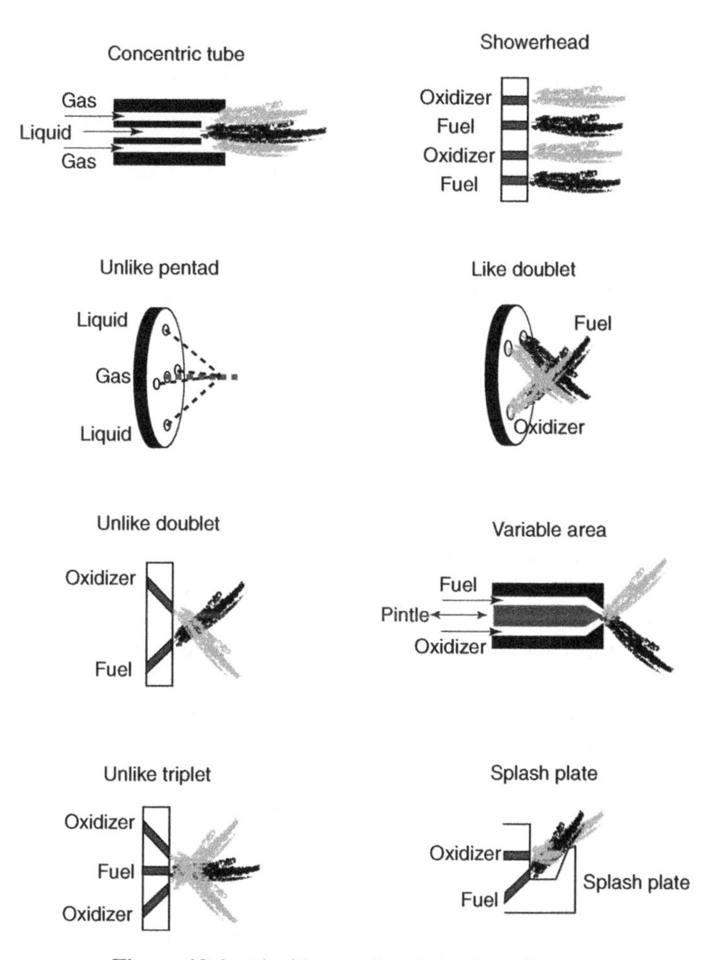

Figure 10.4 Liquid propellant injection schemes.

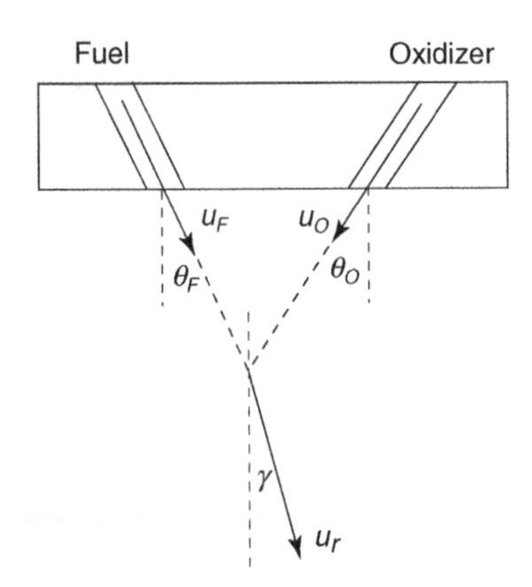

Figure 10.5 Angular deflection in impinging streams.

an angle that can be estimated from the momentum of the impinging streams, as shown in Figure 10.5. Momentum balances in the x- and y-directions are

$$\dot{m}_F u_F \cos\theta_F + \dot{m}_O u_O \cos\theta_O = (\dot{m}_F + \dot{m}_O) u_r \cos\gamma \tag{10.29}$$

$$\dot{m}_F u_F \sin\theta_F - \dot{m}_O u_O \sin\theta_O = (\dot{m}_F + \dot{m}_O) u_r \sin\gamma \tag{10.30}$$

The above two equations can be used to solve for the resultant angle, γ.

$$\tan\gamma = \frac{\dot{m}_F u_F \sin\theta_F - \dot{m}_O u_O \sin\theta_O}{\dot{m}_F u_F \cos\theta_F + \dot{m}_O u_O \cos\theta_O} = \frac{MR \cdot VR \sin\theta_F - \sin\theta_O}{MR \cdot VR \cos\theta_F + \cos\theta_O} \tag{10.31}$$

MR = mixture ratio $= \frac{\dot{m}_F}{\dot{m}_O}$

VR = velocity ratio $= \frac{u_F}{u_O}$

The injection assembly in the SSME, for example, is quite elaborate, using internal passages to promote rapid mixing and evaporation of the propellants, prior to injection into the combustion chamber, as shown in Figure 10.6. There are 600 co-axial injection elements fitted onto a 17.74-in. diameter faceplate, to inject 839 lbm/s of oxygen and 239 lbm/s of hydrogen (O/F ratio of 6.03) against the combustion chamber pressure of 2865 psia at the full throttle setting. Co-axial injectors have simpler design, in comparison to impinging or swirl injectors, and also incur less pressure loss along with having better combustion stability. Each of the main injector elements has the geometry shown in Figure 10.6. The recess allows for more rapid mixing of the oxygen and hydrogen streams, which is accomplished in the shear layer between the two streams.

Much know-how in injector designs has been accumulated, where some of the established guidelines are as follows.

1. Durability of the faceplate and injector elements is affected strongly by the local mixture ratio near the contact point, where contact with highly reactive propellants can cause failure of the hardware due to chemical attack or erosion.

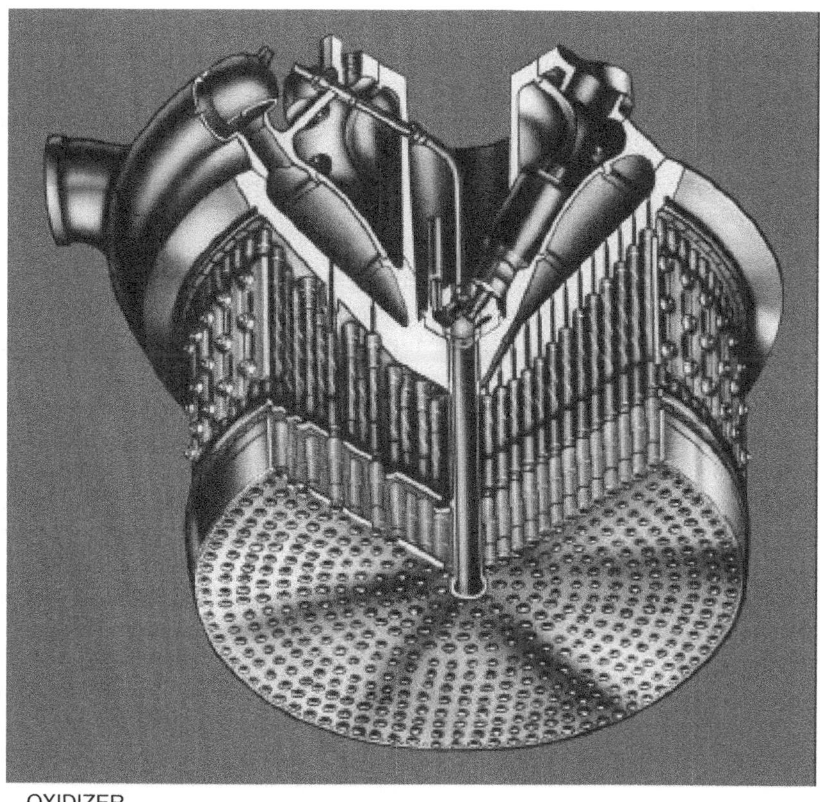

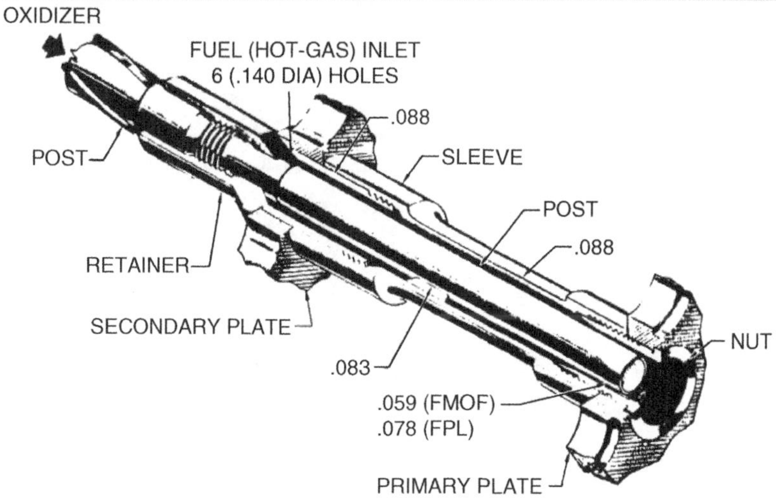

Figure 10.6 SSME injector assembly and injector geometry. Courtesy of NASA.

2. The injector elements, their location and orientation on the injector face, determine the propellant mixing, atomization and the mixture ratio of the propellants.
3. Performance of co-axial injectors depends on the individual element design, not on inter-element placements, while relative orientations and placements of impingement injectors are the determining factors.
4. High combustion rates, sought in rocket injection system design, are accompanied by high heat transfer rates to the walls.

The combustor chamber is where the propellants are injected into and burned. In some instances the convergent-divergent geometry starts at the downstream end of the combustion chamber, where the nozzle is attached for further expansion. The combustion chamber obviously needs to be designed so that stable and complete combustion can occur over continuous and repetitive operations. For material durability, under high temperature and pressure conditions, regenerative cooling is often used, which loops the propellant across the outside of the combustor walls. Thus, the combustor walls lose heat, while the propellant gains it to accelerate its evaporation prior to being injected. This kind of combustor thus has a liner, casing and coolant (propellant) passages and a manifold. The space shuttle MCC, for example, has 430 vertical milled slots through which to send the liquid hydrogen as coolant.

The MCC, as shown in Figure 10.7, is relatively compact, considering the mass flow rates of the propellants, and contains the combustion chamber, throat and initial divergent section, where the expansion ratio (the ratio of MCC exit area to the throat area) is 4.48, with a throat area of 93 in^2. A throat ring is welded to the outside of the casing at that point to strengthen the structure. The liner material is the North American Rockwell alloy (NARloy Z), which consists mostly of copper, with silver and zirconium added. The main nozzle is attached with a flange.

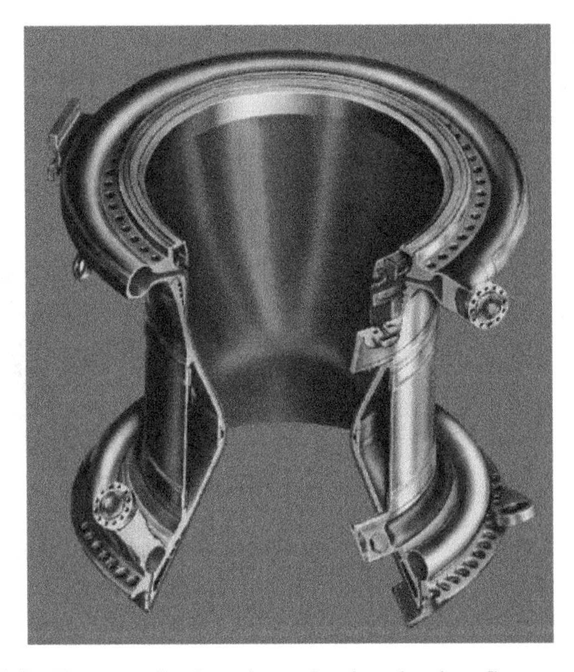

Figure 10.7 The space shuttle main combustion chamber. Courtesy of NASA.

While accurate and detailed analysis of the combustion dynamics, including injection, mixing and combustion, is difficult even using the state-of-the-art CFD methods, we can estimate the chamber length required to fully burn a liquid fuel droplet in an oxidizing environment. It involves numerical integration of the dynamical equation for the drop, coupled with the drop size decrease due to evaporation and combustion of the fuel.

$$\frac{\pi}{6}D^3\rho_L\frac{du}{dt} = -C_D\frac{\pi}{4}D^2\rho_g\frac{(u-u_g)^2}{2} \tag{10.32}$$

D = drop diameter
u = drop velocity
u_g = gas velocity
ρ_L = drop density
ρ_g = gas density
C_D = drag coefficient

Since the distance of the drop motion is of more interest, we can use the fact that $\frac{du}{dt} = \frac{dx}{dt}\frac{du}{dx} = u\frac{du}{dx}$ to rewrite Eq. (10.31).

$$\frac{\pi}{6}D^3\rho_L u\frac{du}{dx} = -C_D\frac{\pi}{4}D^2\rho_g\frac{(u-u_g)^2}{2} \tag{10.33}$$

This is coupled to the so-called d-square law for the drop diameter change, that is, the drop diameter squared decreases linearly in time for most combustion processes.

$$u\frac{d(D^2)}{dx} = -K \tag{10.34}$$

K = drop burning rate

Equations (10.33) and (10.34) can be numerically integrated. Some typical history of the drop size and velocity is shown in Figure 10.8.

10.2.3.1 Combustion Instability

Combustion instability in rockets and other combustion chambers refers to undesirable pressure oscillations resulting from feedback mechanisms between pressure and heat release. Certain frequency components then may resonate with the natural frequencies of the combustion chamber, with high-amplitude oscillations. Due to the high temperatures and pressures involved in combustion chambers, combustion instabilities can easily damage or destroy the chamber if unchecked. The Rayleigh criterion for combustion instability is

$$\text{If } \frac{1}{T}\int_T \dot{Q}'(t)p'(t)dt > 0, \text{ then the combustion is unstable.} \tag{10.35}$$

$\dot{Q}'(t)$ = fluctuation in heat release rate
$p'(t)$ = fluctuation in the chamber pressure

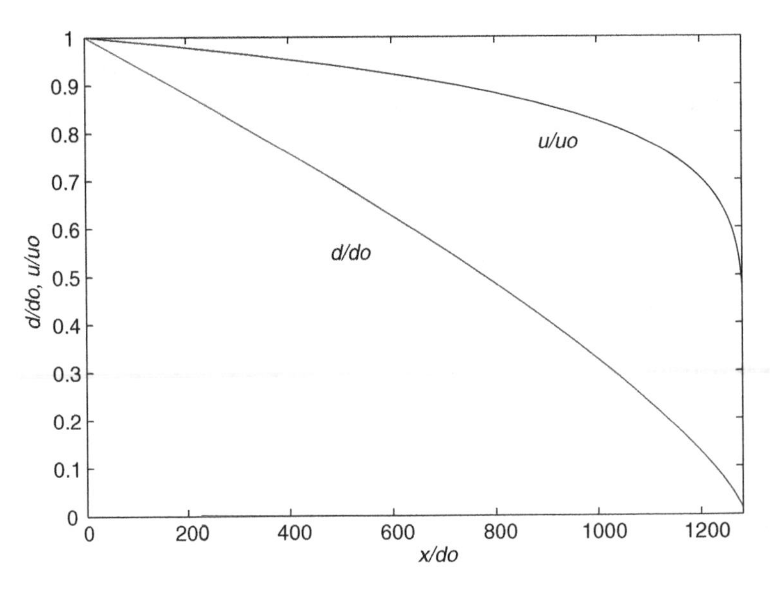

Figure 10.8 The drop size and velocity history.

For example, if there is a drop in combustion pressure, which leads to an increase in the injected propellant mass, then this added mass will release additional heat at a time lag of, say, τ. This time lag effect can be assessed through the integral in Eq. (10.35), by assuming sinusoidal variation for the pressure and heat release fluctuations, with the latter with a time lag of τ.

$$p'(t) = \bar{p}\sin\frac{2\pi t}{T} \tag{10.36a}$$

$$\dot{Q}'(t) = \bar{Q}\sin\frac{2\pi(t-\tau)}{T} \tag{10.36b}$$

$$\int_T \dot{Q}'(t)p'(t)dt = \bar{p}\bar{Q}\int_t^{t+T}\sin\frac{2\pi t'}{T}\sin\frac{2\pi(t'-\tau)}{T}dt' = \bar{p}\bar{Q}\frac{T}{2}\cos\frac{2\pi\tau}{T} \tag{10.37}$$

Thus, combustion instability is expected if the time lag lies between

$$0 < \tau < \frac{T}{4}, \frac{3T}{4} < \tau < \frac{5T}{4}, \ldots \tag{10.38}$$

This kind of relatively simple model of combustion instability is called time-lag analysis.

We can look at an example of a time-lag model for combustion instability. Let us consider the rate of conversion of liquid propellant to gas mass, ω_p. This conversion is due to the mass of the propellant injected into the combustion chamber at time, $t-\tau$, during a time interval $d(t-\tau)$.

$$\omega_p dt = \dot{m}_p(t-\tau)d(t-\tau) \tag{10.39}$$

$\dot{m}_p(t-\tau)$ = mass flow rate of the propellant at time, $t-\tau$.

We can neglect the high-frequency effects of the mass flow rate fluctuations, and approximate the mass flow rate term as its value at some mean time lag, $\bar{\tau}$, so that

$$\dot{m}_p(t-\tau) \approx \dot{m}_p(t-\bar{\tau}) = \bar{\omega}_p \tag{10.40}$$

Substitution of Eqs. (10.39) leads to

$$\omega_p = \bar{\omega}_p\left(1 - \frac{d\tau}{dt}\right) \tag{10.41}$$

One of the ideas behind the time-lag theory is that the time delay is due to the finite time required for the propellant to vaporize and cause heat release. This integrated effect of this time lag can be assessed through an integral over the time lag for some function f that relates the effects of pressure, temperature and other variables on the vaporization and combustion processes. For now, we will just focus on the effect of pressure.

$$\int_{t-\tau}^{t} f(p,T,\ldots)dt' \approx \int_{t-\tau}^{t} f(p)dt' = F = \text{const.} \tag{10.42}$$

If we use a Taylor expansion for $f(p)$, and also a model for f in the form, $f = Cp^n$, then we have

$$f(p) = f(\bar{p}) + \left(\frac{df}{dp}\right)_{\bar{p}} p' = f(\bar{p})\left[1 + n\frac{p'}{\bar{p}}\right] \tag{10.43}$$

since $\frac{df}{dp} = \frac{n}{p}$.

Differentiating Eq. (10.42) requires taking into account that τ is a function of time and that F is some constant.

$$\frac{dF}{dt} = \frac{dF}{dt}\frac{dt}{dt} + \frac{dF}{d(t-\tau)}\frac{d(t-\tau)}{dt} = f(p(t)) - f(p(t-\tau))\left(1 - \frac{d\tau}{dt}\right) = 0 \tag{10.44}$$

Substituting this in Eq. (10.43), and using $\frac{1}{1+x} \approx 1 - x$ for $x \ll 1$, we can solve for $1 - d\tau/dt$.

$$1 - \frac{d\tau}{dt} \approx 1 + n\left[\frac{p'(t)}{\bar{p}} - \frac{p'(t-\tau)}{\bar{p}}\right] \tag{10.45}$$

Use of this result back in Eq. (10.41) results in

$$\omega_p = \bar{\omega}_p + \omega_p' = \bar{\omega}_p\left(1 - \frac{d\tau}{dt}\right) = \bar{\omega}_p + \bar{\omega}_p n\left[\frac{p'(t)}{\bar{p}} - \frac{p'(t-\tau)}{\bar{p}}\right] \tag{10.46}$$

In Eq. (10.46), ω_p has been decomposed into the time-mean and fluctuating terms, and comparison of the terms reveals that the fluctuating term is

$$\omega_p' = \bar{\omega}_p n\left[\frac{p'(t)}{\bar{p}} - \frac{p'(t-\tau)}{\bar{p}}\right] \tag{10.47}$$

We expect that the heat release has the same form as the mass conversion rate term, so we can write

$$\dot{Q}' = \bar{\dot{Q}}n\left[\frac{p'(t)}{\bar{p}} - \frac{p'(t-\tau)}{\bar{p}}\right] \tag{10.48}$$

Thus, the Rayleigh criterion can be evaluated using this time-lag model.

$$\frac{1}{T}\int_{t-T}^{t} \dot{Q}'(t)p'(t)dt = \frac{\bar{\dot{Q}}n}{\bar{p}T}\int_{t-T}^{t} \left[(p'(t))^2 - p'(t)p'(t-\tau)\right]dt \tag{10.49}$$

For $n > 0$, there will be an unstable component in the rocket combustion process due to the squared pressure fluctuation term, while the pressure correlation term in the integral can either contribute to or diminish this effect depending on the time lag, τ.

This time-lag analysis is zero-dimensional, and also assumes some constant n and time lag τ; however, in real combustion chambers both can be a strong function of the oscillation frequency and spatial location. For more accurate, detailed analyses of combustion instabilities, a set of full thermo-acoustic equations is numerically solved. For example, longitudinal, radial and tangential pressure oscillations can exist, with many individual and coupled modes. Figure 10.9a shows the basic pressure wave patterns, while Figure 10.9b includes more complex modes. The frequency of the longitudinal modes can be estimated by

$$f = k\frac{c}{2x_c} \tag{10.50}$$

$k = 1, 2, 3, \ldots$ = mode number for longitudinal waves
c = average speed of sound in the combustion chamber
x_c = effective acoustic length (taken as the distance between the injector faceplate and the nozzle throat, minus approximately ½ of the convergent nozzle length)

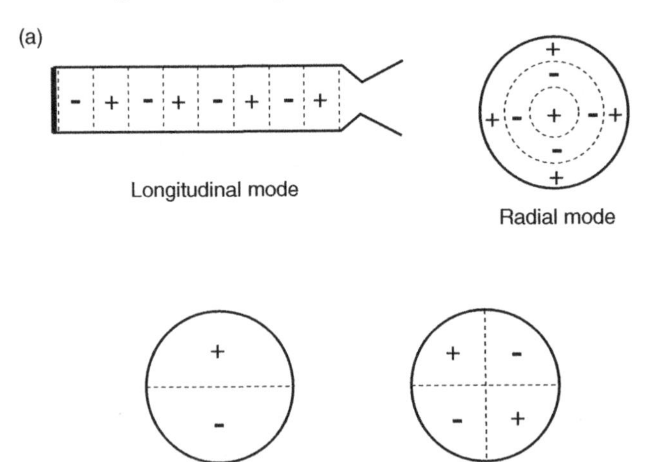

Figure 10.9 (a) Basic pressure oscillation modes. (b) More complex pressure oscillation modes.

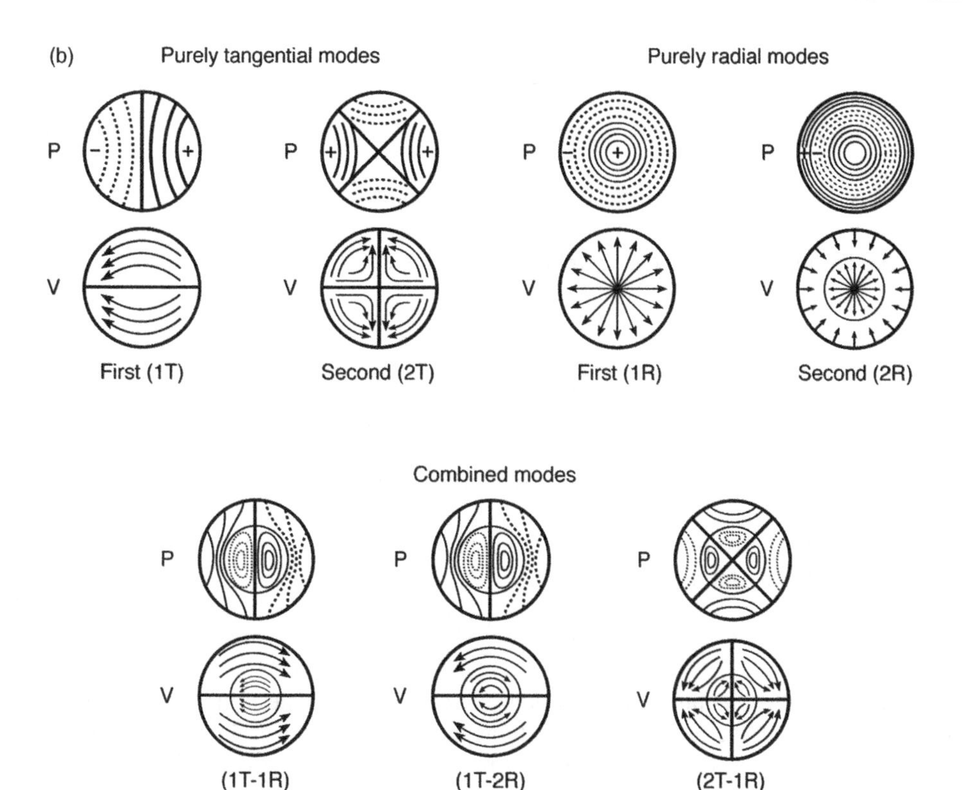

Figure 10.9 *(Continued)*

Similarly, radial and tangential frequencies are estimated by

$$f = \frac{S_{mn}c}{2\pi r_c} \tag{10.51}$$

$m, n = 1, 2, 3, \ldots$ = tangential and radial mode numbers
r_c = radius of the combustion chamber

Detailed thermo-acoustic analysis for combustion instability involves solving for the pressure field in the combustor as a function of time, or its frequency-domain equivalent. The acoustic equation in a quiescent ambient (mean velocity and pressure gradient are zero) is relatively simply derived. We can consider the density, velocity, and pressure as a sum of respective time-mean and fluctuating components. For simplicity, we will write our velocity vector simply as v.

$$\rho = \bar{\rho} + \rho' \tag{10.52a}$$

$$v = \bar{v} + v' \tag{10.52b}$$

$$p = \bar{p} + p' \tag{10.52c}$$

With these, we can write the continuity and the momentum equations.

$$\frac{\partial \rho'}{\partial t} + \bar{\rho}\nabla \cdot v' = 0 \quad (10.53)$$

$$\bar{\rho}\frac{\partial v'}{\partial t} + \nabla p' = 0 \quad (10.54)$$

From the definition of the acoustic velocity c, we have

$$c^2 = \left(\frac{\partial p}{\partial s}\right)_s \quad (10.55)$$

$$p - \bar{p} = \left(\frac{\partial p}{\partial s}\right)_{\bar{p},\bar{\rho}} (\rho - \bar{\rho}) \rightarrow p' = c_o^2 \rho' \quad (10.56)$$

c_o = acoustic velocity at the mean pressure and density.

Taking the time derivative of Eq. (10.53) and taking the gradient of Eq. (10.54) allows us to eliminate the density term from both.

$$\nabla^2 p' - \frac{1}{c_o^2}\frac{\partial p'}{\partial t} = 0 \quad (10.57)$$

Inside a combustion chamber, the conditions are certainly not quiescent, and gas and liquid velocity terms (both the mean and fluctuations) need to be included. In addition, the above acoustic equation is "fed" by the source terms on the right-hand side, which are typically classified into four categories.

$$\nabla^2 p' - \frac{1}{c_o^2}\frac{\partial p'}{\partial t} = h_I + h_{II} + h_{III} + h_{IV} \quad (10.58)$$

h_I = linear coupling of pressure and velocity fluctuations (e.g. $\frac{1}{c_o^2}\frac{\partial}{\partial t}(\bar{u} \cdot \nabla p')$).

h_{II} = non-linear coupling of pressure and velocity fluctuations (e.g. $\frac{1}{c_o^2}\frac{\partial}{\partial t}(u' \cdot \nabla p')$).

h_{III} = entropy fluctuations (e.g. $\nabla\left[\frac{\bar{p}s'}{c_p}(\bar{u} \cdot \nabla)\bar{u}\right]$).

h_{IV} = combustion heat release and viscous dissipation terms

Computational solutions of the above equation with appropriate combustor geometry and boundary conditions require complex algorithms, but can provide a detailed time- and space-dependent pressure field. This kind of data are analyzed for determination of unstable oscillation modes, detection of their onsets and possible remedies. Remedies involve suppression of problematic oscillation modes and treatment of the combustor walls to dissipate acoustic energy. Suppression devices can simply be baffles, such as the ones shown in Figure 10.10, to block the radial, tangential and longitudinal modes. Since the direction of the flow of propellants and combustion products is in the axial direction, radial and tangential baffles will have small effects on the rest of the operation of the combustor. However, suppression of the longitudinal modes can only involve baffles that impede the flow. In some cases, the baffles are made of combustible materials so that, over time, they serve as the suppression devices as well as additional propellants.

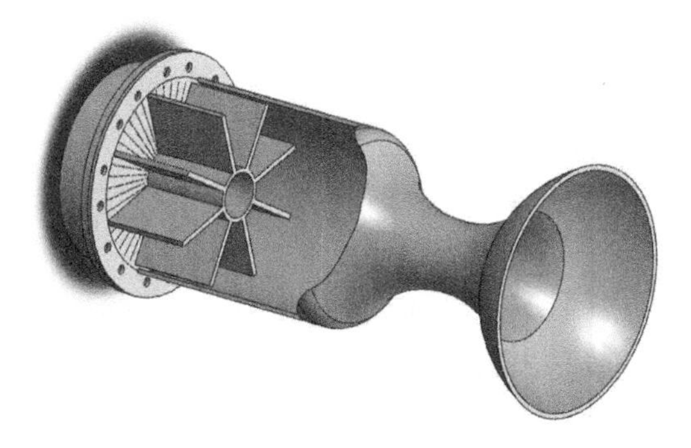

Figure 10.10 Baffles for suppression of acoustic waves.

Resonators and acoustic liners at the combustor wall are used to dissipate acoustic energy, as shown in Figure 10.11. For the resonator geometry shown in Figure 10.11, the natural, undamped frequency is

$$f = c_o\sqrt{\frac{A}{lV_R}} \tag{10.59}$$

Equation (10.59) shows that wall resonators are only effective for high-frequency components, since at low frequency the resonator has to be excessively large. Combustor liners are more often used in gas-turbine combustors than rockets, due to cooling requirements.

10.2.3.2 Propellant Feed Systems

Due to the high power required for the pumps to feed the propellants into the combustion chamber at high pressures, gas generators, gas expanders or staged combustion are often used.

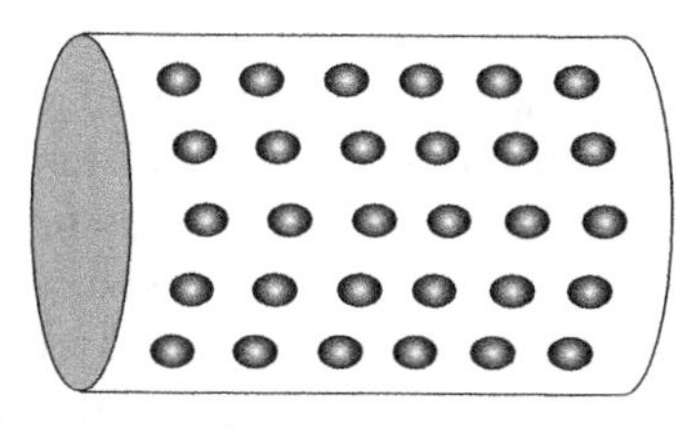

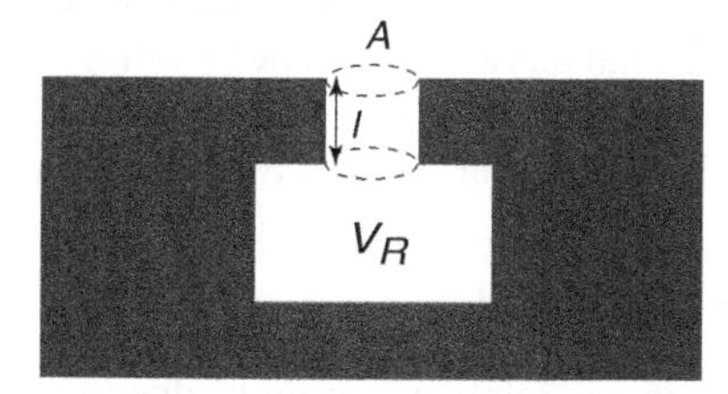

Figure 10.11 Combustor liners and wall resonators.

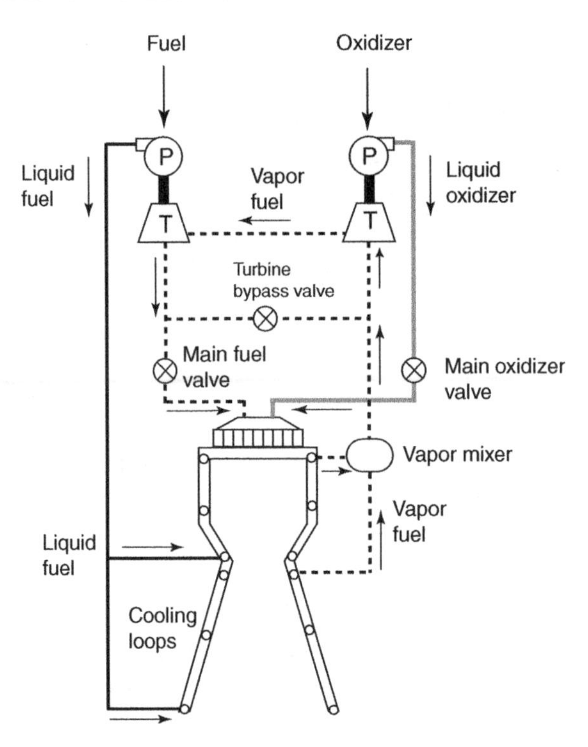

Figure 10.12 Gas expander flow diagram.

In gas generators, a small portion of the propellants is used in a small combustor to power the turbine-pump assembly. The exhaust gas is typically dumped overboard, so that the operation of gas generators subtracts from the available energy. Gas expanders use one of the liquid propellants in a regenerative cooling loop around the combustion chambers and/or the nozzle. The heated and vaporized propellant then drives the turbine-pump. In practice, only hydrogen has sufficiently high cooling capabilities for gas expander cycle, but the low density of hydrogen requires large turbine elements or a series of turbines. Figure 10.12 shows a schematic of the gas expander cycle to power the turbo-pumps.

The SSME, as shown in Figure 10.3, makes use of staged combustion, where pre-burners on both the fuel and oxidizer feed streams power the respective turbine-pump combinations. The pre-burners are fed with a controlled mixture of fuel and oxidizers, and increased enthalpy of the combustion products drives the turbine, which in turn powers the pump. In SSME, the mixture going into the two pre-burners is extremely fuel-rich with 76% of the total hydrogen and 11% of the total oxygen flow. The resulting gas is thus also fuel-rich and goes into the main combustion chamber, after being merged with the remaining hydrogen for the fuel feed and with the remaining oxygen for the oxidizer feed.

10.3 Solid Propellant Combustion

Solid propellants are advantageous, relative to liquids, due to the ease of storage and their stability. However, the combustion cannot be throttled, other than to a limited extent with the cross-sectional geometry, and certainly it cannot be shut off once ignited. Figure 10.13

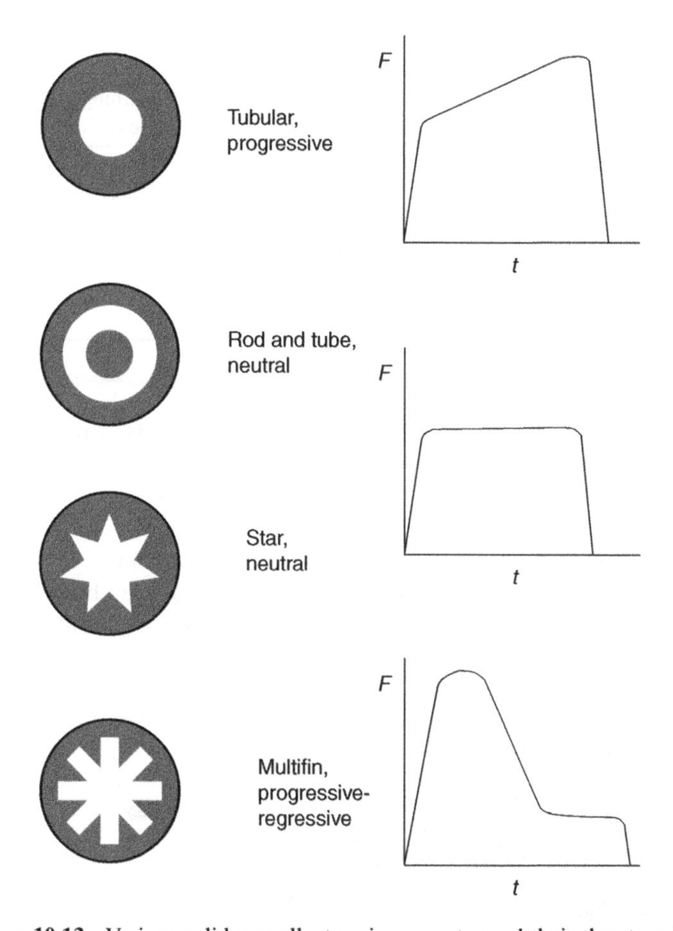

Figure 10.13 Various solid propellant grain geometry and their thrust profiles.

shows various solid propellant "grain" geometry and their thrust profiles. Since the burning rate depends on the exposed surface area, the different cross-sectional shapes, and thus the area, determine the burning rate at a given time. Depending on the thrust requirements, the grain is designed for regressive, neutral or progressive burning.

Table 10.10 shows some of the typical solid propellants. Some propellants (e.g. DB = double base, HMX, NEPE) are used alone, or with metal (aluminum) particles, while others (PVC, PU, PBAN, etc.) are always used with an oxidizer (AP). The former type of solid propellant is called double-base (DB), and these contain in their molecular structure both fuel and oxidizer components. In that sense, they are analogous to monopropellants. The latter type involves physical combinations of fuel (such as PVC, PU, PBAN) and oxidizer (AP) components, and are referred to as composite propellants. Double-base propellants, upon their decomposition and volatilization, undergo premixed combustion, while composite propellants at the microscopic level will generate many "flamelets", consisting of premixed and non-premixed combustion zones, as shown in Figure 10.14. The generic DB (double-base) refers to a combination of nitrocellulose and nitroglycerin, in which the liquid-propellant nitroglycerin also acts as the "plasticizer" to gelatinize nitrocellulose to form

Table 10.10 Commonly-used solid propellants.

Propellant	Density [lbm/in^3]	Flame Temperature [°F]	Burning rate [in/s]	Pressure exponent, n
PVC/AP/Al	0.064	6260	0.45	0.35
PS/AP/Al	0.062	5460	0.31	0.33
PBAN/AP/Al	0.064	6260	0.32	0.35
CTPB/AP/Al	0.064	6160	0.45	0.40
HTPB/AP/Al	0.065	6160	0.28	0.30
PBAA/AP/Al	0.064	6260	0.35	0.35

shapes. Nitrocellulose has a somewhat complex chemical formula, $C_6H_{7.55}O_5(NO_2)_{2.45}$. During combustion, nitrocellulose breaks down into C/H and C/H/O fragments that act as fuel, and also NO_2 separates to act as the oxidizer.

Composite propellants contain fuel and oxidizer components that are physically mixed. Oxidizers are typically fine crystalline particles of ammonium perchlorate, ammonium nitrate, potassium perchlorate, and so on. Ammonium perchlorate, for example, has a chemical formula, NH_4ClO_4, where the chlorine atom is surrounded by four oxygen atoms and an NH_4 ion is attached to one of the oxygen atoms. Upon heating, ammonium perchlorate

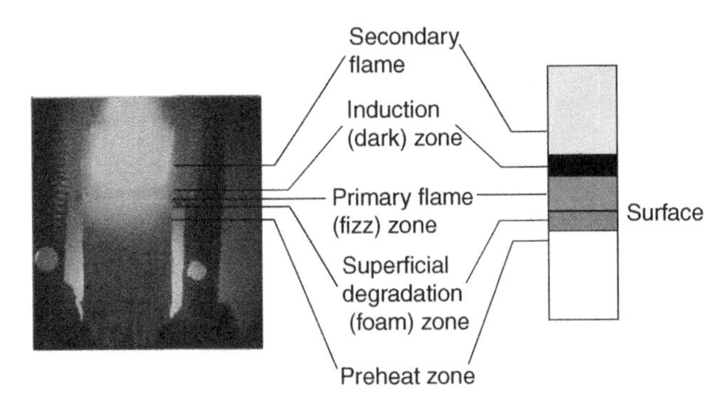

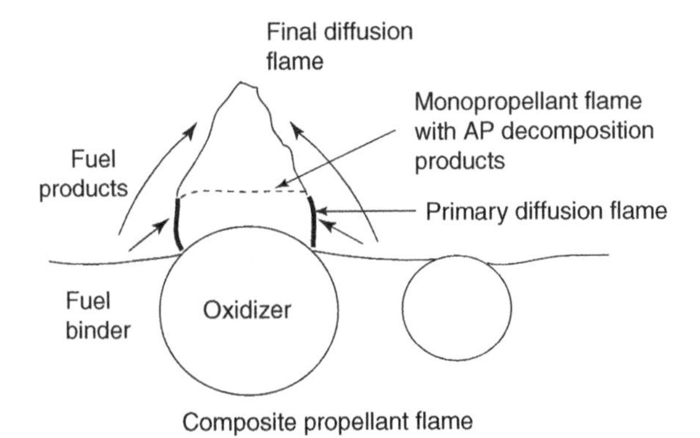

Figure 10.14 Flame shapes for double-base and composite propellants. Courtesy of NASA.

undergoes the following decomposition reaction to produce chlorine and oxygen.

$$2NH_4ClO_4 \leftrightarrow Cl_2+N_2+2O_2+4H_2O \tag{10.60}$$

Ammonium nitrate (AN) and potassium perchlorate (PP) follow similar decomposition to produce oxidizers.

$$\text{AN}: \quad 2NH_4NO_3 \leftrightarrow N_2O + 2H_2O \tag{10.61}$$

$$\text{PP}: \quad KClO_3 \leftrightarrow 2KCl + 3O_2 \tag{10.62}$$

The oxidizer particles are embedded in hydrocarbon binders (also called plasticizers) such as polyurethane (PU), polyvinyl chloride (PVC), polybutadiene (PB) shown in Table 10.10, which also act as the fuel. Acronyms such as PVC and PB refer to familiar plastic materials, which have other more mundane uses. Plasticizers can easily be molded into desired shapes, and have superior stability characteristics. In addition to the main oxidizer/binder, solid propellants may contain metalized particles, stabilizers, burning rate catalysts and anti-aging agents.

10.3.1 Burning Rate Analysis

The solid propellant combustion can be analyzed in a simple way, by considering a burning rate as function of the combustion chamber pressure.

$$\text{Burning rate} = S_b = a(p_2^o)^n \quad [\text{cm/s}]\text{or}[\text{in/s}] \tag{10.63}$$

a, n = empirical constants

To the first order, this is how the burning rate is empirically determined through measurements of the surface regress speed of the propellant and fitting the data against the chamber stagnation pressure. However, in most instances the empirical parameters a, n are constant only for some limited pressure ranges, so that we would use

$$\begin{aligned} S_b &= a_1(p_2^o)^{n1} \quad \text{for} \quad p_2^o < p_1 \\ S_b &= a_2(p_2^o)^{n2} \quad \text{for} \quad p_1 < p_2^o < p_2 \\ &\ldots \end{aligned} \tag{10.64}$$

The empirical constants are determined by measuring the burning rate directly as a function of pressure, in a pressurized strand burner. The pressure is varied in the burner, and the burning rate can be measured optically or using a series of trip wires in the solid propellant. The trip wires are melted away as the burning surface reaches these wires, to trigger a signal which can be measured in time. Typically S_b ranges from 0.3 to 25 cm/s.

The burning rate can be converted to the rate of gas production, analogous to the mass flow rate expression.

$$\dot{m}_g = \rho_p A_b S_b = \rho_p A_b a(p_2^o)^n \tag{10.65}$$

ρ_p = propellant density
A_b = burning surface area

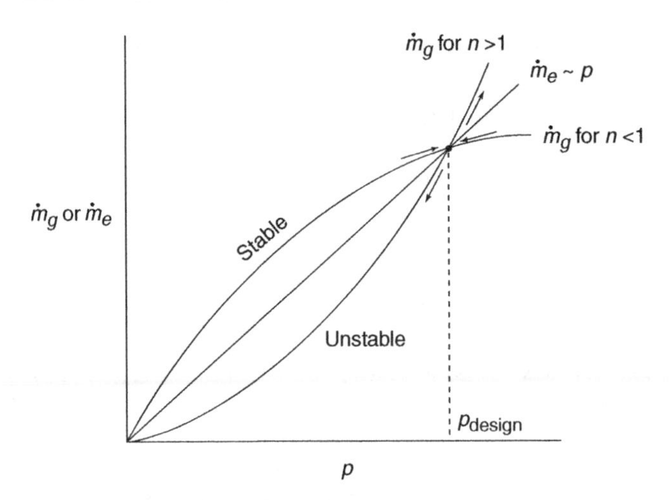

Figure 10.15 A plot of the mass flow rates in a solid rocket combustion chamber.

We can compare the rate of gas production to the mass flow rate ejected through the nozzle in a choked flow.

$$\dot{m}_n = \dot{m}_e = \frac{A * p_2^o}{\sqrt{RT_2^o}} \left(\frac{2\gamma}{\gamma + 1} \right)^{\frac{\gamma+1}{2(\gamma-1)}} \tag{10.66}$$

Comparing Eqs. (10.65) and (10.66), the nozzle flow rate increases linearly, while the gas production rate depends on the n-th power of the chamber pressure. We can determine the basic stability of the solid propellant through these two expressions, with the aid of Figure 10.15. If $n > 1$, any positive departure from the chamber pressure will send the gas production rate above that of the nozzle flow rate, which will have the effect of increasing the pressure yet again. Thus, this is a positive feedback loop leading to an indefinite increase in chamber pressure. Conversely, if $n < 1$ the positive departure will be suppressed through an increase in the nozzle flow rate, bringing the pressure to the design point. We can easily see a similar effect for negative departures, where the chamber pressure will spiral down for $n > 1$, while chamber pressure will be restored for $n < 1$. Therefore, one of the fundamental requirements of solid propellants is the burning rate whose pressure is of $n < 1$ exponential dependence. Typically, we find $n < 0.8$ in most solid propellants.

We can further examine the effect of the burning rate on the solid rocket operation, with a simplified solid rocket motor geometry, as shown in Figure 10.16. The time rate of change of

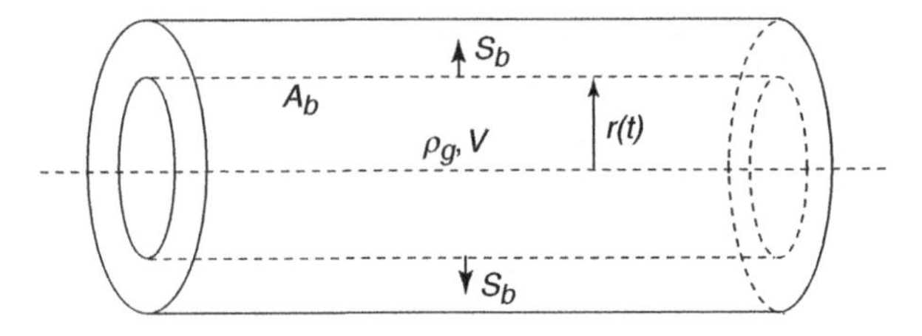

Figure 10.16 A simplified solid rocket motor geometry, for burning rate analysis.

gas mass (M_g) in the combustion chamber is given by

$$\frac{dM_g}{dt} = \frac{d}{dt}(\rho_g V) = \rho_g \frac{dV}{dt} + V\frac{d\rho_g}{dt} \tag{10.67}$$

Since the volume change is due to the surface regression of the solid propellant grain, we can write

$$\frac{dV}{dt} = S_b A_b \tag{10.68}$$

Also, assuming ideal-gas behavior of the combustion product gas,

$$p^o = \rho_g RT^o \rightarrow \frac{d\rho_g}{dt} = \frac{1}{RT^o}\frac{dp^o}{dt} \tag{10.69}$$

We have dropped the subscript "2" for the stagnation pressure in the combustion chamber, for brevity. So Eq. (10.67) becomes

$$\frac{dM_g}{dt} = \rho_g S_b A_b + \frac{V}{RT^o}\frac{dp^o}{dt} \tag{10.70}$$

Now, we write the conservation of mass,

$$\frac{dM_g}{dt} = \dot{m}_g - \dot{m}_n \tag{10.71}$$

Equating the right-hand sides of Eqs. (10.70) and (10.71), we have the following result.

$$(\rho_p - \rho_g)A_b a(p^o)^n = \frac{V}{RT^o}\frac{dp^o}{dt} + \dot{m}_n \tag{10.72}$$

The steady-state combustion chamber pressure can be found by setting the time-rate term in Eq. (10.71) to zero, and solving for p^o, while once again using the expression for the nozzle flow rate (Eq. (10.65)).

$$(p^o)_{steady-state} = \left[\left(\frac{\gamma+1}{2}\right)^{\frac{\gamma+1}{2(\gamma-1)}} \frac{(\rho_p - \rho_g)a}{\gamma}\frac{A_b}{A*}\sqrt{\gamma RT^o}\right]^{\frac{1}{1-n}} \tag{10.73}$$

This expression requires the burning surface area, A_b, so implicitly assumes that the $dA_b/dt \ll 1$. The ratio, $K = A_b/A*$ is sometimes referred to as kleming, meaning pinching, and the typical range is from 200 to 1000 during combustion. We can see from Eq. (10.73) that the steady-state pressure is linear with K for $n = 0$, more than linear for $0 < n < 1$. The steady-state pressure literally blows up when $n = 1$. The dependence of the steady-state pressure change on the dK can be found by taking the logarithmic differentials of both sides of Eq. (10.73).

$$\frac{dp^o}{p^o} = \frac{1}{1-n}\frac{dK}{K} \tag{10.74}$$

The steady-state is achieved after the initial (startup) or any transient periods. For the transient combustion, the pressure fluctuation is governed by Eq. (10.72), which is a first-order non-linear ODE of the form

$$\frac{dp^o}{dt} + Ap^o - B(p^o)^n = 0 \tag{10.75}$$

This has a solution of the form

$$\frac{p^o}{(p^o)_{t=0}} = \exp\left(\frac{n-1}{\tau}t\right) \tag{10.76}$$

The time constant, τ, is

$$\tau \approx \frac{\left(\frac{\gamma+1}{2}\right)^{\frac{\gamma+1}{2(\gamma-1)}}}{\sqrt{\gamma RT^o}} \frac{V_o}{A*} \tag{10.77}$$

V_o = initial volume

Thus, the burning rate exponent n again has an important influence on the transient pressure, where the pressure will grow indefinitely if $n > 1$ or converge to a steady-state value if $n < 1$ in Eq. (10.76). Again, in most propellants $n < 0.8$.

The combustion and the induced gas motion in a solid rocket motor are quite complex. The flame structure is also different between double-base and composite propellants. Composite propellants have more complex flame structure, as shown in Figure 10.16. AP, or similar, oxidizer particles decompose into fragments, which are capable of burning by themselves, so that a premixed flame resides over the AP particle. The excess oxidizer components diffuse to the side to form the primary diffusion flame, with the fuel vaporized from the binder. There are also excess oxidizer components downstream of the premixed flame, to allow the final diffusion flame.

Double-base propellants have both the fuel and oxidizer components in the chemical formulation, so that a nearly homogeneous premixed flame burns over the propellant surface, as depicted in Figure 10.14. However, the vertical structure is divided into distinct zones. The heat from the flame supplies the sensible enthalpy to the combustion product side, but also to the propellant itself. This heat vaporizes the propellants in the preheated zone (~10 μm), also called the foam zone. When the temperature of the vaporized propellants becomes sufficiently high, the molecular bonds start to rupture to form decomposition products in the fizz zone. Some of the decomposition products recombine so that the net result is exothermic. Chemical reactions continue in the dark zone, where active combustible products are formed. These in turn are burned in the flame zone, where the adiabatic flame temperature is reached.

This complex process can be simplified for a thermal analysis of a double-base propellant flame, as shown in Figure 10.17, where the structure is reduced to the preheat and the flame zone. The energy equation applicable to any point along the perpendicular direction is:

$$\rho u c_p \frac{dT}{dx} = k\frac{d^2T}{dx^2} + w_g q_r \tag{10.78}$$

Here, k is the thermal conductivity, w_g and q_r are the reaction rate and the heat of combustion, respectively, and u the gas velocity or the solid propellant regression rate. There is no reaction in the solid phase. So, after removing the reaction rate term from Eq. (10.78), and putting

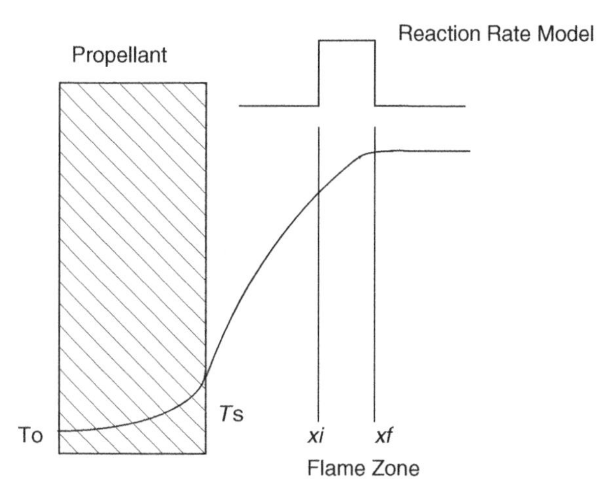

Figure 10.17 Flame structure over double-base solid propellants.

appropriate solid phase properties, we have the following ODE:

$$\rho_p S_b c_p \frac{dT}{dx} = k_p \frac{d^2T}{dx^2} \quad (10.79)$$

In the above equation, S_b is the surface regression speed, and ρu has been replaced by solid-phase mass flux, $\rho_p S_b$. Integration gives the following temperature profile in the solid phase:

$$T(x) - T_o = (T_s - T_o)\exp\left(\frac{\rho_p c_p S_b x}{k_p}\right) \quad (10.80)$$

From this, the heat flux at the surface can be calculated, by differentiating the temperature and multiplying by the propellant thermal conductivity.

$$q^- = k_p \left(\frac{dT}{dx}\right)_s = \rho_p S_b c_p (T_s - T_o) \quad (10.81)$$

In the gas phase, we need to retain all of the terms and, of course, use the gas phase properties. Using an integration factor, $\exp\left(-\frac{\rho_g u_g c_g x}{k_g}\right)$, Eq. (10.80) can also be integrated to give the heat flux at the surface ($x = 0$).

$$q^+ = k_g \left(\frac{dT}{dx}\right)_{x=0} = q_r \int_0^\infty \exp\left(-\frac{\rho_g u_g c_g x}{k_g}\right) w_g dx \quad (10.82)$$

Evaluation of the integral is difficult, due to the typically non-linear reaction rate term, so we model the reaction rate as shown in Figure 10.17, as a top-hat profile with a mean reaction rate of $\bar{w}_g$ within the [x_i, x_f], and zero everywhere else.

$$k_g \left(\frac{dT}{dx}\right)_{x=0} = \frac{k_g}{\rho_g u_g c_g} \bar{w}_g q_r \left(\exp\left(-\frac{\rho_g u_g c_g x_i}{k_g}\right) - \exp\left(-\frac{\rho_g u_g c_g x_f}{k_g}\right)\right) \quad (10.83)$$

Now if x_i is small, that is, if the flame is very close to the propellant surface, then the first exponential term is close to 1, and the second exponential term is usually small, so we have:

$$k_g \left(\frac{dT}{dx}\right)_{x=0} \approx \frac{k_g}{\rho_g u_g c_g} \bar{w}_g q_r \tag{10.84}$$

We use the heat flux expressions of Eqs. (10.81) and (10.84) and apply this to the propellant surface, where these heat fluxes are balanced by the heat required to vaporize the propellant at the surface, $\rho_p S_b q_s$. Also, q_s is the heat of vaporization.

$$\frac{k_g}{\rho_g u_g c_g} \bar{w}_g q_r = \rho_p S_b c_p (T_s - T_o) - \rho_p S_b q_s \tag{10.85}$$

Solving for the regression rate, S_b, we get

$$S_b = \left[\frac{\lambda_g \bar{w}_g q_r}{\rho_p{}^2 c_p c_g (T_s - T_o - q_s/c_p)}\right]^{1/2} \tag{10.86}$$

10.4 Rocket Nozzles

The nozzle expansion and accompanying flow acceleration can be approximately described by one-dimensional gas dynamics, similar to the gas-turbine nozzles, but the flow in a rocket nozzle can be quite complex, as schematically shown in Figure 10.18. The residual liquid or solid particles from the combustion chamber can form multi-phase flow quite far downstream in the nozzle. High temperature in the nozzle flow does not allow chemical equilibrium, and there may be continuous ionization and recombination reactions. There is a substantial amount of heat and momentum transfer with the wall, leading to interaction of subsonic and supersonic flows in the boundary layer. Due to the large expansion ratio, the flow is not uniform, and may even involve shocks and recirculation in the periphery. In the ideal design conditions, the nozzle expansion ratio would be chosen so that the exit pressure matches that of the ambient, so that only a shear layer exists between the high-speed nozzle flow and the ambient (adapted nozzle). At sea level, the ambient pressure is higher, so that the nozzle flow is over-expanded, and the flow slows down outside of the nozzle through a series of shocks. In more severe cases, the shocks can enter the nozzle, resulting in poor performance. At high altitudes, the flow can be under-expanded and expansion waves deflect the flow outward, resulting in a large "bulge" in the rocket plumes. These factors can hamper the nozzle operation, in particular the non-adapted (under- or overexpansion) nozzle flows, as shown for the SSME and Vulcain (the rocket motor for the European Ariane series) nozzles in Table 10.11.

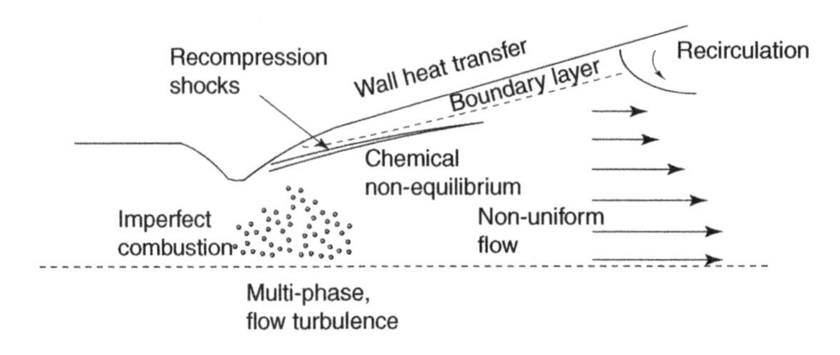

Figure 10.18 Flow processes in a rocket nozzle.

Table 10.11 Loss terms that contribute to percentage reduction in the thrust coefficient (C_F) in conventional rocket nozzles.

Loss term to C_F	Vulcain (%)	SSME (%)
Chemical non-equilibrium	0.2	0.1
Wall friction	1.1	0.6
Divergent flow, non-uniform flow profile	1.2	1.0
Under- or overexpansion	0–15	0–15

Some of the concepts to reduce the effects of non-adapted flow are shown in Figure 10.19. The dual-bell nozzle, for example, induces controlled flow separation at low altitudes, and the inflection point essentially serves as the nozzle exit. At higher altitudes, the flow follows the contour of the second bell and the full nozzle exit area is utilized. Extendible nozzles use a similar idea, where an extension is applied at high altitudes. The so-called plug or aerospike nozzles use a contoured wall to guide the flow continuously to optimum plume condition.

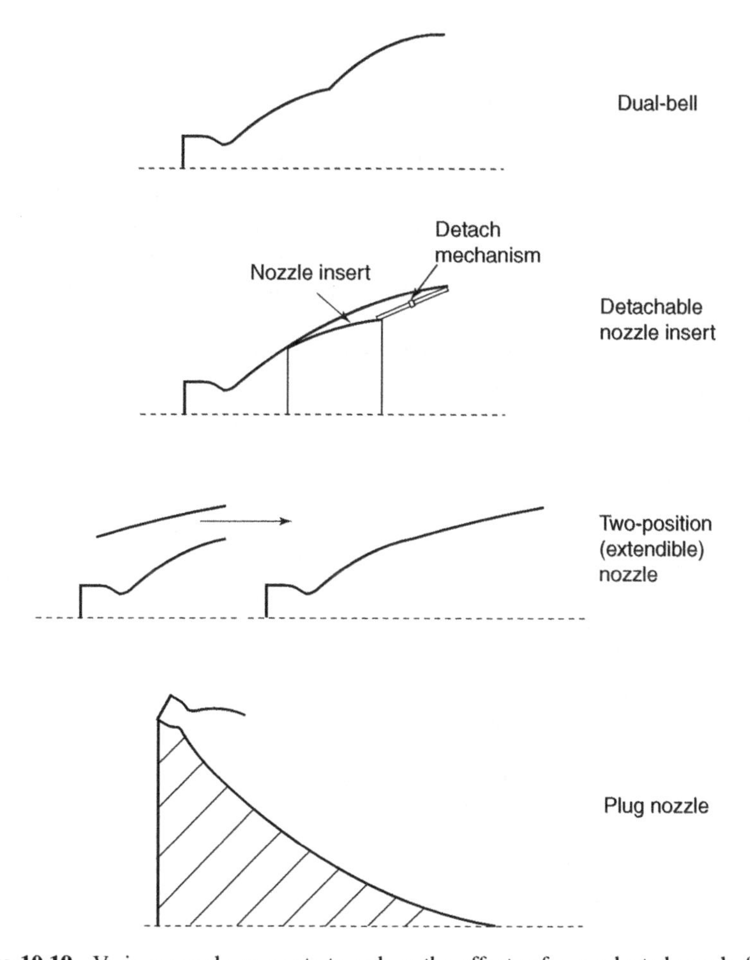

Figure 10.19 Various nozzle concepts to reduce the effects of non-adapted nozzle flows.

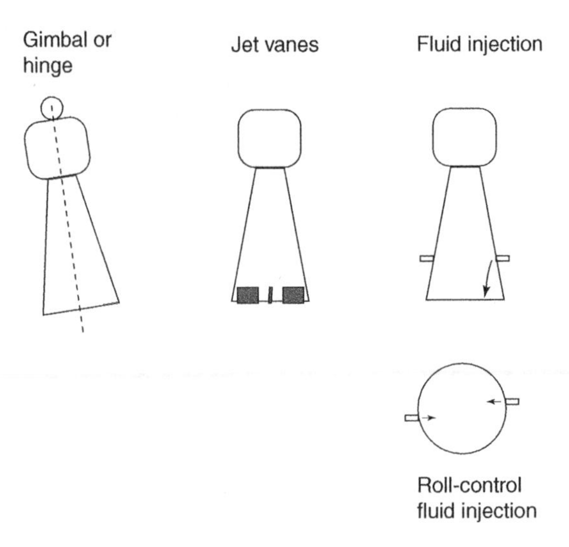

Figure 10.20 Various thrust vector control methods.

10.4.1 Thrust Vector Control

Thrust vector control (TVC) can be achieved with either fixed or movable nozzles. Fixed-nozzle TVC methods include simple jet vanes or tabs at the nozzle exit, which would obviously create non-uniform flow and pressure, as shown in Figure 10.20. The jet vanes can also be angled so that they generate roll motion. These devices, however, impede the flow, and result in lows of thrust by up to 25%. Fluid injection into the nozzle creates local shocks, as the high speed nozzle flow intersects with the slow injected fluid. The asymmetry creates redirection of the thrust vector. The injected fluid can be drawn from the combustion chamber, or from a separate tank of inert liquid or gas. The fluid injection can also generate roll control, as shown in Figure 10.20. Movable nozzles pivot the nozzle section using various geometries.

10.4.2 Nozzle and Combustion Chamber Heat Transfer

The nozzles and combustion chamber are subject to intense heat transfer, through convection, conduction and radiation modes. V-2 rockets, used during World War II, had throat heat flux of 2.5 MW/m^2, while modern rockets such as the SSME are exposed to nearly 100 MW/m^2 at the throat. For material integrity and stability, this heat needs to be dissipated through active or passive cooling methods.

For the gas-side (inside of the nozzle or combustor) heat transfer, the convection effects can be estimated using the Newton's law of cooling.

$$q_{conv} = h(T_\infty - T_w) = \text{heat flux}[\text{W/m}^2] \tag{10.87}$$

h = heat transfer coefficient
T_∞ = combustion product gas temperature
T_w = wall temperature

The heat transfer coefficient, h, in turn, can be found from the Nusselt number.

$$\mathrm{Nu} = \frac{hL}{k} = ARe_D^{4/5}\mathrm{Pr}^{1/3} = \text{Nusselt number} \tag{10.88}$$

L = length scale (the local diameter or the distance from the throat)
k = thermal conductivity of the combustion product gas
Re_D = Reynolds number based on the local diameter, D.
$\mathrm{Pr} = \frac{c_p}{\mu k}$ = Prandtl number

For the combustion chamber and the nozzle section (down to about one throat diameter downstream of the throat), the constant $A = 0.023$ is used with L = local diameter. For the nozzle proper, $A = 0.025 \sim 0.028$ is used with L = distance from the throat.

The radiation heat transfer from the gas to the wall is estimated by assuming that the combustion product gas radiates like a black body with emissivity, $\varepsilon = 1$. This radiation is absorbed by the wall, while the wall itself radiates back some heat. The energy balance using the Stefan–Boltzmann law for radiation is

$$q_{rad} = \alpha_w \sigma T_\infty^4 - \varepsilon_w \sigma T_w^4 = \text{radiation heat flux}[\mathrm{W/m^2}] \tag{10.90}$$

σ = Stafan–Boltzmann constant $= 5.67 \times 10^{-8} \frac{\mathrm{W}}{\mathrm{K^4 m^2}}$
α_w = absorptivity of the wall
ε_w = emissivity of the wall

The wall itself is subject to convection and radiation heat flux on the gas side, while conduction through the wall transfers heat to the ambient and cooling fluids, if any. If we assume a cylindrical shape for the nozzle or the combustion chamber, then the heat transfer rate due to conduction is

$$q'_{cond} = 2\pi k_w \frac{T_w - T_o}{\ln \frac{r_o}{r_w}} = \text{heat rate per unit length}[\mathrm{W/m}] \tag{10.91}$$

k_w = thermal conductivity of the wall material
r_w = inner radius at the wall
r_o = outer radius of the wall
T_o = temperature at the outer side of the wall

The temperature distribution is a logarithmic function of the radius.

$$T(r) = \frac{T_w - T_o}{\ln \frac{r_w}{r_o}} \ln \frac{r}{r_o} + T_o \tag{10.92}$$

For a wall of small thickness, the above temperature profile is linearized.

$$T(r) = T_w + \frac{T_o - T_w}{\delta} r \quad (\text{for } \delta/r_w \ll 1) \tag{10.93}$$

$\delta = r_o - r_w$ = wall thickness

The heat removal rate by the coolant in contact with the outer wall can also be estimated by the Nusselt number. However, since the coolant is typically liquid (one of the propellants), vaporization can occur at the wall surface at high temperatures. If the bubbles thus formed diffuse into the liquid, then heat removal is enhanced due to the absorption of the heat of vaporization. This is called nucleate boiling. At high heat fluxes, however, the boiling can be pervasive and vapor bubbles can essentially cover the hot surfaces, and this then impedes heat transfer into the liquid since vapor thermal conductivity is much lower. This is termed film boiling and can lead to severe loss of cooling effectiveness, and must be avoided.

10.5 MATLAB® Program

```
%MCE101: Plot of normalized exit velocity
R=8314;
g1=1.15; g2=1.25; g3=1.35;
% Start of calculations
for i=1:50;
%
pr(i)=1/(2.5+(i-1)*(25-2.5)/50);
prr(i)=1/pr(i);

A1=1-pr(i)^((g1-1)/g1);
A2=1-pr(i)^((g2-1)/g2);
A3=1-pr(i)^((g3-1)/g3);

U1(i)=sqrt(A1);
U2(i)=sqrt(A2);
U3(i)=sqrt(A3);
end
plot(prr,U1,prr,U2,prr,U3)
```

10.6 Problems

10.1. For a rocket thrust chamber, following specifications are given:

Thrust $= F = 12$ MN
Combustion chamber pressure $= p^o = 12$ MPa
Combustion chamber temperature $= T^o = 2800$ K.

If $\gamma = 1.4$ and $R = 0.287$ kJ/kg K, determine the following ($p_e = p_a = 0.1$ MPa): Determine (a) specific impulse; (b) mass flow rate; (c) throat diameter; (d) exit diameter; (e) characteristic velocity and (f) thrust coefficient.

10.2. a. For $p_e/p_2^o = 0.15$, $\gamma = 1.35$, and molecular weight of 15 kg/kmol, plot the rocket jet exhaust velocity as a function of the stagnation temperature using Equation 10.4.

b. For $p_e/p_2^o = 0.15$, $\gamma = 1.35$, and $T_2^o = 2500$ K, plot U_e as a function of the molecular weight.

10.3. For $\gamma = 1.2$, plot the thrust coefficient, C_F, as a function of the area ratio, $A_e/A*$, for p_2^o/p_o from 2 to 1000 given $p_e/p_2^o = 0.55$.

$$C_F = \frac{F}{A * p_2^o} = \sqrt{\frac{2\gamma^2}{\gamma - 1}\left(\frac{2}{\gamma + 1}\right)^{\frac{\gamma+1}{\gamma-1}}\left[1 - \left(\frac{p_e}{p_2^o}\right)^{\frac{\gamma-1}{\gamma}}\right]} + \left(\frac{p_e}{p_2^o} - \frac{p_o}{p_2^o}\right)\frac{A_e}{A*} \quad (10.8)$$

10.4. For $\gamma = 1.1$, 1.2, 1.25, 1.30 and 1.4, plot the thrust coefficient, C_F, as a function of p_2^o/p_o for area ratio, $A_e/A*$, ranging from 5 to 100 given $p_e/p_2^o = 0.55$.

10.5. For the SSME shown in Figure 10.3, draw a flow chart for all the flows and estimate the volumetric flow rates in each of the flow lines.

10.6. Search the internet for data on the Gibbs function and tabulate as a function of temperature for some commonly encountered species: O_2, O, H_2, H, H_2O, CO_2, CO, N_2, N and OH.

10.7. What are the equilibrium concentrations of O_2 and O at $p = 1$, 0.5 and 0.1 atm, for $T = 2000$, 3000 and 4000 K? The oxygen dissociation reaction is: $O_2 \leftrightarrow 2O$

10.8. Plot the equilibrium concentrations of CO_2, CO and O_2, as a function of pressure at $T = 2500$K.

$$CO_2 \leftrightarrow CO + \frac{1}{2}O_2$$

10.9. Use an online equilibrium calculator, such as http://cearun.grc.nasa.gov/, to compute the equilibrium concentrations of reaction between liquid H_2 and liquid O_2 at an equivalence ratio of 1.0 for pressures of 0.1, 1, 10 and 100 atm.

10.10. Use an online equilibrium calculator, such as http://cearun.grc.nasa.gov/, to plot the equilibrium concentrations of the 10 most abundant species, as a function of the equivalence ratio, during chemical reaction of liquid H_2 and liquid O_2, at a pressure of 10 atm.

10.11. Use an online equilibrium calculator such as http://cearun.grc.nasa.gov/, to plot the equilibrium concentrations of solid-propellant UDMH as a function of the oxidizer–fuel mass ratio, for pressures of 1, 10 and 100 atm.

10.12. Estimate the Sauter mean diameter (D32) of the droplets from a pressure-atomized injector, for a liquid propellant of density 450 kg/m^3, surface tension of 8 N/m, and injection velocity of $u_{inj} = 120$ m/s. The mean speed of the droplets is measured to be 100 m/s. Neglect the viscous effects.

10.13. The length, L, required for complete evaporation of monopropellant droplets can be estimated as follows.

$$L = \frac{r_o^2\left(\frac{u_o}{\sqrt{\gamma RT}} + \frac{9}{34}\frac{\Pr}{B}\right)}{2 + \frac{9}{2}\frac{\Pr}{B}}\frac{\sqrt{\gamma RT}}{\nu}\frac{\rho_L}{\rho}\frac{\Pr}{\ln(1 + B)}, \text{ where } B = \frac{c_p(T - T_o)}{h_{fg}}$$

r_o = initial droplet radius = 50 μm
u_o = initial droplet velocity
T = combustion chamber temperature = 3000 K
Pr = Prandtl number of the ambient gas = 0.7

ν = kinematic viscosity of the ambient gas = $2 \times 10^{-6} \frac{m^2}{s}$
R = specific gas constant = 0.280 kJ/kg K
ρ_L = liquid density = 700 kg/m^3
ρ = gas density
T_o = initial droplet temperature = 300 K
h_{fg} = heat of vaporization = 750 kJ/kg
c_p = 1.025 kJ/kg K

Plot L as a function of the combustion chamber pressure p.

10.14. A rocket nozzle has the following specifications. $A* = 0.3m^2; \frac{A_e}{A*} = 5$. The ambient pressure is 0 MPa (vacuum). For the combustion chamber, the stagnation temperature is 3000 K, with the fluid of $\gamma = 1.4$ and molecular weight of 20 kg/kmol. If the stagnation pressure varies as a function of time (e.g. in solid rocket motors) as follows, find the values as a function of time of (a) the exhaust velocity, U_e; (b) the mass flow rate and (c) the thrust.

$$p_2^o = 20 - 0.04t \quad [\text{MPa}], t[\text{s}].$$

10.15. Plot the steady-state pressure in a solid-propellant motor as a function of $A_b/A*$. $S_b = 0.12(p_2^o)^{0.6}$ [cm/s], where pressure is in [atm]. $\gamma = 1.2$, $R = 0.255$ kJ/kg K, $T^o = 2500$ K, and $\rho_p = 779$ kg/m^3. The gas density may be neglected.

10.16. A hollow-cylinder solid-propellant grain burns only on the interior surface, at a burning rate of $S_b = 1.2(p^o)0.6$ cm/s where p^o is in MPa. At some point, interior diameter (d) of the grain is measured to 0.4 m when p^o is 8 MPa. The outer diameter of the grain is 1 m, while the length is 6 m. The propellant density 2500 kg/m^3. Estimate the rate of change of the combustion pressure if the combustion temperature is constant at 2800 K for $\gamma = 1.24$.

10.17. A solid rocket propellant has the following burning properties:

$$S_b = 0.06 \text{ cm/s at 5 atm}$$
$$S_b = 0.36 \text{ cm/s at 100 atm}$$

a. Find a and n in $S_b = ap_o^n$.
b. The propellant density is 1900 kg/m^3, $p_o = 7$ MPa, and $T_o = 2500$ K. What must be the steady-state ratio of the burning surface area, A_b, to the throat area, $A*$, if $\gamma = 1.25$ and molecular weight is 20.5?
c. If $A_b/A*$ increases at 0.7%/sec from the steady-state value, estimate the time that it would take the chamber pressure to double.

Bibliography

Hill, P. and Petersen, C. (1992) *Mechanics and Thermodynamics of Propulsion*, 2nd edn, Addison-Wesley Publishing Company.

Cornelisse, J.W., Schoyer, H.F.R. and Wakker, K.F. (1979) *Rocket Propulsion and Spaceflight Dynamics*, Pitman.

11

Non-Chemical Rockets

In electric propulsion devices, ionized mass is accelerated in an electromagnetic field. Other conceptual non-rocket systems include solar sails and nuclear propulsion. Aero-braking, tethers, rail-gun launch or gravity assist may also be classified as non-chemical propulsion methods. Most of the non-chemical rockets thus use the momentum principle of one form or another where high momentum ejection results in net thrust. Here, we will mostly deal with ion propulsion devices, which are currently in use, and await some breakthrough physics and engineering for more novel propulsion concepts.

Electric propulsion can be divided into electrothermal, electrostatic and electromagnetic. As the name suggests, electrothermal devices such as resistojet or arcjet use electrical heating of propellants to increase their thermal energy, as shown in Figure 11.1. The propellant is then accelerated in a nozzle, similar to chemical rockets, so that only the means of elevating the thermal energy of the propellant differs. The electrostatic and electromagnetic propulsion involves ionization of the propellants and acceleration of the ions in electric or electromagnetic fields as shown in Figure 11.1, which produces very high specific impulse. Thus, the primary advantage of ion propulsion is its high specific impulse. Table 11.1 shows some of the basic parameters of these devices, in comparison to monopropellant thrusters often used for similar purposes.

The high specific impulse, I_{sp}, of ion rockets means that a relatively small propellant mass is required. The propellant mass required to achieve thrust, F, is given by

$$m_p = \dot{m}_p \Delta t = \frac{F\Delta t}{U_e} = \frac{F\Delta t}{g_o I_{sp}} \tag{11.1}$$

The reduction of the propellant mass comes with the cost of high mass needed for the electric power supply system. We can estimate the power supply mass as being a linear function of the output power, P. The power conversion efficiency, η, may be defined as the ratio of propellant kinetic energy to the electric power output P. Then we have

$$m_S = \alpha P = \frac{\alpha \dot{m}_p U_e^2}{2\eta} = \frac{\alpha F U_e}{2\eta} = \frac{\alpha g_o F I_{sp}}{2\eta} \tag{11.2}$$

Aerospace Propulsion, First Edition. T.-W. Lee.
© 2014 John Wiley & Sons, Ltd. Published 2014 by John Wiley & Sons, Ltd.
Companion Website: www.wiley.com/go/aerospaceprop

Table 11.1 Typical performance of electric propulsion systems, in comparison to monopropellants.

Propulsion system	F [mN]	I_{sp} [s]	η [%]	Burn time	Propellants
Electrothermal	200∼1000	200∼1000	30∼90	Months	NH_3, N_2, H_2, N_2H_4, NH_3 $N_2H_4H_2$
Electrostatic	0.01∼200	1500∼5000	60∼80	Months	Xe, Kr, Ar
Electromagnetic	0.01∼2000	600∼5000	30∼50	Weeks ∼ Years	Teflon, Ar, Xe, H_2, Li
Monopropellant	30∼100 000	200∼250	80∼90	Hours	N_2H_4, etc.

Thus, the high specific impulse requires high system mass. An optimum can be found by differentiating $m_S + m_p$ with respect to I_{sp}.

$$\left(I_{sp}\right)_{opt} = \frac{1}{g_o}\left(\frac{2\eta\Delta t}{\alpha}\right)^{1/2} \tag{11.3}$$

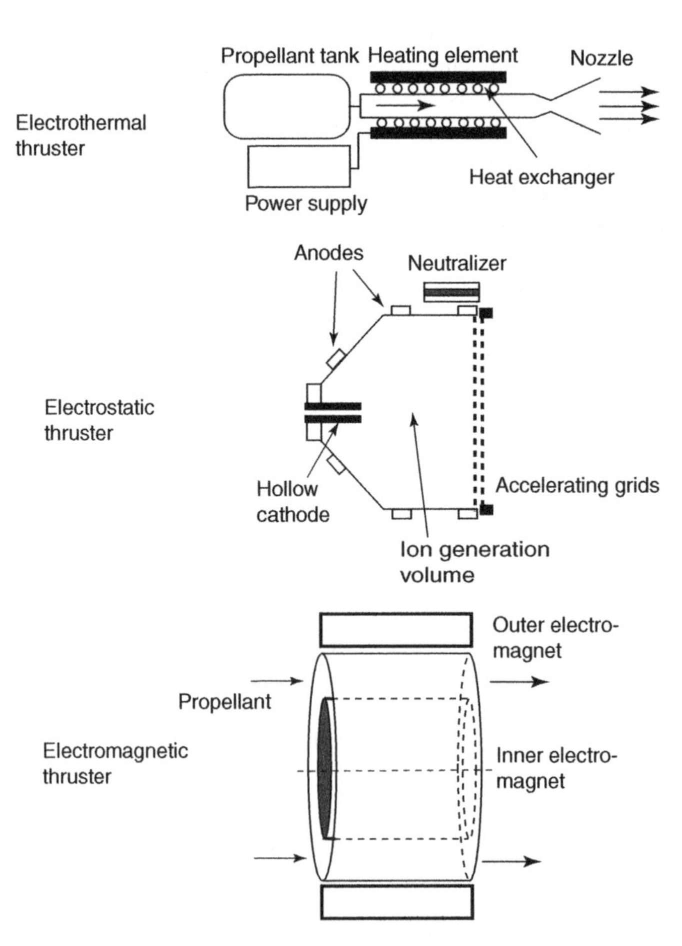

Figure 11.1 Electric rockets.

Substitution of this optimum I_{sp} in Eq. (11.2) then gives the required power supply system mass. For realistic values of α, ion propulsion systems are limited to small thrust applications according to Eq. (11.2).

11.1 Electrothermal Devices

Electrothermal rockets deposit energy to the propellant by either resistive heating (resistojet) or electrical sparks (arcjet). The resistive heating can be quantified by

$$P = I^2 R \quad [\mathrm{W}] \tag{11.4}$$

I = current in amps [A]
R = resistance in ohms [Ω]

The resistance is a function of the material resistivity (ρ_e), length (l) and cross-sectional area (A) of the heating element.

$$R = \frac{\rho_e l}{A} \quad [\Omega] \tag{11.5}$$

The surface heating of the propellant can be estimated assuming that the propellant is radiating like a blackbody. The steady-state energy balance for the propellant gas is

$$h(T_w - T_g) + \sigma\varepsilon(T_w^4 - T_g^4) = 0 \tag{11.6}$$

h = heat transfer coefficient
T_w = surface temperature of the heating element
T_g = propellant gas temperature
σ = Stefan–Boltzmann constant $= 5.670 \times 10^{-8} \frac{\mathrm{J}}{\mathrm{K^4 m^2 s}}$
ε = emissivity of the surface

To maximize the heat transfer to the propellant and to minimize the radiation loss to the surroundings, various geometries for electrothermal heating can be envisioned as shown in Figure 11.2.

The propellant gas absorbs the heat, in principle according to its specific heat. High specific heat of the propellant is directly translated into the high exhaust velocity, since the propellant contains a large amount of thermal energy that can be converted to kinetic energy. At the molecular level, at high temperatures, the propellant gas absorbs the heat in several different ways. The first obvious effect of heating is an increase in the propellant gas temperature, which is indicative of the kinetic energy of the random motion of the molecules. The energy is also called the translational energy. At high temperatures, the molecular vibration, rotation and electronic energy may become excited, so that the molecules contain more energy than indicated by the translational energy, or the gas temperature. At yet higher temperatures, the molecules can dissociate into separate atoms, and this again requires heat input (dissociation energy) being given to the propellant. Ionization can also occur for both the molecules and the atoms, if they receive ionization

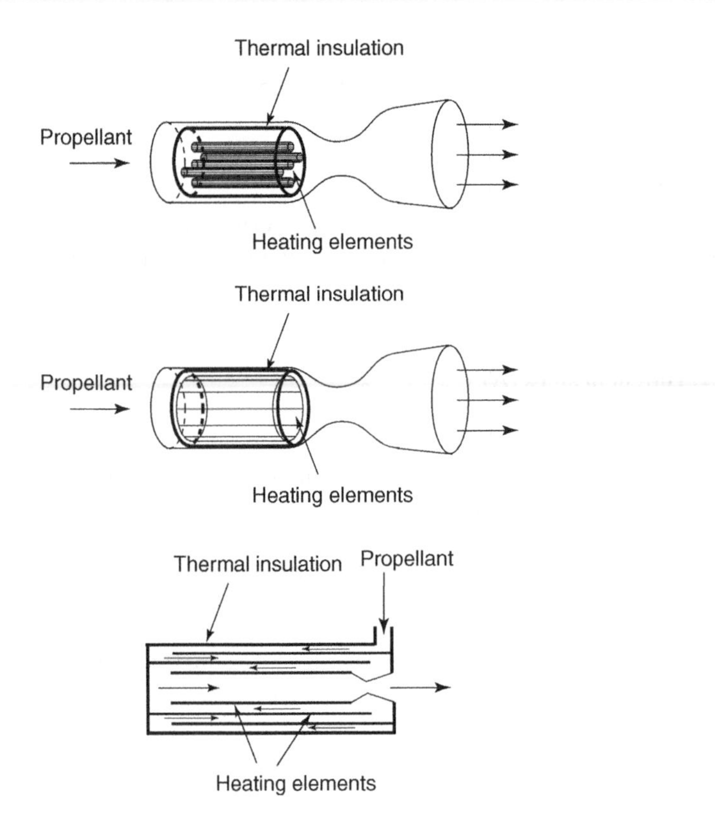

Figure 11.2 Various electrothermal rockets.

energy. Therefore, the net content of the energy consists of translational, rotational, vibrational, electronic, dissociation and ionization. In theory, this energy content can be calculated from the mole fraction of constituent particles (molecules, atoms, ions and electrons) and their energy contents. To start, the individual mole fractions, α's, give the number concentrations.

$$N_2 = \alpha_2 N_o = \text{number concentration of neutral molecules} \tag{11.7a}$$

$$N_1 = \alpha_1 N_o = \text{number concentration of neutral atoms} \tag{11.7b}$$

$$N_{2+} = \alpha_{2+} N_o = \text{number concentration of molecular ions} \tag{11.7c}$$

$$N_{1+} = \alpha_{1+} N_o = \text{number concentration of atomic ions} \tag{11.7d}$$

$$N_e = \alpha_e N_o = \text{number concentration of electrons} \tag{11.7e}$$

N_o = initial number concentration of molecules

The mole fractions are not independent, but are governed by the conservation of atomic particles and conservation of electric charge.

$$\alpha_2 + \alpha_{2+} + \frac{1}{2}\alpha_1 + \frac{1}{2}\alpha_{1+} = 1 \tag{11.8}$$

$$\alpha_{2+} + \alpha_{1+} = \alpha_e \tag{11.9}$$

For molecules (neutral and ionic), the energy consists of translational, rotational, vibrational and electronic.

$$e_{2or2+} = \alpha_{2or2+}N_o\left(\frac{3}{2}kT + \beta_{rot}kT + \beta_{vib}kT + \sum_i \beta_i\varepsilon_i\right) \tag{11.10}$$

k = Boltzmann's constant = $1.3806503 \times 10^{-23}\ \frac{\text{m}^2\text{kg}}{\text{s}^2\text{K}}$
β_{rot} = fraction of rotationally excited molecules
β_{vib} = fraction of vibrationally excited molecules
β_i = fraction of electronically excited molecules to the i-th state
ε_i = energy of the i-th electronic state

We can write similar expressions for atoms and electrons, except that they can only have translational and electronic energies.

$$e_{1or1+} = \alpha_{1or1+}N_o\left(\frac{3}{2}kT + \sum_i \beta_i\varepsilon_i\right) \tag{11.11}$$

$$e_e = \alpha_e N_o \frac{3}{2}kT \tag{11.12}$$

Figuring out all the α's and β's is a lengthy process, containing many approximations. In any event, the important aspect is that the energy contained in all of the other modes can be transferred back to thermal energy through molecular collisions and recombination reactions (back to neutral molecules). If there is sufficient time for the energy transfer to occur ($\xi = 0$), then all of the energy contained in these modes will be added to the propellant enthalpy. At the other extreme, if the flow is too fast, and there is no time for any of these energy transfer to occur ($\xi = 1$, or "frozen flow"), then there is only the thermal enthalpy that can be converted to the kinetic energy of the nozzle flow. The relaxation time for the excited energy modes to return to ground states depends on the propellant pressure and temperature, so that it is one of the strong considerations for choosing a propellant for a given chamber conditions, so that for our purposes we can simply express the energy content as

$$e_{\max} = e_{trans} + (1 - \xi)\Delta e \tag{11.13}$$

e_{max} = total energy content
e_{trans} = thermal energy (due to translational motion of molecules)
Δe = energy contained in all other modes
ξ = frozen flow fraction

Thus, we want propellants that would have minimal frozen flow fraction at the operating conditions.

Arcjets use gaseous discharges, or electric sparks, to impart energy to the propellant gas. The high-enthalpy propellant gas then undergoes gas-dynamic expansion in a nozzle to generate thrust. If the electric potential, or voltage, is sufficiently high to overcome the initial electric resistance, then current can be established through a stream of electrons (from cathode to anode) through a discharge plasma. The discharge occurs through the following sequence of events. First, a small number of electrons and ionized particles can exist in the gap between the cathode and anode. These particles are created by stray radiation such as cosmic rays or simply through thermal effects, albeit at very low probabilities. These particles would not last very long, except that, in a strong electric field, they can accelerate toward respective electrodes, that is, electrons toward the anode and ions toward the cathode. These accelerated particles collide with the gas molecules, causing an avalanche of electrons and excited ions. The excited ions emit radiation when they return to the ground states, hence the bluish glow from electrical sparks. When the ions accelerate and collide with the cathode surface, further electrons are produced. With this supply of electrons and further collisions to produce yet more ions, the discharge becomes a self-sustaining process as long as the electrical potential exists. These collisions also increase the thermal energy of the plasma, which is how the energy is imparted to the propellant in an arcjet. To concentrate the energy of the electrical spark to a small volume of propellants, a fluid vortex or a physical constrictor may be used as shown in Figure 11.3.

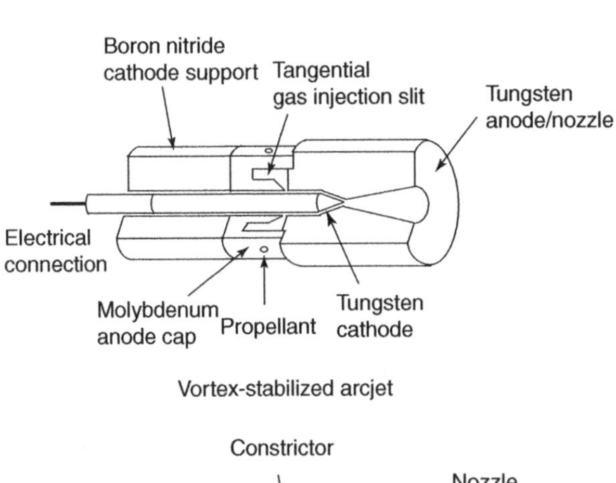

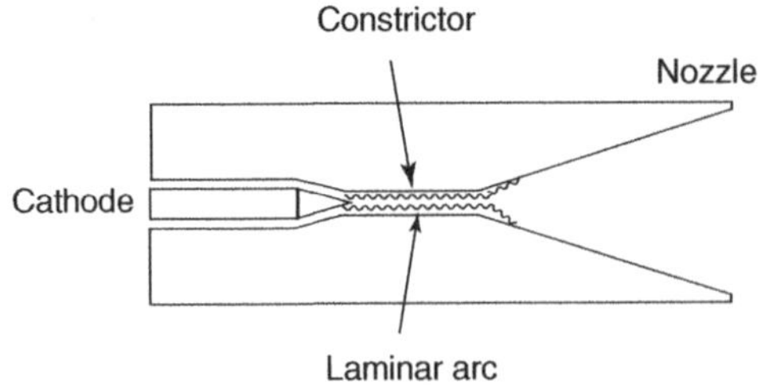

Figure 11.3 Arcjet thrusters.

The plasma temperature can be quite high, and arcjets can thus generate higher specific impulse than resistojets. For example, at 1 atm pressure a peak temperature of up to 60 000 K can be achieved with electrical potential of 250 V/cm for a current of 150 A.

11.2 Ion Thrusters

Ion thrusters work by generating ions from the propellant, and then accelerating these ions in an electric or electromagnetic field. The ion generation also produces excess electrons, which need to be discharged from the spacecraft to avoid accumulating negative charges. Otherwise, the negative charges will attract the ions back to the spacecraft to negate the momentum that was produced.

The operation of the ion thruster is schematically shown in Figure 11.4. The propellant is typically an inert gas, such as xenon, which is injected from a downstream location toward the upstream end of the ionization chamber. The ion generation is achieved through electron bombardment where a discharge cathode located centrally at the upstream end of the chamber produces electrons, which accelerate toward the anode chamber wall. The electron motion is also modified by a magnetic field in order to maximize the residence time of the electrons, which in turn increases the number of collisions between the electrons and the propellant atoms. An alternative means of ion generation is the so-called electron cyclone resonance, which uses electromagnetic radiation (typically microwave) coupled with a

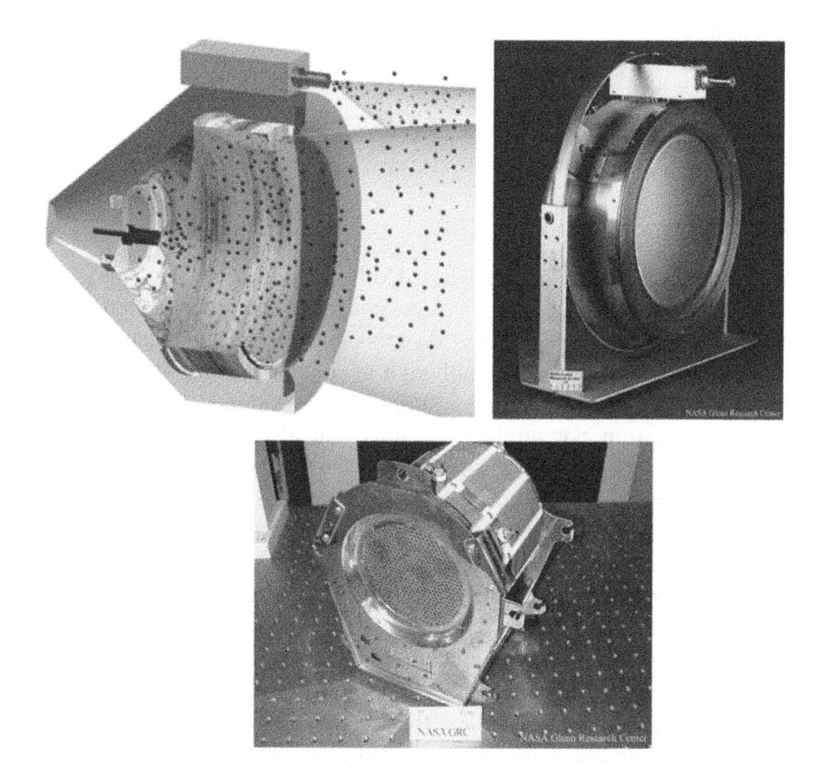

Figure 11.4 Ion thruster. Courtesy of NASA.

magnetic field raising the electronic energy states to the point of ionization. As noted above, the excess electrons are dispelled out of the spacecraft through a second hollow cathode called a neutralizer. The ions are accelerated in an electric or electromagnetic field. In the electric acceleration, the anode grid with thousands of co-axial apertures draw the electrons and then send them through the apertures.

A much simpler form of electrostatic propulsion is through the "field emission" effect. The separate ion generator of conventional ion thrusters is replaced by a capillary slit in field emission electric propulsion (FEEP) devices. The FEEP thruster consists of a capillary slit through which propellants such as cesium flow, and an electric field between the slit and the accelerating grid. If the electric field close to the slit exit reaches the field emission levels (10^9 V/m), then electrons are emitted from the slit surface to cause ionization of the propellant. Ions are then accelerated by the same electric field, between the anode (slit) and the cathode grid, close to the thruster exit. Again the ions discharged from the thruster are neutralized by the electron beam, as in conventional ion thrusters.

For an ion with an electrical charge of q and mass m in an electric potential difference (voltage) of V_a, the conversion of the potential energy to kinetic energy generates the exhaust velocity.

$$\frac{mU_e^2}{2} = qV_a \rightarrow U_e = \sqrt{\frac{2qV_a}{m}} \tag{11.14}$$

The specific impulse is

$$I_{sp} = \frac{1}{g_o}\sqrt{\frac{2qV_a}{m}} \tag{11.15}$$

11.2.1 Ion Generation

Ions can be produced in a number of ways. Collision with particles (molecules, atoms or electrons) with high kinetic energy, electromagnetic radiation (photon absorption), strong electric field, surface-contact electron transfer or chemical reactions with another molecule or atom can all produce ions, although for ion propulsion the electron collisions are primarily used. The electrical power supplied to accelerate electrons is consumed to overcome the ionization potential of the ions, but also is lost through thermal and radiative losses. The ion generator should also be designed to convert a very high percentage of the propellant into ions, emitting few neutral atoms into the acceleration chamber. Neutral atoms in the acceleration chamber can interfere with the motion of ions and also erode the accelerating electrodes. Ion propulsion systems typically operate for a mission duration of the order of 10 000 hours or more, so that a stable ion source is an important element.

In surface-contact ionization sources, alkali atoms such as cesium adhere to metal surfaces due to the negative charges of the conducting electrons of the metal. If the ionization potential is lower than the work function of the surface, then the electrons of the alkali atoms are yielded to the metal. The resulting alkali ions are bounced from the surface at high metal surface temperatures, and return to the flow. Thus, surface-contact ionization requires a heated

Example 11.1 Specific impulse of ion thruster using mercury (Hg)

For singly ionized mercury, $m = 3.33 \times 10^{-25}$ kg and $q = 1.598 \times 10^{-19}$ C. For an electrical potential of $V_a = 1000$ V, the exhaust velocity is given by Eq. (11.14).

$$U_e = \sqrt{\frac{2qV_a}{m}} = \sqrt{\frac{2(1.598 \times -10^{-19}\,\mathrm{C})(1000\,\mathrm{V})}{3.33 \times 10^{-25}\,\mathrm{kg}}} = 30,980\,\mathrm{m/s}$$

Here, we have used the unit conversion of 1 J = (1 V) × (1 C). And, $I_{sp} = 3161\ \mathrm{s}^{-1}$ from Eq. (11.15).

The beam current, I, representing the charge flow rate of the ions is

$$I = \frac{q}{m}\dot{m}_p \tag{11.16}$$

The beam power P_b can be calculated by the product of the current and voltage.

$$P_b = IV_a = \frac{q}{m}\dot{m}_p V_a \tag{11.17}$$

Combining this result with Eq. (10.14) shows that the beam power is

$$P_b = \dot{m}_p \frac{U_e^2}{2} \tag{11.18}$$

A power efficiency can be defined as

$$\eta = (\text{beam power})/(\text{total electrical power}) = \frac{P_b}{P_b + \varepsilon_i + \varepsilon_l} \tag{11.19}$$

ξ_i = ionization energy
ξ_l = energy loss

chamber to vaporize the propellant and a porous metal ionizer which is also heated to a temperature of 1300–1500 K for tungsten.

The electron bombardment to produce the ions involves motion of electrons from the central cathode to the anode wall in a magnetic field. The rate of production of ions during electron bombardment can be express as

$$\frac{dn_i}{dt} = Q_i \dot{N}_e n_A \tag{11.20}$$

n_i = number density of ions
Q_i = ionization cross-section
$\dot{N}_e$ = number of electrons per unit volume per unit time
n_A = number density of propellant atoms

The rate of ion production depends on the number density of electrons and propellant atoms, and also on the ionization cross-section, which is the probability that the electron–atom collision would produce ions. Thus, the ionization cross-section is related to the kinetic energy of the electrons and atoms, as well as to the ionization energy. The ions produced can also recombine to return to the neutral state similar to a chemical reaction, so that the equilibrium concentration of ions can be estimated.

$$\mathrm{A} \leftrightarrow \mathrm{A}^+ + \mathrm{e} \tag{11.21}$$

A = neutral atom
A^+ = ion
e = electron

The equilibrium constant K_n can be written in a similar manner as for chemical reactions.

$$K_n = \frac{n_i n_e}{n_A} = \frac{F_i F_e}{F_A} \tag{11.22}$$

n_e = number density of electrons

In Eq. (11.22), the F's are the so-called partition functions which represent the probability of a particular particle existing in its state. For example, for the neutral atom the energy states include the translational kinetic energy and electronic energy levels, so that the total partition functions, F_A, can be decomposed into translational (f_{At}) and electronic (f_{Ae}) partition functions.

$$F_A = f_{A,t} f_{Ae} \tag{11.23}$$

The translational particle function represents the probability (or number fraction) of finding the neutral atom A with a certain kinetic energy expressed in terms of temperature, T. From statistical mechanics, this partition function is

$$f_{At} = \frac{(2\pi m_A kT)^{3/2}}{h^3} \tag{11.24}$$

m_A = mass of the atom
k = Boltzmann's constant (1.38×10^{-23} J/K)
h = Planck's constant (6.62×10^{-34} J-s)

The partition function of the electron energy states can be written in a similar manner.

$$f_{Ae} = \sum_j g_j \exp\left(-\frac{\varepsilon_j}{kT}\right) \tag{11.25}$$

g_j = degeneracy of the state, j (number of possible states with electronic energy ξ_j)
ξ_j = energy level of state j

The partition function for the ions is similar, except that we use the different ion electronic energy levels, ε'_j.

$$F_i = f_{i,t} f_{i,e} = \frac{(2\pi m_i kT)^{3/2}}{h^3} \sum_j g'_j \exp\left(-\frac{\varepsilon'_j}{kT}\right) \quad (11.26)$$

The electron partition function is simpler, since it carries only the kinetic energy with a spin degeneracy of 2 (i.e. there are two types of electrons with different spins).

$$F_e = 2\frac{(2\pi m_e kT)^{3/2}}{h^3} \quad (11.27)$$

We can now define the degree of ionization, α.

$$\alpha = \frac{n_i}{n_o} = \frac{n_i}{n_A + n_i} \quad (11.28)$$

n_o = total number of atoms available for ionization

We can also relate the number of electrons and ions, through the conservation of electrical charge.

$$n_i = n_e \quad (11.29)$$

The combined plasma can be assumed to obey the ideal gas equation of state, so that

$$p = (n_e + n_i + n_A)kT = (1+\alpha)n_o kT \quad (11.30)$$

Using these results to rewrite K_n, we have

$$K_n = \frac{n_i n_e}{n_A} = \frac{n_i^2}{n_A} = \frac{\alpha^2 n_o}{(1-\alpha)} \quad (11.31)$$

Substitution of n_o from Eq. (11.30) into Eq. (11.31) and then (11.22) leads to an expression for the degree of ionization.

$$\frac{\alpha^2}{1-\alpha^2} = \frac{2(2\pi m_e kT)^{5/2}}{ph^3} \frac{f_{i,e}}{f_{Ae}} \exp\left(-\frac{\varepsilon_j}{kT}\right) \quad (11.32)$$

This result is called the Saha equation, and shows that the degree of ionization increases with increasing temperature and decreasing pressure.

The motion of electrons during ion generation can be estimated using Newton's second law. The actual motion is much more complex, with interaction between particles, and ongoing ionization and recombination reactions. For preliminary analysis, however, a simple solution as presented below can give some idea as to the required dimensions and fields for ion generators.

$$m\frac{d\vec{u}}{dt} = \vec{F} \tag{11.33}$$

For a charged particle, such as an electron, the force acting upon the particle is

$$\vec{F} = q\vec{E} + q(\vec{u} \times \vec{B}) \tag{11.34}$$

q = electrical charge
$\vec{E}$ = electrical field strength
$\vec{B}$ = magnetic field strength

For an axisymmetric geometry with a constant magnetic field as shown in Figure 11.5, the radial component of Newton's second law is

$$m_e\left(\frac{du_r}{dt} - u_\theta\frac{d\theta}{dt}\right) = -e\left(-\frac{dV}{dr}\right) + e\left(r\dot{\theta}\right)B \tag{11.35}$$

$$E_r = -\frac{dV}{dr}$$

$B_r = B_r = B$ = magnetic field strength
e = electron electrical charge (−)

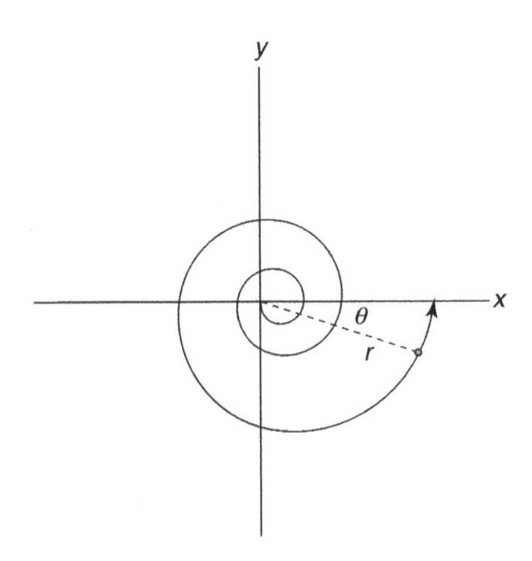

Figure 11.5 Motion of a "free" electron in an ion generator.

Using the dot notation for the time derivative throughout, this becomes

$$m_e(\ddot{r} - r\dot{\theta}^2) = e\frac{dV}{dr} + er\dot{\theta}B \tag{11.36}$$

The tangential component for symmetric electric field is

$$m_e(2\dot{r}\dot{\theta} + r\ddot{\theta}) = -e\dot{r}B \tag{11.37a}$$

$$\text{Or,}\quad m_e\frac{d}{dt}(r^2\dot{\theta}) = -r(e\dot{r}B) = -eB\frac{d}{dt}\left(\frac{1}{2}r^2\right) \tag{11.37b}$$

This is integrable.

$$m_e r^2\dot{\theta} = \frac{-eB}{2}r^2 + C \tag{11.38}$$

The constant may be evaluated using the boundary condition

$$\dot{\theta} \approx 0 \text{ at } r = r_c \tag{11.39}$$

This gives

$$\dot{\theta} = \frac{-eB}{2m_e}\left(1 - \frac{r_c^2}{r^2}\right) \tag{11.40}$$

The angular speed increases with increasing magnetic strength B and increasing distance from the core, so that electrons travel in spirals (as opposed to a straight line from the core to the chamber wall) resulting in much higher probability of high-energy collisions with propellant atoms.

11.2.2 Acceleration of Ions

The ions generated in the ion generator need to be accelerated to high speeds in order to produce thrust. In a simple electrostatic thruster, the cathode placed at the downstream end serves this function, as shown in Figure 11.6.

In order to analyze the ion acceleration, let us consider the electrostatic and electromagnetic forces on a charged particle. The force between two charged particles is given by the Coulomb's law.

$$\vec{F}_{ij} = \frac{1}{4\pi\varepsilon_o}\frac{q_i q_j}{r_{ij}^2}\vec{e}_{ij} \tag{11.41}$$

ξ_o = permittivity of free space = $8.85 \times 10^{-12}\ \frac{C^2}{Nm^2}$
r_{ij} = distance between the two particles

Example 11.2 Find the motion of a particle in a non-uniform magnetic field of $B = B_a\left(\frac{r}{r_a}\right)^2$.

Substituting the above magnetic field in Eq. (11.37b), we have

$$m_e\frac{d}{dt}(r^2\dot{\theta}) = -r(e\dot{r}B) = -eB_a\left(\frac{r}{r_a}\right)^2 r\dot{r} = -e\frac{B_a}{r_a^2}\frac{d}{dt}\left(\frac{r^4}{4}\right) \tag{E11.2.1}$$

Integration gives

$$m_e r^2\dot{\theta} = -e\frac{B_a}{r_a^2}\left(\frac{r^4}{4}\right) + C \tag{E11.2.2}$$

At the cathode, the angular velocity is small at $r = r_c \ll 1$, so that C is approximately zero.

The equation of motion is then

$$\dot{\theta} = -e\frac{B_a}{4m_e}\left(\frac{r}{r_a}\right)^2 \tag{E11.2.3}$$

$$\text{At } r = r_a, \dot{\theta}_a = -e\frac{B_a}{4m_e} \text{ or } (u_\theta)_a = r_a\dot{\theta}_a = -e\frac{r_aB_a}{4m_e} \tag{E11.2.4}$$

q_i, q_j = electrical charge of particle i and j

$\vec{e}_{ij}$ = unit vector connecting the line between the two charged particles

This can be extended to a medium consisting of many charged particles, by superposing the Coulomb's law. Also, in a medium with a finite conductivity, the permittivity of free space is modified by the dielectric constant, κ. The dielectric constant for air, for example, is 1.0006,

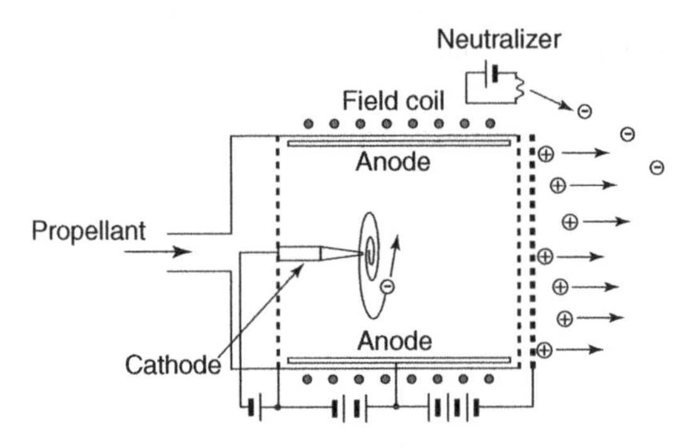

Figure 11.6 An electrostatic thruster.

80 for water, 11.68 for silicon, and so on.

$$\vec{F}_{ij} = \frac{1}{4\pi\kappa\varepsilon_o}\sum_j \frac{q_i q_j}{r_{ij}^2}\vec{e}_{ij} = \frac{1}{4\pi\varepsilon}\sum_j \frac{q_i q_j}{r_{ij}^2}\vec{e}_{ij} \tag{11.42}$$

$$\varepsilon = \kappa\varepsilon_o$$

In some instances, as in the ion acceleration chamber, we are interested in the force acting on a particular particle, so that the electrical field becomes a useful concept. The electrical field is defined as the electrical force per electrical charge.

$$\vec{F}_{ij} = q_i\vec{E}_i \tag{11.43}$$

Then, the electrical field can computed as

$$\vec{E}_i = \frac{1}{4\pi\varepsilon}\sum_j \frac{q_j}{r_{ij}^2}\vec{e}_{ij} \tag{11.44}$$

For a single particle of charge q, enclosed in a spherical volume, the divergence theorem for the electrical field is

$$\int \vec{E}\cdot d\vec{A} = \int \nabla\cdot\vec{E}\,dV \tag{11.45a}$$

$$\frac{1}{4\pi\varepsilon}\frac{q}{R^2}4\pi R^2 = \frac{q}{\varepsilon} \approx (\nabla\cdot\vec{E})V \tag{11.45b}$$

$$\text{So that } \nabla\cdot\vec{E} = \frac{q/V}{\varepsilon} \tag{11.45c}$$

This can be extended to multiple charges by introducing the charge density, which replaces the numerator in Eq. (11.45c).

$$\rho_e = \frac{dq}{dV} = (\text{charge})/(\text{volume}) \tag{11.46}$$

Then, the current density (flux), j, may be defined as

$$j = \rho_e u = (\text{charge density})(\text{mean speed of the particles}) \tag{11.47}$$

Equation (11.45c) becomes

$$\nabla\cdot\vec{E} = \frac{\rho_e}{\varepsilon} \tag{11.48}$$

This is the Gauss's law.

Now, let us consider a simple ion acceleration chamber where a voltage difference of $(V_a - V)$ is applied over a distance of Δx, as shown in Figure 11.7. The electrical potential energy and kinetic energy of the ions are conserved.

$$qV_a = qV + \frac{mu_e^2}{2} \tag{11.49}$$

This gives us the ion velocity that is achieved in the electric field.

$$u_e = \sqrt{\frac{2q(V_a - V)}{m}} = \frac{j}{\rho_e} \tag{11.50}$$

We can use this relationship to express the charge density, ρ_e.

$$\rho_e = \frac{j}{\sqrt{\frac{2q}{m}}}(V_a - V)^{-1/2} \tag{11.51}$$

For the one-dimensional electric field as shown in Figure 11.7$E = -dV/dx$, so that Eq. (11.48) becomes

$$\frac{d^2V}{dx^2} = -\frac{\rho_e}{\varepsilon} = -\frac{j}{\varepsilon\sqrt{2q/m}}(V_a - V)^{-1/2} \tag{11.52}$$

This can be solved conveniently by substituting $V^* = Va - V$, and multiplying by $2dV^*/dx$.

$$2\frac{dV^*}{dx}\frac{d^2V^*}{dx^2} = 2\alpha\frac{dV^*}{dx}(V^*)^{-1/2} \tag{11.53}$$

$$\alpha = \frac{j}{\varepsilon\sqrt{2q/m}}$$

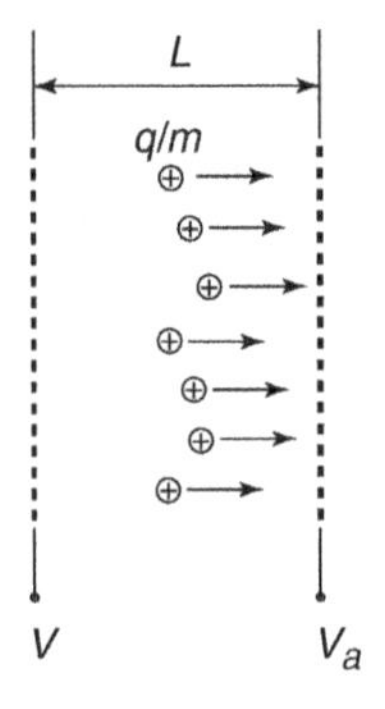

Figure 11.7 An ion accelerator.

$$d\left[\left(\frac{dV^*}{dx}\right)^2\right] = 2\alpha(V^*)^{-1/2}dV^* \tag{11.54}$$

This form is twice integrated, along with the boundary condition of $V^* = 0$ and $dV^*/dx = 0$ at $x = 0$, to give the following solution.

$$\frac{4}{3}(V^*)^{3/4} = 2\sqrt{\alpha}x \tag{11.55}$$

$$\text{Or, } V = V_a - \left(\frac{q}{4}\frac{j}{\varepsilon\sqrt{2q/m}}\right)^{2/3} x^{4/3} \tag{11.56}$$

For chamber length of L and $|V - V_a| = V_a$,

$$j = \frac{4}{9}\varepsilon\sqrt{\frac{2q}{m}}\frac{V_a^{3/2}}{L^2} \tag{11.57}$$

We can relate the mass flux with the current density, j.

$$\frac{\dot{m}}{A} = \rho u_e = \rho_e\frac{m}{q}u_e = \frac{\rho_e u_e}{q/m} = \frac{j}{q/m} \tag{11.58}$$

ρ = ion mass density

For this electric field, the ion velocity from Eq. (11.50) is

$$u_e = \sqrt{\frac{2qV_a}{m}} = \frac{j}{\rho_e} \tag{11.59}$$

So, the thrust per unit area of the thruster is

$$\frac{F}{A} = \frac{\dot{m}u_e}{A} = \frac{\dot{m}}{A}u_e = \frac{j}{q/m}\sqrt{\frac{2qV_a}{m}} = \sqrt{\frac{2}{q/m}}jV_a^{1/2} \tag{11.60}$$

Using Eq. (11.57),

$$\frac{F}{A} = \frac{8}{9}\varepsilon\left(\frac{V_a}{L}\right)^2 \tag{11.61}$$

11.2.3 Electromagnetic Thrusters

Electromagnetic thrusters typically use either self-generated or applied magnetic fields. In magnetoplasmadynamic (MPD) thrusters, as shown in Figure 11.8, a continuous electrical

Example 11.3 A simple ion thruster

Let us compute the requirements for a 0.5 N ($I_{sp} = 3000$s) thruster with propellant that has $q/m = 500$ C/kg. The applied electric field is $(V_a/L)_{max} = 10^5$ V/cm.

From Eq. (11.15), we obtain $V_a = \frac{(g_o I_{sp})^2}{q/m} = 866000$ V.

Then, Eq. (11.59) gives $j = \frac{4}{9}\varepsilon\sqrt{\frac{2q}{m}}\frac{V_a^{3/2}}{L^2} = 13.37$ A/m^2.

Equation (11.63) can then be used to find the required cross-sectional area to generate 0.5 of thrust.

$$A = \frac{F}{\frac{8}{9}\varepsilon_o\left(\frac{V_a}{L}\right)^2} = 0.000\,626\text{ m}^2$$

field generates a magnetic field, while in pulsed plasma thrusters (PPT), a series of short arcs is used. In PPTs, the strong arcs can ablate solid propellants such as Teflon so that complexities of gas propellant feed are removed. In these devices, the strong electrical current produces the ions, and also generates sufficient magnetic field to propel these ions through the Lorentz force, expressed by

$$\vec{F} = q\vec{E} + q(\vec{u} \times \vec{B}) \tag{11.34}$$

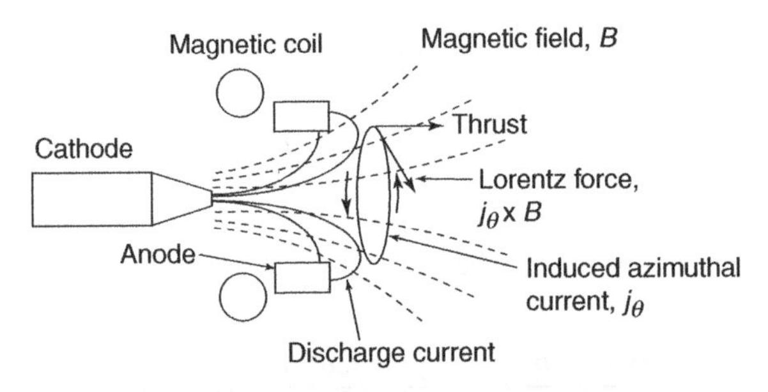

Figure 11.8 A magnetoplasmadynamic thruster. Courtesy of NASA.

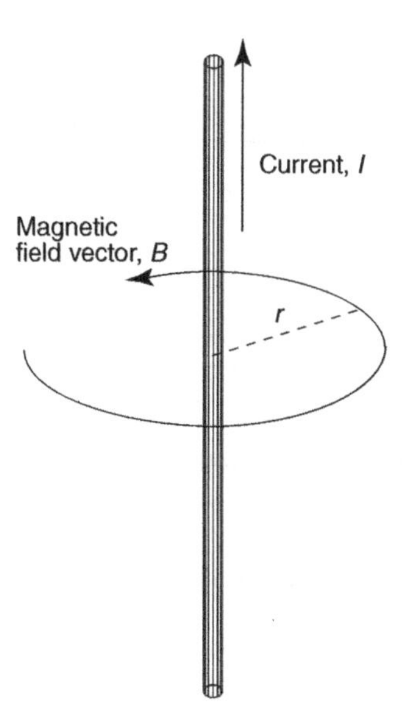

Figure 11.9 Ampere's law.

According to Ampere's law, a current generates a magnetic field, as shown in Figure 11.9.

$$\oint \vec{B} \cdot d\vec{s} = \frac{4\pi}{c} I_{encl.} \tag{11.62}$$

For example, if we consider a circular path around a straight wire, as shown in Figure 11.9, then the path integral in Eq. (11.62) becomes

$$B(r)2\pi r = \frac{4\pi}{c} I \tag{11.63}$$

So the magnetic field generated by the current through the wire, I, is

$$B(r) = \frac{2I}{rc} \quad [T, \text{ tesla}] \tag{11.64}$$

The unit for the magnetic field is tesla.

$$1\,\mathrm{T} = 1\,\frac{\mathrm{N}}{\mathrm{C(m/s)}} \tag{11.65}$$

Due to the speed of light in the denominator of Eq. (11.64), a large current is required to generate even a moderate magnetic field. The magnetic field is also a vector, which follows

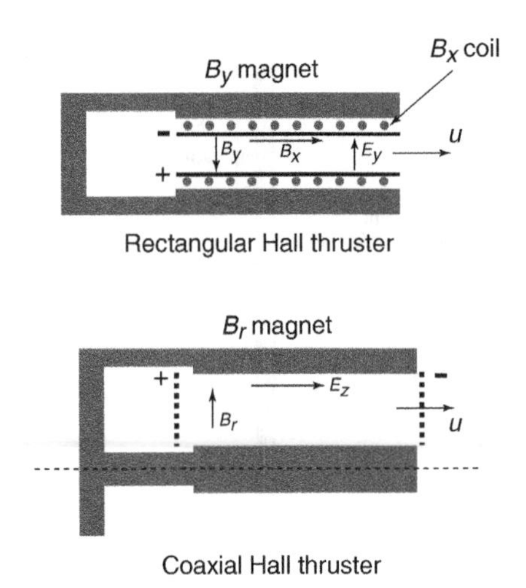

Figure 11.10 Hall effect thrusters.

the right-hand rule, so that for the upward current in Figure 11.9, the direction of the magnetic field vector is counter-clockwise.

Hall effect thrusters (HET) use applied magnetic field at fairly low plasma density or strong magnetic field. The electrical current associated with the ion generation produces the current density, or the movement of charges. This current density interacts with the magnetic field, according to the second term in Eq. (11.34) to produce ion acceleration, which is referred to as the Hall effect. The relative orientations of the electrical and magnetic fields need to produce the force component in the azimuthal direction, as shown in Figure 11.10.

Under simple configurations such as the one shown in Figure 11.11, we may be able estimate the acceleration of ions in a magnetic field. In addition, we consider isothermal flows, for which a simple solution is found. The governing equations of motion follow the three conservation laws of mass, momentum and energy, along with some constitutive relationships.

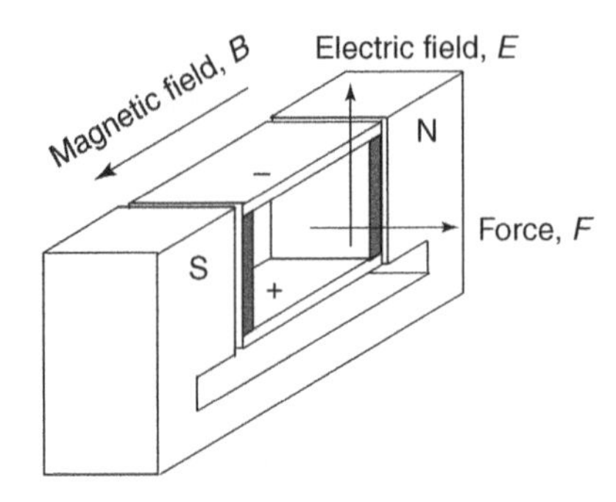

Figure 11.11 Ion acceleration in a simple electromagnetic field.

$$\rho u = M = const. \tag{11.66}$$

$$\rho u \frac{du}{dx} = -\frac{dp}{dx} + jB \tag{11.67}$$

$$\rho u \frac{d}{dx}\left(c_p T + \frac{u^2}{2}\right) = jE \tag{11.68}$$

$$j = \sigma(E - uB) \tag{11.69}$$

$$p = \rho RT \tag{11.70}$$

For isothermal flows, $dT/dx = 0$, so that Eqs. (11.67) and (11.68) can be combined.

$$\begin{aligned} Mu\frac{du}{dx} &= \xi\left(M\frac{du}{dx} + \frac{dp}{dx}\right) \\ \xi &= \frac{E}{B} \end{aligned} \tag{11.71}$$

From the ideal gas equation of state (11.70) and conservation of mass (11.66), we also have

$$\frac{dp}{dx} = RT\frac{d\rho}{dx} = -MRT\frac{1}{u^2}\frac{du}{dx} \tag{11.72}$$

Inserting this result in Eq. (11.71) gives us

$$\xi = \frac{u^3}{u^2 - RT} \tag{11.73}$$

We could define the speed of sound in the plasma as $a_o = (RT)^{1/2}$, in which case ξ will have a singularity when $u = a_o$. To avoid this problem, we will only consider supersonic speeds, where the initial velocity is at least sonic or greater. Using Eq. (11.69) and ξ, we can rewrite the energy equation.

$$Mu\frac{du}{dx} = \sigma E^2\left(1 - \frac{u}{\xi}\right) = \sigma B^2\left(\xi^2 - u\xi\right) \tag{11.74}$$

Substituting Eq. (11.74) the results from the second and third part of the equation, respectively, in

$$Mu^3\frac{du}{dx} = \sigma E^2 a_o^2 \tag{11.75}$$

$$M\frac{du}{dx} = \sigma B^2 a_o^2 \frac{u^3}{(u^2 - a_o^2)^2} \tag{11.76}$$

Now, Eq. (11.75) can be integrated under constant σE^2 in the x-direction.

$$u^* = \left[1 + 4\frac{\sigma E^2 L}{M u_o^2}\left(\frac{a_o}{u_o}\right)^2 x^*\right]^{1/4} \tag{11.77}$$

u_o = initial speed
$u^* = u/u_o$
$x^* = x/L$
L = chamber length

Similarly, Eq. (11.76) can be solved if σB^2 is kept constant in the x-direction.

$$\left(\frac{u_o}{a_o}\right)^2 (u^{*2} - 1) - 4\ln u^* + \left(\frac{a_o}{u_o}\right)^2 \left[1 - \left(\frac{1}{u^*}\right)^2\right] = 2\frac{\sigma B^2 L}{M} x^* \tag{11.78}$$

Another solution is for a constant applied power for which jE = P = constant.

$$u^* = \left(1 + 2\frac{PL}{M u_o^2} x^*\right)^{1/2} \tag{11.79}$$

11.3 Problems

11.1. Plot $m_S + m_p$ using Eqs. (11.1) and (11.2), for $\alpha = 10$, 100, 1000 kg/kW, and $\eta = 0.1$, 0.2 and 0.4. Compare the trace of $(I_{sp})_{opt}$ with Eq. (11.3).

11.2. Plot the propellant gas temperature in electrothermal thrust chamber, using Eq. (11.6). $h = 10$, 50 and 100 W/m^{2}K; $T_w = 1000$, 2000 and 3000 K. $\varepsilon = 0.2$.

11.3. Estimate the specific impulse for a nitrogen arcjet, operating at 7350 K and pressure of 1 atm. At this temperature, nitrogen is assumed to be completely dissociated to atomic nitrogen, which expands in an isentropic, frozen flow through a nozzle of area ratio $A_e/A^* = 100$, into vacuum. Use $\gamma = 5/3$.

11.4. If hydrogen molecules are heated at 50 atm pressure from 300 K to 6000 K, near-complete dissociation will occur. If the equilibrium constant is approximated as $\ln(K_p) = \ln\left(\frac{p_H}{p_{H2}}\right) = 14.901 - \frac{55870}{T}$ [atm], calculate the fraction of the energy that is lost to dissociation, provided that no recombination has taken place. Use average specific heats of atomic and molecular hydrogen of 25 kJ/kmol-K and 38 kJ/kmol-K, respectively. The heat of formation of H is +217 kJ/gmol at 300 K.

11.5. Tabulate the charge-to-mass ratio, q/m, for singly ionized noble gases, Hg, Ar and Xe, and plot the exhaust velocity and specific impulse as a function of the applied voltage, V_a, using Eqs. (11.14) and (11.15).

11.6. Using the Saha equation, plot the degree of ionization as a function of pressure, for argon. The degeneracies, g_j, can be assumed to be 1, while atomic and ionic electronic energy levels, ε_j, ε_j', need to be looked up.

$$\frac{\alpha^2}{1-\alpha^2} = \frac{2(2\pi m_e kT)^{5/2}}{ph^3}\frac{f_{i,e}}{f_{Ae}}\exp\left(-\frac{\varepsilon_j}{kT}\right) \quad (11.32)$$

11.7. Plot the motion of an electron in a non-uniform magnetic field of $B = B_a\left(\frac{r}{r_a}\right)^2$, using Eq. (E11.2.3).

$$\dot{\theta} = -e\frac{B_a}{4m_e}\left(\frac{r}{r_a}\right)^2 \quad (E11.2.3)$$

Calculate the current density, j [mA/cm^2], the thrust per unit area, N/cm^2, for cesium ions in an electric field of $E = -500$ V/0.2 mm.
q/m for cesium is 7.25×10^5 C/kg and the electric permittivity can be taken as $\varepsilon = 8.85 \times 10^{-3}\ \frac{\text{A·s}}{\text{kg·m}}$.

11.8. For a cesium atom ionization, the following reaction can be written.

$$Cs \leftrightarrow Cs^+ + e^-$$

Write an equation for the degree of ionization, α, where $\alpha = 0$ for zero ionization and 1 for full ionization, in terms of the equilibrium constant, K_i and p/p_{REF}.

11.9. For a xenon ion thruster, a thrust of 5 N at a specific impulse of 5000s is required, for accelerating electrode spacing of 5 mm. What is (a) the required voltage difference between the electrodes; and (2) the area of the thruster exit plane?

11.10. Plot the current density, mass flux, exhaust speed and F/A as a function of the applied voltage, V_a, for xenon ions, with $L = 0.1$ m.

11.11. An ion thruster using an electrostatic field needs to be designed to produce $I_{sp} = 3500$s, using propellant that can be ionized to $q/m = 500$ C/kg. For an electric field of 10^5 V/cm, what is the required cross-sectional area to produce 1 N of thrust?

11.12. An ion thruster with 5 N thrust, using mercury as propellant, uses an electrical potential of 5000 V. What is the required beam power? If the electrode distance is 5 mm, what is the current density and the jet cross-section?

11.13. For the ion beam deflection device shown below, the doubly charged particles move at a speed of $u_2 = \sqrt{2}u_1$. Calculate the ratio of deflection angles, θ_2/θ_1 for doubly and singly charged particles for the following two cases: (a) a constant vertical electric field of 1000 V/mm; (b) a constant magnetic field out of the paper at 1.2 tesla and (c) a combined electric and magnetic fields given in (a) and (b), respectively. The units (tesla, V/m) are self-consistent SI units (no unit conversion is needed).

u_1 = velocity of singly charged ion; u_2 = velocity of doubly charged ion.
θ_1 = deflection angle for singly charged ion; θ_2 = deflection angle for doubly charged ion.

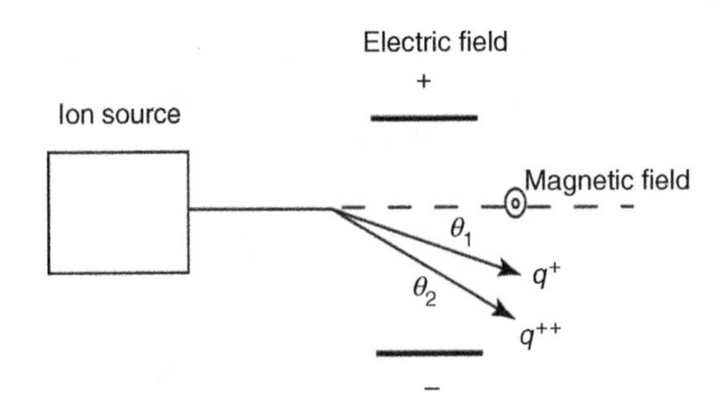

Bibliography

Hill, P. and Petersen, C. (1992) *Mechanics and Thermodynamics of Propulsion*, 2nd edn, Addison-Wesley Publishing Company.

Jahn, R.G. (1968) *Physics of Electric Propulsion*, McGraw-Hill.

Appendices

Appendix A: Standard Atmospheric Air Properties

$$p_{REF} = 101325\frac{\mathrm{N}}{\mathrm{m}^2} = 2116\frac{\mathrm{lbf}}{\mathrm{ft}^2}$$

$$T_{REF} = 288.15K = 518.7°\mathrm{R}$$

$$\rho_{REF} = 1.225\frac{\mathrm{kg}}{\mathrm{m}^3} = 0.076\,47\frac{\mathrm{lbm}}{\mathrm{ft}^3}\cdot$$

$$a_{REF} = 340.294\frac{\mathrm{m}}{\mathrm{s}} = 1116\frac{\mathrm{ft}}{\mathrm{s}}\cdot$$

h [km]	ρ/ρ_{REF}	p/p_{REF}	T/T_{REF}	a [m/s]	$\mu \times 10^6 \frac{\mathrm{kg}}{\mathrm{m}\cdot\mathrm{s}}$
−2	1.2067E + 00	1.2611E + 00	1.0451	347.9	18.51
0	1.0000E + 00	1.0000E + 00	1.0000	340.3	17.89
2	8.2168E − 01	7.8462E − 01	0.9549	332.5	17.26
4	6.6885E − 01	6.0854E − 01	0.9088	324.6	16.61
6	5.3887E − 01	4.6600E − 01	0.8648	316.5	15.95
8	4.2921E − 01	3.5185E − 01	0.8198	308.1	15.27
10	3.3756E − 01	2.6153E − 01	0.7748	299.5	14.58
12	2.5464E − 01	1.9146E − 01	0.7519	295.1	14.22
14	1.8600E − 01	1.3985E − 01	0.7519	295.1	14.22
16	1.3589E − 01	1.0217E − 01	0.7519	295.1	14.22
18	9.9302E − 02	7.4662E − 02	0.7519	295.1	14.22
20	7.2578E − 02	5.4569E − 02	0.7519	295.1	14.22
22	5.2660E − 02	3.9945E − 02	0.7585	296.4	14.32
24	3.8316E − 02	2.9328E − 02	0.7654	297.7	14.43
26	2.7964E − 02	2.1597E − 02	0.7723	299.1	14.54
28	2.0470E − 02	1.5950E − 02	0.7792	300.4	14.65

(continued)

Aerospace Propulsion, First Edition. T.-W. Lee.
© 2014 John Wiley & Sons, Ltd. Published 2014 by John Wiley & Sons, Ltd.
Companion Website: www.wiley.com/go/aerospaceprop

(*Continued*)

h [km]	ρ/ρ_{REF}	p/p_{REF}	T/T_{REF}	a [m/s]	$\mu \times 10^6 \frac{kg}{m \cdot s}$
30	1.5028E − 02	1.1813E − 02	0.7861	301.7	14.75
32	1.1065E − 02	8.7740E − 03	0.7930	303.0	14.86
34	8.0709E − 03	6.5470E − 03	0.8112	306.5	15.14
36	5.9245E − 03	4.9198E − 03	0.8304	310.1	14.43
38	4.3806E − 03	3.7218E − 03	0.8496	313.7	15.72
40	3.2615E − 03	2.8337E − 03	0.8688	317.2	16.01
42	2.4445E − 03	2.1708E − 03	0.8880	320.7	16.29
44	1.8438E − 03	1.6727E − 03	0.9072	324.1	16.57
46	1.3992E − 03	1.2961E − 03	0.9263	327.5	16.85
48	1.0748E − 03	1.0095E − 03	0.9393	329.8	17.04
50	8.3819E − 04	7.8728E − 04	0.9393	329.8	17.04
52	6.5759E − 04	6.1395E − 04	0.9336	328.8	16.96
54	5.2158E − 04	4.7700E − 04	0.9145	325.4	16.68
56	4.1175E − 04	3.6869E − 04	0.8954	322.0	16.40
58	3.2344E − 04	2.8344E − 04	0.8763	318.6	16.12
60	2.5276E − 04	2.1668E − 04	0.8573	315.1	15.84
62	1.9647E − 04	1.6468E − 04	0.8382	311.5	15.55
64	1.5185E − 04	1.2439E − 04	0.8191	308.0	15.26
66	1.1668E − 04	9.3354E − 05	0.8001	304.4	14.94
68	8.9101E − 05	6.9593E − 05	0.7811	300.7	14.67
70	6.7601E − 05	5.1515E − 05	0.7620	297.1	14.38
72	5.0905E − 05	3.7852E − 05	0.7436	293.4	14.08
74	3.7856E − 05	2.7635E − 05	0.7300	290.7	13.87
76	2.8001E − 05	2.0061E − 05	0.7164	288.0	13.65
78	2.0597E − 05	1.4477E − 05	0.7029	285.3	13.43
80	1.5063E − 05	1.0384E − 05	0.6893	282.5	13.21
82	1.0950E − 05	7.4002E − 06	0.6758	279.7	12.98
84	7.9106E − 06	5.2391E − 06	0.6623	276.9	12.76
86	5.6777E − 06	3.6835E − 06	0.6488	274.1	12.53

h [kft]	ρ/ρ_{REF}	p/p_{REF}	T/T_{REF}	a [ft/s]	$\mu \times 10^6 \frac{lbf \cdot s}{ft^2}$
−1	1.0296	1.0367	1.0069	1120.3	0.376
0	1.0000	1.0000	1.0000	1116.5	0.374
1	0.9711	0.9644	0.9931	1112.6	0.372
2	0.9428	0.9298	0.9863	1108.7	0.370
3	0.9151	0.8963	0.9794	1104.9	0.368
4	0.8881	0.8637	0.9725	1101.0	0.366
5	0.8617	0.9321	0.9656	1097.1	0.364
6	0.8359	0.8014	0.9588	1093.2	0.362
7	0.8107	0.7717	0.9519	1089.3	0.360
8	0.7861	0.7429	0.9450	1085.3	0.358
9	0.7621	0.7149	0.9381	1081.4	0.355
10	0.7386	0.6878	0.9313	1077.4	0.353

(*Continued*)

h [kft]	ρ/ρ_{REF}	p/p_{REF}	T/T_{REF}	a [ft/s]	$\mu \times 10^6 \frac{lbf \cdot s}{ft^2}$
11	0.7157	0.6616	0.9244	1073.4	0.351
12	0.6933	0.6362	0.9175	1069.4	0.349
13	0.6715	0.6115	0.9107	1065.4	0.347
14	0.6502	0.5877	0.9038	1061.4	0.345
15	0.6295	0.5646	0.8969	1057.4	0.343
16	0.6092	0.5422	0.8901	1053.3	0.341
17	0.5895	0.5206	0.8832	1049.2	0.339
18	0.5702	0.4997	0.8763	1045.1	0.337
19	0.5514	0.4795	0.8695	1041.0	0.335
20	0.5332	0.4599	0.8626	1036.9	0.332
21	0.5153	0.4410	0.8558	1032.8	0.330
22	0.4980	0.4227	0.8489	1028.6	0.328
23	0.4811	0.4051	0.8420	1024.5	0.326
24	0.4646	0.3880	0.8352	1020.3	0.324
25	0.4486	0.3716	0.8283	1016.1	0.322
26	0.4330	0.3557	0.8215	1011.9	0.319
27	0.4178	0.3404	0.8146	1007.7	0.317
28	0.4031	0.3256	0.8077	1003.4	0.315
29	0.3887	0.3113	0.8009	999.1	0.313
30	0.3747	0.2975	0.7940	994.8	0.311
31	0.3611	0.2843	0.7872	990.5	0.308
32	0.3480	0.2715	0.7803	986.2	0.306
33	0.3351	0.2592	0.7735	981.9	0.304
34	0.3227	0.2474	0.7666	977.5	0.302
35	0.3106	0.2360	0.7598	973.1	0.300
36	0.2988	0.2250	0.7529	968.7	0.297
37	0.2852	0.2145	0.7519	968.1	0.297
38	0.2719	0.2044	0.7519	968.1	0.297
39	0.2592	0.1949	0.7519	968.1	0.297
40	0.2471	0.1858	0.7519	968.1	0.297
41	0.2355	0.1771	0.7519	968.1	0.297
42	0.2245	0.1688	0.7519	968.1	0.297
43	0.2140	0.1609	0.7519	968.1	0.297
44	0.2040	0.1534	0.7519	968.1	0.297
45	0.1945	0.1462	0.7519	968.1	0.297
46	0.1854	0.1394	0.7519	968.1	0.297
47	0.1767	0.1329	0.7519	968.1	0.297
48	0.1685	0.1267	0.7519	968.1	0.297
49	0.1606	0.1208	0.7519	968.1	0.297
50	0.1531	0.1151	0.7519	968.1	0.297
51	0.1460	0.1097	0.7519	968.1	0.297
52	0.1391	0.1046	0.7519	968.1	0.297
53	0.1326	0.0997	0.7519	968.1	0.297
54	0.1264	0.0951	0.7519	968.1	0.297
55	0.1205	0.0906	0.7519	968.1	0.297

(*continued*)

(*Continued*)

h [kft]	ρ/ρ_{REF}	p/p_{REF}	T/T_{REF}	a [ft/s]	$\mu \times 10^6 \frac{\text{lbf} \cdot \text{s}}{\text{ft}^2}$
56	0.1149	0.0864	0.7519	968.1	0.297
57	0.1096	0.0824	0.7519	968.1	0.297
58	0.1044	0.0785	0.7519	968.1	0.297
59	0.0996	0.0749	0.7519	968.1	0.297
60	0.0949	0.0714	0.7519	968.1	0.297
61	0.0905	0.0680	0.7519	968.1	0.297
62	0.0863	0.0649	0.7519	968.1	0.297
63	0.0822	0.0618	0.7519	968.1	0.297
64	0.0784	0.0590	0.7519	968.1	0.297
65	0.0747	0.0562	0.7519	968.1	0.297

Appendix B: Specific Heats for Air as a Function of Temperature

T (K)	C_p [kJ/kg · K]	C_v [kJ/kg · K]	γ
100	1.032	0.745	1.385
150	1.012	0.725	1.396
200	1.007	0.720	1.399
250	1.006	0.719	1.399
300	1.007	0.720	1.399
350	1.009	0.722	1.398
400	1.014	0.727	1.395
450	1.021	0.734	1.391
500	1.030	0.743	1.386
550	1.040	0.753	1.381
600	1.051	0.764	1.376
650	1.063	0.776	1.370
700	1.075	0.788	1.364
750	1.087	0.800	1.359
800	1.099	0.812	1.353
850	1.110	0.823	1.349
900	1.121	0.834	1.344
950	1.131	0.844	1.340
1000	1.141	0.854	1.336
1100	1.159	0.872	1.329
1200	1.175	0.888	1.323
1300	1.189	0.902	1.318
1400	1.207	0.920	1.312
1500	1.230	0.943	1.304
1600	1.248	0.961	1.299
1700	1.267	0.980	1.293
1800	1.286	0.999	1.287
1900	1.307	1.020	1.281
2000	1.337	1.050	1.273

(Continued)

T (K)	C_p [kJ/kg · K]	C_v [kJ/kg · K]	γ
2100	1.372	1.085	1.265
2200	1.417	1.130	1.254
2300	1.478	1.191	1.241
2400	1.558	1.271	1.226
2500	1.665	1.378	1.208
3000	2.726	2.439	1.118

Source: Incropera, F.P., and DeWitt, D.P., Fundamentals of Heat and Mass Transfer, 3rd Ed.,1990 John Wiley and Sons, Inc.

Appendix C: Normal Shock Properties

M_1	p_2/p_1	ρ_2/ρ_1	T_2/T_1	p_{02}/p_{01}	p_{02}/p_1	M_2
1.0000E + 00	1.0000E + 00	1.0000E + 00	1.0000E + 00	1.0000E + 00	1.8930E + 00	1.0000E + 00
1.0200E + 00	1.0471E + 00	1.0334E + 00	1.0132E + 00	9.9999E − 01	1.9380E + 00	9.8052E − 01
1.0400E + 00	1.0952E + 00	1.0671E + 00	1.0263E + 00	9.9992E − 01	1.9840E + 00	9.6203E − 01
1.0600E + 00	1.1442E + 00	1.1009E + 00	1.0393E + 00	9.9975E − 01	2.0320E + 00	9.4445E − 01
1.0800E + 00	1.1941E + 00	1.1349E + 00	1.0522E + 00	9.9943E − 01	2.0820E + 00	9.2771E − 01
1.1000E + 00	1.2450E + 00	1.1691E + 00	1.0649E + 00	9.9893E − 01	2.1330E + 00	9.1177E − 01
1.1200E + 00	1.2968E + 00	1.2034E + 00	1.0776E + 00	9.9821E − 01	2.1850E + 00	8.9656E − 01
1.1400E + 00	1.3495E + 00	1.2378E + 00	1.0903E + 00	9.9726E − 01	2.2390E + 00	8.8204E − 01
1.1600E + 00	1.4032E + 00	1.2723E + 00	1.1029E + 00	9.9605E − 01	2.2940E + 00	8.6816E − 01
1.1800E + 00	1.4578E + 00	1.3069E + 00	1.1154E + 00	9.9457E − 01	2.3500E + 00	8.5488E − 01
1.2000E + 00	1.5133E + 00	1.3416E + 00	1.1280E + 00	9.9280E − 01	2.4080E + 00	8.4217E − 01
1.2200E + 00	1.5698E + 00	1.3764E + 00	1.1405E + 00	9.9073E − 01	2.4660E + 00	8.2999E − 01
1.2400E + 00	1.6272E + 00	1.4112E + 00	1.1531E + 00	9.8836E − 01	2.5260E + 00	8.1830E − 01
1.2600E + 00	1.6855E + 00	1.4460E + 00	1.1657E + 00	9.8568E − 01	2.5880E + 00	8.0709E − 01
1.2800E + 00	1.7448E + 00	1.4808E + 00	1.1783E + 00	9.8268E − 01	2.6500E + 00	7.9631E − 01
1.3000E + 00	1.8050E + 00	1.5157E + 00	1.1909E + 00	9.7937E − 01	2.7140E + 00	7.8596E − 01
1.3200E + 00	1.8661E + 00	1.5505E + 00	1.2035E + 00	9.7575E − 01	2.7780E + 00	7.7600E − 01
1.3400E + 00	1.9282E + 00	1.5854E + 00	1.2162E + 00	9.7182E − 01	2.8440E + 00	7.6641E − 01
1.3600E + 00	1.9912E + 00	1.6202E + 00	1.2290E + 00	9.6758E − 01	2.9120E + 00	7.5718E − 01
1.3800E + 00	2.0551E + 00	1.6549E + 00	1.2418E + 00	9.6304E − 01	2.9800E + 00	7.4829E − 01
1.4000E + 00	2.1200E + 00	1.6897E + 00	1.2547E + 00	9.5819E − 01	3.0490E + 00	7.3971E − 01
1.4200E + 00	2.1858E + 00	1.7243E + 00	1.2676E + 00	9.5306E − 01	3.1200E + 00	7.3144E − 01
1.4400E + 00	2.2525E + 00	1.7589E + 00	1.2807E + 00	9.4765E − 01	3.1910E + 00	7.2345E − 01
1.4600E + 00	2.3202E + 00	1.7934E + 00	1.2938E + 00	9.4196E − 01	3.2640E + 00	7.1574E − 01
1.4800E + 00	2.3888E + 00	1.8278E + 00	1.3069E + 00	9.3600E − 01	3.3380E + 00	7.0829E − 01
1.5000E + 00	2.4583E + 00	1.8621E + 00	1.3202E + 00	9.2979E − 01	3.4130E + 00	7.0109E − 01
1.5200E + 00	2.5288E + 00	1.8963E + 00	1.3336E + 00	9.2332E − 01	3.4890E + 00	6.9413E − 01
1.5400E + 00	2.6002E + 00	1.9303E + 00	1.3470E + 00	9.1662E − 01	3.5670E + 00	6.8739E − 01
1.5600E + 00	2.6725E + 00	1.9643E + 00	1.3606E + 00	9.0970E − 01	3.6450E + 00	6.8087E − 01
1.5800E + 00	2.7458E + 00	1.9981E + 00	1.3742E + 00	9.0255E − 01	3.7240E + 00	6.7455E − 01
1.6000E + 00	2.8200E + 00	2.0317E + 00	1.3880E + 00	8.9520E − 01	3.8050E + 00	6.6844E − 01

(continued)

(*Continued*)

M_1	p_2/p_1	ρ_2/ρ_1	T_2/T_1	p_{02}/p_{01}	p_{02}/p_1	M_2
1.6200E+00	2.8951E+00	2.0653E+00	1.4018E+00	8.8765E−01	3.8870E+00	6.6251E−01
1.6400E+00	2.9712E+00	2.0986E+00	1.4158E+00	8.7992E−01	3.9690E+00	6.5677E−01
1.6600E+00	3.0482E+00	2.1318E+00	1.4299E+00	8.7201E−01	4.0530E+00	6.5119E−01
1.6800E+00	3.1261E+00	2.1649E+00	1.4440E+00	8.6394E−01	4.1380E+00	6.4579E−01
1.7000E+00	3.2050E+00	2.1977E+00	1.4583E+00	8.5572E−01	4.2240E+00	6.4054E−01
1.7200E+00	3.2848E+00	2.2304E+00	1.4727E+00	8.4736E−01	4.3110E+00	6.3545E−01
1.7400E+00	3.3655E+00	2.2629E+00	1.4873E+00	8.3886E−01	4.3990E+00	6.3051E−01
1.7600E+00	3.4472E+00	2.2952E+00	1.5019E+00	8.3024E−01	4.4880E+00	6.2570E−01
1.7800E+00	3.5298E+00	2.3273E+00	1.5167E+00	8.2151E−01	4.5780E+00	6.2104E−01
1.8000E+00	3.6133E+00	2.3592E+00	1.5316E+00	8.1268E−01	4.6700E+00	6.1650E−01
1.8200E+00	3.6978E+00	2.3909E+00	1.5466E+00	8.0376E−01	4.7620E++00	6.1209E−01
1.8400E+00	3.7832E+00	2.4224E+00	1.5617E+00	7.9476E−01	4.8550E+00	6.0780E−01
1.8600E+00	3.8695E+00	2.4537E+00	1.5770E+00	7.8569E−01	4.9500E+00	6.0363E−01
1.8800E+00	3.9568E+00	2.4848E+00	1.5924E+00	7.7655E−01	5.0450E+00	5.9957E−01
1.9000E+00	4.0450E+00	2.5157E+00	1.6079E+00	7.6736E−01	5.1420E+00	5.9562E−01
1.9200E+00	4.1341E+00	2.5463E+00	1.6236E+00	7.5812E−01	5.2390E+00	5.9177E−01
1.9400E+00	4.2242E+00	2.5767E+00	1.6394E+00	7.4884E−01	5.3380E+00	5.8802E−01
1.9600E+00	4.3152E+00	2.6069E+00	1.6553E+00	7.3954E−01	5.4380E+00	5.8437E−01
1.9800E+00	4.4071E+00	2.6369E+00	1.6713E+00	7.3021E−01	5.5390E+00	5.8082E−01
2.0000E+00	4.5000E+00	2.6667E+00	1.6875E+00	7.2087E−01	5.6400E+00	5.7735E−01
2.0500E+00	4.7363E+00	2.7400E+00	1.7285E+00	6.9751E−01	5.9000E+00	5.6906E−01
2.1000E+00	4.9783E+00	2.8119E+00	1.7705E+00	6.7420E−01	6.1650E+00	5.6128E−01
2.1500E+00	5.2263E+00	2.8823E+00	1.8132E+00	6.5105E−01	6.4380E+00	5.5395E−01
2.2000E+00	5.4800E+00	2.9512E+00	1.8569E+00	6.2814E−01	6.7160E+00	5.4706E−01
2.2500E+00	5.7396E+00	3.0186E+00	1.9014E+00	6.0553E−01	7.0020E+00	5.4055E−01
2.3000E+00	6.0050E+00	3.0845E+00	1.9468E+00	5.8329E−01	7.2940E+00	5.3441E−01
2.3500E+00	6.2763E+00	3.1490E+00	1.9931E+00	5.6148E−01	7.5920E+00	5.2861E−01
2.4000E+00	6.5533E+00	3.2119E+00	2.0403E+00	5.4014E−01	7.8970E+00	5.2312E−01
2.4500E+00	6.8363E+00	3.2733E+00	2.0885E+00	5.1931E−01	8.2080E+00	5.1792E−01
2.5000E+00	7.1250E+00	3.3333E+00	2.1375E+00	4.9901E−01	8.5260E+00	5.1299E−01
2.5500E+00	7.4196E+00	3.3919E+00	2.1875E+00	4.7928E−01	8.8500E+00	5.0831E−01
2.6000E+00	7.7200E+00	3.4490E+00	2.2383E+00	4.6012E−01	9.1810E+00	5.0387E−01
2.6500E+00	8.0263E+00	3.5047E+00	2.2902E+00	4.4156E−01	9.5190E+00	4.9965E−01
2.7000E+00	8.3383E+00	3.5590E+00	2.3429E+00	4.2359E−01	9.8620E+00	4.9563E−01
2.7500E+00	8.6563E+00	3.6119E+00	2.3966E+00	4.0623E−01	1.0210E+01	4.9181E−01
2.8000E+00	8.9800E+00	3.6636E+00	2.4512E+00	3.8946E−01	1.0570E+01	4.8817E−01
2.8500E+00	9.3096E+00	3.7139E+00	2.5067E+00	3.7330E−01	1.0930E+01	4.8469E−01
2.9000E+00	9.6450E+00	3.7629E+00	2.5632E+00	3.5773E−01	1.1300E+01	4.8138E−01
2.9500E+00	9.9863E+00	3.8106E+00	2.6206E+00	3.4275E−01	1.1680E+01	4.7821E−01
3.0000E+00	1.0333E+01	3.8571E+00	2.6790E+00	3.2834E−01	1.2060E+01	4.7519E−01
3.0500E+00	1.0686E+01	3.9025E+00	2.7383E+00	3.1450E−01	1.2450E+01	4.7230E−01
3.1000E+00	1.1045E+01	3.9466E+00	2.7986E+00	3.0121E−01	1.2850E+01	4.6953E−01
3.1500E+00	1.1410E+01	3.9896E+00	2.8598E+00	2.8846E−01	1.3250E+01	4.6689E−01
3.2000E+00	1.1780E+01	4.0315E+00	2.9220E+00	2.7623E−01	1.3660E+01	4.6435E−01
3.2500E+00	1.2156E+01	4.0723E+00	2.9851E+00	2.6451E−01	1.4070E+01	4.6192E−01
3.3000E+00	1.2538E+01	4.1120E+00	3.0492E+00	2.5328E−01	1.4490E+01	4.5959E−01
3.3500E+00	1.2926E+01	4.1507E+00	3.1142E+00	2.4252E−01	1.4920E+01	4.5735E−01

(*Continued*)

M_1	p_2/p_1	ρ_2/ρ_1	T_2/T_1	p_{02}/p_{01}	p_{02}/p_1	M_2
3.4000E+00	1.3320E+01	4.1884E+00	3.1802E+00	2.3223E−01	1.5350E+01	4.5520E−01
3.4500E+00	1.3720E+01	4.2251E+00	3.2472E+00	2.2237E−01	1.5790E+01	4.5314E−01
3.5000E+00	1.4125E+01	4.2609E+00	3.3151E+00	2.1295E−01	1.6240E+01	4.5115E−01
3.5500E+00	1.4536E+01	4.2957E+00	3.3839E+00	2.0393E−01	1.6700E+01	4.4925E−01
3.6000E+00	1.4953E+01	4.3296E+00	3.4537E+00	1.9531E−01	1.7160E+01	4.4741E−01
3.6500E+00	1.5376E+01	4.3627E+00	3.5245E+00	1.8707E−01	1.7620E+01	4.4565E−01
3.7000E+00	1.5805E+01	4.3949E+00	3.5962E+00	1.7919E−01	1.8100E+01	4.4395E−01
3.7500E+00	1.6240E+01	4.4262E+00	3.6689E+00	1.7166E−01	1.8570E+01	4.4231E−01
3.8000E+00	1.6680E+01	4.4568E+00	3.7426E+00	1.6447E−01	1.9060E+01	4.4073E−01
3.8500E+00	1.7126E+01	4.4866E+00	3.8172E+00	1.5760E−01	1.9550E+01	4.3921E−01
3.9000E+00	1.7578E+01	4.5156E+00	3.8928E+00	1.5103E−01	2.0050E+01	4.3774E−01
3.9500E+00	1.8036E+01	4.5439E+00	3.9694E+00	1.4475E−01	2.0560E+01	4.3633E−01
4.0000E+00	1.8500E+01	4.5714E+00	4.0469E+00	1.3876E−01	2.1070E+01	4.3496E−01
4.0500E+00	1.8970E+01	4.5983E+00	4.1254E+00	1.3303E−01	2.1590E+01	4.3364E−01
4.1000E+00	1.9445E+01	4.6245E+00	4.2048E+00	1.2756E−01	2.2110E+01	4.3236E−01
4.1500E+00	1.9926E+01	4.6500E+00	4.2852E+00	1.2233E−01	2.2640E+01	4.3113E−01
4.2000E+00	2.0413E+01	4.6749E+00	4.3666E+00	1.1733E−01	2.3180E+01	4.2994E−01
4.2500E+00	2.0906E+01	4.6992E+00	4.4489E+00	1.1256E−01	2.3720E+01	4.2878E−01
4.3000E+00	2.1405E+01	4.7229E+00	4.5322E+00	1.0800E−01	2.4270E+01	4.2767E−01
4.3500E+00	2.1910E+01	4.7460E+00	4.6165E+00	1.0364E−01	2.4830E+01	4.2659E−01
4.4000E+00	2.2420E+01	4.7685E+00	4.7017E+00	9.9481E−02	2.5390E+01	4.2554E−01
4.4500E+00	2.2936E+01	4.7904E+00	4.7879E+00	9.5501E−02	2.5960E+01	4.2453E−01
4.5000E+00	2.3458E+01	4.8119E+00	4.8751E+00	9.1698E−02	2.6540E+01	4.2355E−01
4.5500E+00	2.3986E+01	4.8328E+00	4.9632E+00	8.8062E−02	2.7120E+01	4.2260E−01
4.6000E+00	2.4520E+01	4.8532E+00	5.0523E+00	8.4586E−02	2.7710E+01	4.2168E−01
4.6500E+00	2.5060E+01	4.8731E+00	5.1424E+00	8.1263E−02	2.8310E+01	4.2079E−01
4.7000E+00	2.5605E+01	4.8926E+00	5.2334E+00	7.8086E−02	2.8910E+01	4.1992E−01
4.7500E+00	2.6156E+01	4.9116E+00	5.3254E+00	7.5047E−02	2.9520E+01	4.1908E−01
4.8000E+00	2.6713E+01	4.9301E+00	5.4184E+00	7.2140E−02	3.0130E+01	4.1826E−01
4.8500E+00	2.7276E+01	4.9482E+00	5.5124E+00	6.9359E−02	3.0750E+01	4.1747E−01
4.9000E+00	2.7845E+01	4.9659E+00	5.6073E+00	6.6699E−02	3.1380E+01	4.1670E−01
4.9500E+00	2.8420E+01	4.9831E+00	5.7032E+00	6.4153E−02	3.2010E+01	4.1595E−01
5.0000E+00	2.9000E+01	5.0000E+00	5.8000E+00	6.1716E−02	3.2650E+01	4.1523E−01
5.1000E+00	3.0178E+01	5.0326E+00	5.9966E+00	5.7151E−02	3.3950E+01	4.1384E−01
5.2000E+00	3.1380E+01	5.0637E+00	6.1971E+00	5.2966E−02	3.5280E+01	4.1252E−01
5.3000E+00	3.2605E+01	5.0934E+00	6.4014E+00	4.9126E−02	3.6630E+01	4.1127E−01
5.4000E+00	3.3853E+01	5.1218E+00	6.6097E+00	4.5600E−02	3.8010E+01	4.1009E−01
5.5000E+00	3.5125E+01	5.1489E+00	6.8218E+00	4.2361E−02	3.9410E+01	4.0897E−01
5.6000E+00	3.6420E+01	5.1749E+00	7.0378E+00	3.9383E−02	4.0840E+01	4.0791E−01
5.7000E+00	3.7738E+01	5.1998E+00	7.2577E+00	3.6643E−02	4.2300E+01	4.0690E−01
5.8000E+00	3.9080E+01	5.2236E+00	7.4814E+00	3.4120E−02	4.3780E+01	4.0594E−01
5.9000E+00	4.0445E+01	5.2464E+00	7.7091E+00	3.1795E−02	4.5280E+01	4.0503E−01
6.0000E+00	4.1833E+01	5.2683E+00	7.9406E+00	2.9651E−02	4.6820E+01	4.0416E−01
6.1000E+00	4.3245E+01	5.2893E+00	8.1760E+00	2.7672E−02	4.8370E+01	4.0333E−01
6.2000E+00	4.4680E+01	5.3094E+00	8.4153E+00	2.5845E−02	4.9960E+01	4.0254E−01
6.3000E+00	4.6138E+01	5.3287E+00	8.6584E+00	2.4156E−02	5.1570E+01	4.0179E−01

(*continued*)

(*Continued*)

M_1	p_2/p_1	ρ_2/ρ_1	T_2/T_1	p_{02}/p_{01}	p_{02}/p_1	M_2
6.4000E+00	4.7620E+01	5.3473E+00	8.9055E+00	2.2594E−02	5.3200E+01	4.0107E−01
6.5000E+00	4.9125E+01	5.3651E+00	9.1564E+00	2.1148E−02	5.4860E+01	4.0038E−01
6.6000E+00	5.0653E+01	5.3822E+00	9.4113E+00	1.9808E−02	5.6550E+01	3.9972E−01
6.7000E+00	5.2205E+01	5.3987E+00	9.6700E+00	1.8566E−02	5.8260E+01	3.9909E−01
6.8000E+00	5.3780E+01	5.4145E+00	9.9326E+00	1.7414E−02	6.0000E+01	3.9849E−01
6.9000E+00	5.5378E+01	5.4298E+00	1.0199E+01	1.6345E−02	6.1760E+01	3.9791E−01
7.0000E+00	5.7000E+01	5.4444E+00	1.0469E+01	1.5351E−02	6.3550E+01	3.9736E−01
7.1000E+00	5.8645E+01	5.4586E+00	1.0744E+01	1.4428E−02	6.5370E+01	3.9683E−01
7.2000E+00	6.0313E+01	5.4722E+00	1.1022E+01	1.3569E−02	6.7210E+01	3.9632E−01
7.3000E+00	6.2005E+01	5.4853E+00	1.1304E+01	1.2769E−02	6.9080E+01	3.9583E−01
7.4000E+00	6.3720E+01	5.4980E+00	1.1590E+01	1.2023E−02	7.0970E+01	3.9536E−01
7.5000E+00	6.5458E+01	5.5102E+00	1.1879E+01	1.1329E−02	7.2890E+01	3.9491E−01
7.6000E+00	6.7220E+01	5.5220E+00	1.2173E+01	1.0680E−02	7.4830E+01	3.9447E−01
7.7000E+00	6.9005E+01	5.5334E+00	1.2471E+01	1.0075E−02	7.6800E+01	3.9405E−01
7.8000E+00	7.0813E+01	5.5443E+00	1.2772E+01	9.5102E−03	7.8800E+01	3.9365E−01
7.9000E+00	7.2645E+01	5.5550E+00	1.3077E+01	8.9819E−03	8.0820E+01	3.9326E−01
8.0000E+00	7.4500E+01	5.5652E+00	1.3387E+01	8.4878E−03	8.2870E+01	3.9289E−01
9.0000E+00	9.4333E+01	5.6512E+00	1.6693E+01	4.9639E−03	1.0480E+02	3.8980E−01
1.0000E+01	1.1650E+02	5.7143E+00	2.0388E+01	3.0448E−03	1.2920E+02	3.8758E−01
1.1000E+01	1.4100E+02	5.7619E+00	2.4471E+01	1.9451E−03	1.5630E+02	3.8592E−01
1.2000E+01	1.6783E+02	5.7987E+00	2.8943E+01	1.2866E−03	1.8590E+02	3.8466E−01
1.3000E+01	1.9700E+02	5.8276E+00	3.3805E+01	8.7709E−04	2.1810E+02	3.8368E−01
1.4000E+01	2.2850E+02	5.8507E+00	3.9055E+01	6.1380E−04	2.5280E+02	3.8289E−01
1.5000E+01	2.6233E+02	5.8696E+00	4.4694E+01	4.3953E−04	2.9020E+02	3.8226E−01
1.6000E+01	2.9850E+02	5.8851E+00	5.0722E+01	3.2119E−04	3.3010E+02	3.8174E−01
1.7000E+01	3.3700E+02	5.8980E+00	5.7138E+01	2.3899E−04	3.7260E+02	3.8131E−01
1.8000E+01	3.7783E+02	5.9088E+00	6.3944E+01	1.8072E−04	4.1760E+02	3.8095E−01
1.9000E+01	4.2100E+02	5.9180E+00	7.1139E+01	1.3865E−04	4.6530E+02	3.8065E−01
2.0000E+01	4.6650E+02	5.9259E+00	7.8722E+01	1.0777E−04	5.1550E+02	3.8039E−01
2.2000E+01	5.6450E+02	5.9387E+00	9.5055E+01	6.7414E−05	6.2360E+02	3.7997E−01
2.4000E+01	6.7183E+02	5.9484E+00	1.1294E+02	4.3877E−05	7.4210E+02	3.7965E−01
2.6000E+01	7.8850E+02	5.9559E+00	1.3239E+02	2.9534E−05	8.7090E+02	3.7940E−01
2.8000E+01	9.1450E+02	5.9620E+00	1.5339E+02	2.0460E−05	1.0100E+03	3.7920E−01
3.0000E+01	1.0498E+03	5.9669E+00	1.7594E+02	1.4531E−05	1.1590E+03	3.7904E−01
3.2000E+01	1.1945E+03	5.9708E+00	2.0006E+02	1.0548E−05	1.3190E+03	3.7891E−01
3.4000E+01	1.3485E+03	5.9742E+00	2.2572E+02	7.8044E−06	1.4890E+03	3.7880E−01
3.6000E+01	1.5118E+03	5.9769E+00	2.5294E+02	5.8738E−06	1.6690E+03	3.7871E−01
3.8000E+01	1.6845E+03	5.9793E+00	2.8172E+02	4.4885E−06	1.8600E+03	3.7864E−01
4.0000E+01	1.8665E+03	5.9813E+00	3.1206E+02	3.4771E−06	2.0610E+03	3.7857E−01
4.2000E+01	2.0578E+03	5.9830E+00	3.4394E+02	2.7271E−06	2.2720E+03	3.7852E−01
4.4000E+01	2.2585E+03	5.9845E+00	3.7739E+02	2.1630E−06	2.4930E+03	3.7847E−01
4.6000E+01	2.4685E+03	5.9859E+00	4.1239E+02	1.7332E−06	2.7250E+03	3.7842E−01
4.8000E+01	2.6878E+03	5.9870E+00	4.4894E+02	1.4019E−06	2.9670E+03	3.7839E−01
5.0000E+01	2.9165E+03	5.9880E+00	4.8706E+02	1.1438E−06	3.2190E+03	3.7835E−01

Appendix D: Oblique Shock Angle Chart

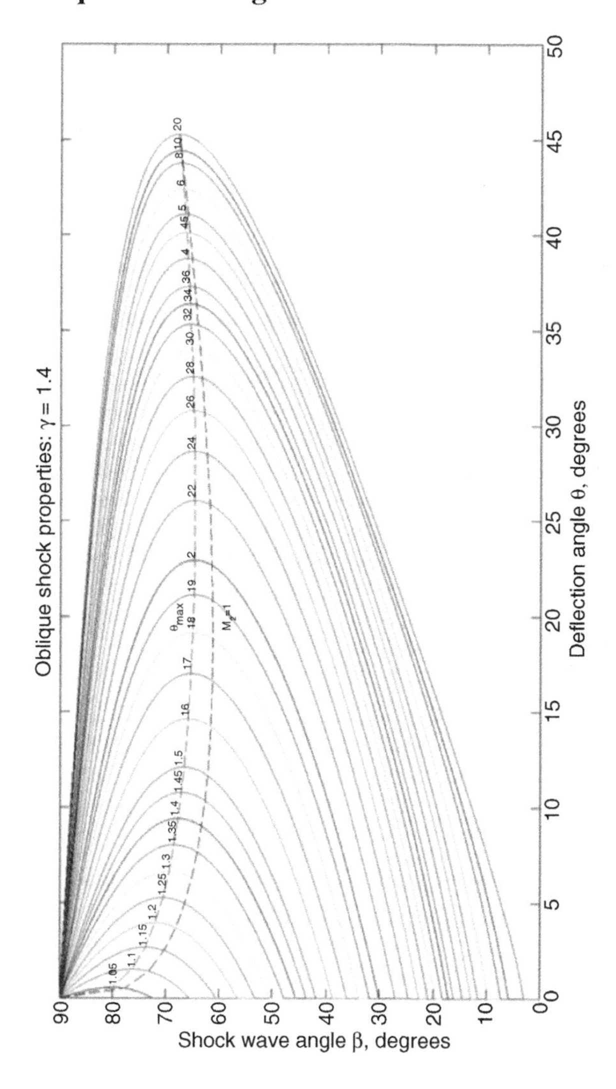

Appendix E: Polynomial Coefficients for Specific Heat of Selected Gases

$$\bar{C}_p/R_u = a_1 + a_2T + a_3T^2 + a_4T^3 + a_5T^4$$

	T (K)	a_1	a_2	a_3	a_4	a_5
CO	300–1000	3.26245165E + 00	1.51194085E − 03	−3.88175522E − 06	5.58194424E − 09	−2.47495123E − 12
	1000–3500	3.02507806E + 00	1.44268852E − 03	−5.63082779E − 07	1.01858133E − 10	−6.91095156E − 15
CO_2	300–1000	2.27572465E + 00	9.92207229E − 03	−1.04091132E − 05	6.86668678E − 09	−2.11728009E − 12
	1000–3500	4.45362282E + 00	3.14016873E − 03	−1.27841054E − 06	2.39399667E − 10	−1.66903319E − 14
H_2	300–1000	3.29812431E + 00	8.24944174E − 04	−8.14301529E − 07	−9.47543433E − 11	4.13487224E − 13
	1000–3500	2.99142337E + 00	7.00064411E − 04	−5.63382869E − 08	−9.23157818E − 12	1.58275179E − 15
H_2O	300–1000	3.38684249E + 00	3.47498246E − 03	−6.35469633E − 06	6.96858127E − 09	−2.506588470E − 12
	1000–3500	2.67214561E + 00	3.056292890E − 03	−8.73026011E − 07	1.20099639E − 10	−6.39161787E − 15
N_2	300–1000	3.29867697E + 00	1.40824041E − 03	−3.96322230E − 06	5.64151526E − 09	−2.44485487E − 12
	1000–3500	2.92664003E + 00	1.48797676E − 03	−5.68476082E − 07	1.00970378E − 10	−6.75335134E − 15
O_2	300–1000	3.21293640E + 00	1.12748635E − 03	−5.75615047E − 07	1.31387723E − 09	−8.76855392E − 13
	1000–3500	3.69757819E + 00	6.13519689E − 04	−1.25884199E − 07	1.77528148E − 11	−1.13643531E − 15
CH_4	300–1000	7.78741479E + 01	1.74766835E − 02	−2.78340904E − 05	3.04970804E − 08	−1.22393068E − 11
	1000–3500	1.68347883E + 00	1.02372356E − 02	−3.87512864E − 06	6.78558487E − 10	−4.50342312E − 14
C_2H_2	300–1000	2.01356220E + 00	1.51904458E − 02	−1.61631888E − 05	9.07899178E − 09	−1.91274600E − 12
	1000–3500	4.43677044E + 00	5.37603907E − 03	−1.91281674E − 06	3.28637895E − 10	−2.15670953E − 14
C_2H_4	300–1000	−8.61487985E − 01	2.79616285E − 02	−3.38867721E − 05	2.78515220E − 08	−9.73787891E − 12
	1000–3500	3.52841878E + 00	1.14851845E − 02	4.41838528E − 06	7.84460053E − 10	−5.26684849E − 14

Source: Kee, R.J., Rupley, F.M., and Miller, J.A., Sandia Report, SAND87-8215, April 1987, "The Chemkin Thermodynamic Data Base"

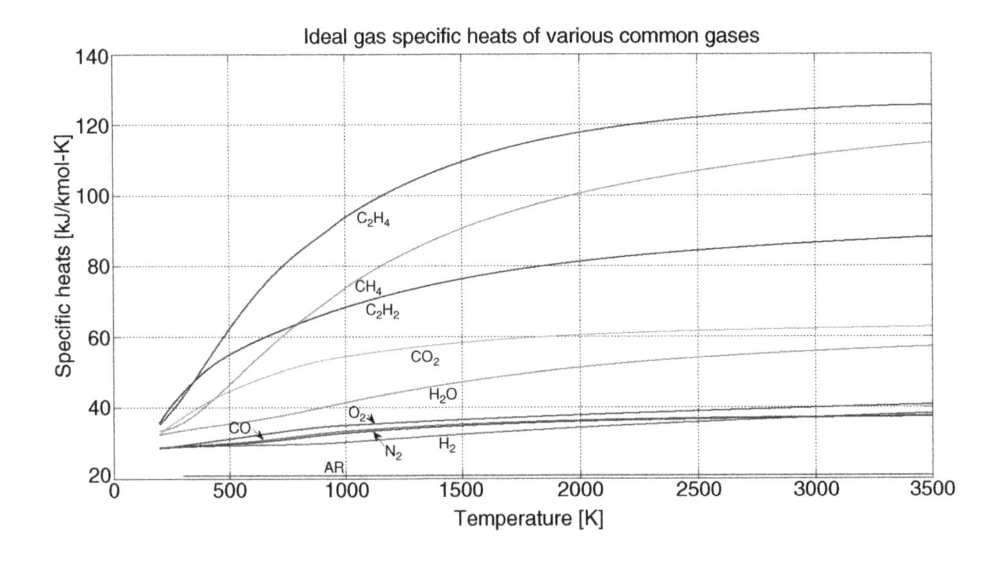

Appendix F: Standard state Gibbs free energy ($T = 298.15$K, $P = 1$ atm) $\bar{g}_f^o(T)[^{kj}/_{kmol}]$

T [K]	CO	CO_2	H_2	H_2O	N_2	NO_2	O_2
200	−128 532	−394 126	0	−232 779	0	45 453	0
298	−137 163	−394 428	0	−228 608	0	51 291	0
300	−137 328	−394 433	0	−228 526	0	51 403	0
400	−146 332	−394 718	0	−223 929	0	57 602	0
500	−155 403	−394 983	0	−219 085	0	63 916	0
600	−164 470	−395 226	0	−214 049	0	70 285	0
700	−173 499	−395 443	0	−208 861	0	76 679	0
800	−182 473	−395 635	0	−203 550	0	83 079	0
900	−191 386	−395 799	0	−198 141	0	89 476	0
1000	−200 238	−395 939	0	−192 652	0	95 864	0
1100	−209 030	−396 056	0	−187 100	0	102 242	0
1200	−217 768	−396 155	0	−181 497	0	108 609	0
1300	−226 453	−396 236	0	−175 852	0	114 966	0
1400	−235 087	−396 301	0	−170 172	0	121 313	0
1500	−243 674	−396 352	0	−164 464	0	127 651	0
1600	−252 214	−396 389	0	−158 733	0	133 981	0
1700	−260 711	−396 414	0	−152 983	0	140 303	0
1800	−269 164	−396 425	0	−147 216	0	146 620	0
1900	−277 576	−396 424	0	−141 435	0	152 931	0
2000	−285 948	−396 410	0	−135 643	0	159 238	0
2100	−294 281	−396 384	0	−129 841	0	165 542	0
2200	−302 576	−396 346	0	−124 030	0	171 843	0
2300	−310 835	−396 294	0	−118 211	0	178 143	0
2400	−319 057	−396 230	0	−112 386	0	184 442	0
2500	−327 245	−396 152	0	−106 555	0	190 742	0
2600	−335 399	−396 061	0	−100 719	0	197 042	0
2700	−343 519	−395 957	0	−94 878	0	203 344	0
2800	−351 606	−395 840	0	−89 031	0	209 648	0
2900	−359 661	−395 708	0	−83 181	0	215 955	0
3000	−367 684	−395 562	0	−77 326	0	222 265	0
3100	−375 677	−395 403	0	−71 467	0	228 579	0
3200	−383 639	−395 229	0	−65 604	0	234 898	0
3300	−391 571	−395 041	0	−59 737	0	241 221	0
3400	−399 474	−394 838	0	−53 865	0	247 549	0
3500	−407 347	−394 620	0	−47 990	0	253 883	0
3600	−415 192	−394 388	0	−42 110	0	260 222	0
T[K]	CO	CO_2	H_2	H_2O	N_2	NO_2	O_2
3700	−423 008	−394 141	0	−36 226	0	266 567	0
3800	−430 796	−393 879	0	−30 338	0	272 918	0
3900	−438 557	−393 602	0	−24 446	0	279 276	0
4000	−446 291	−393 311	0	−18 549	0	285 639	0
4100	−453 997	−393 004	0	−12 648	0	292 010	0
4200	−461 677	−392 683	0	−6742	0	298 387	0

(*continued*)

(Continued)

T [K]	CO	CO_2	H_2	H_2O	N_2	NO_2	O_2
4300	−469 330	−392 346	0	−831	0	304 772	0
4400	−476 957	−391 995	0	5085	0	311 163	0
4500	−484 558	−391 629	0	11 005	0	317 562	0
4600	−492 134	−391 247	0	16 930	0	323 968	0
4700	−499 684	−390 851	0	22 861	0	330 381	0
4800	−507 210	−390 440	0	28 796	0	336 803	0
4900	−514 710	−390 014	0	34 737	0	343 232	0
5000	−522 186	−389 572	0	40 684	0	349 670	0

Source: Kee, R.J., Rupley, F.M., and Miller, J.A., Sandia Report, SAND87-8215B, March 1991, "The Chemkin Thermodynamic Data Base"

Index

Note: Figures are indicated by italic page numbers, Example boxes and Tables by bold numbers

Aerospace Propulsion, First Edition. T.-W. Lee.

© 2014 John Wiley & Sons, Ltd. Published 2014 by John Wiley & Sons, Ltd.

Companion Website: www.wiley.com/go/aerospaceprop

Printed and bound by CPI Group (UK) Ltd, Croydon, CR0 4YY

12/01/2025

14624500-0002